Lecture Notes in Computer Science 6850

Commenced Publication in 1973
Founding and Former Series Editors:
Gerhard Goos, Juris Hartmanis, and Jan van Leeuwen

Editorial Board

FoLLI Publications on Logic, Language and Information

Subline of Lectures Notes in Computer Science

Subline Editors-in-Chief

Subline Area Editors

Richard Moot Christian Retoré

The Logic of Categorial Grammars

A Deductive Account of Natural Language Syntax and Semantics

Springer

Authors

Richard Moot
CNRS
LaBRI
Domaine Universitaire, 351, Cours de la Libération Bat A30
33405 Talence Cedex, France
E-mail: richard.moot@labri.fr

Christian Retoré
Université de Bordeaux
LaBRI
Domaine Universitaire, 351, Cours de la Libération Bat A30
33405 Talence Cedex, France
E-mail: christian.retore@labri.fr

ISSN 0302-9743 e-ISSN 1611-3349
ISBN 978-3-642-31554-1 e-ISBN 978-3-642-31555-8
DOI 10.1007/978-3-642-31555-8
Springer Heidelberg Dordrecht London New York

Library of Congress Control Number: 2012940852

CR Subject Classification (1998): F.3, F.4, I.2.3, I.2, F.1, F.2.2

LNCS Sublibrary: SL 1 – Theoretical Computer Science and General Issues

Typesetting: Camera-ready by author, data conversion by Scientific Publishing Services, Chennai, India

Printed on acid-free paper

Springer is part of Springer Science+Business Media (www.springer.com)

Voor Tessa
À Antoine, Mathilde et Garance

Pour le moment, la linguistique générale m'apparaît comme un système de géométrie. On aboutit à des théorèmes qu'il faut démontrer.

de Saussure (1911)
in Godel (1957)

I think it is correct to say that the difference between the structural linguist and the formal logician is one of stress and degree rather than of kind.

Bar-Hillel (1954)

The correct way to use the insights and techniques of logic is in formulating a general theory of linguistic structure.

Chomsky (1955)

La logique contemporaine est robuste et élégante. Mais elle est aussi maigre. Les logiciens contemporains sont tout à fait conscients de ce fait, et ils s'efforcent d'engraisser un peu la bête. Aristote et sa théorie des catégories auront peut-être toujours quelques conseils à offrir.

Barnes (2005)

Preface

Audience

This book is intended for students (from the third year) in computer science, formal linguistics, and mathematical logic as well as for colleagues interested in categorial grammars and their logical foundations. Though the logical viewpoint of categorial grammars is well-established, until now there has been no textbook available to students and researchers who want to discover all the main results for categorial grammars. It is the goal of this book to fill this gap.

The book has few formal prerequisites; nevertheless, notions of mathematical logic will help.

In addition, some familiarity with proof theory and the typed λ-calculus will help, and for readers who feel they need additional background we recommend (Girard et al, 1988; Girard, 2011), and (Girard, 1995, 2011) for more details on linear logic.

Occasionally, we will assume the reader knows what context-free grammar is, and is aware of their elementary properties such as Chomsky and Greibach normal forms. If some background on formal language theory is needed, we recommend (Hopcroft and Ullman, 1979; Harrison, 1978)

Finally, for some background in linguistics, we recommend (Akmajian et al, 2001).

Nevertheless we have made every effort to make this book as self-contained as possible, and none of this additional background knowledge is strictly mandatory.

The exercises at the end of every chapter, sometimes with hints, should help the reader assimilate the new notions and check that they are well understood.

Contents

These lecture notes present categorial grammars as deductive systems, in the approach called parsing-as-deduction, and we include detailed proofs of their main properties. We provide background and motivation for all results, modern proofs of many of the classical results for categorial grammars, together with our own results, in particular on proof nets.

Chapter 1: Our book starts with AB grammars (Ajdukiewicz-Bar-Hillel) also called classical categorial grammars, basic Lambek grammars... We establish their correspondence with context-free grammars, and briefly present their learnability properties.

Chapter 2: Lambek's syntactic calculus is at the center of this book, and despite all the years it is still a pleasure to read Lambek's original article (Lambek, 1958), written with an elegance which is rarely encountered. In this chapter, we present the correspondence between Lambek grammars and context-free grammars following Pentus, (Pentus, 1993b), and study the structure of the parse-structures/deductions: normalization, sub formula property, decidability, interpolation.

Chapter 3: Here, we turn our attention to one of the important applications of categorial grammar: how to automatically turn a syntactic analysis, that is a deduction, into a lambda term, which combined with lexical lambda terms, produces a logical formula corresponding to the meaning of the analyzed sentence. As we will see, the answer to this question is very simple and builds upon a fundamental result from logic and lambda-calculus: the Curry–Howard isomorphism. We also discuss several applications of this correspondence in the context of Montague semantics.

Chapter 4: The non-associative Lambek calculus NL is a variant of the Lambek calculus where the objects of study are *trees* of formulas instead of lists of formulas. We also discuss Kripke models and polynomial parsing algorithms for NL.

Chapter 5: The multimodal Lambek calculus extends the non-associative Lambek calculus by introducing controlled versions of the structural rules of associativity and commutativity to the logic. Some illustrative examples show how multimodality allows us to move beyond context-free languages.

Chapter 6: In this chapter, we present proof nets for the Lambek calculus, graphs that better describe a deduction than the usual trees, since no two proof nets correspond to the same linguistic analysis. They also enable different parsing techniques, and they can even be used to measure the complexity of human processing.

Chapter 7: In the final chapter, we combine ideas from Chaps. 5 and 6 to give proof nets for the multimodal Lambek calculus, emphasizing their usefulness for parsing categorial grammars. We also discuss Grail, an implementation of multimodal grammars based on proof nets.

How to Read This Book

As with most books, the simplest way to read this one is from start to finish. However, to help the reader who is eager to get to one of the later chapters as soon as possible or the teacher who has only time for a limited amount of course material, Fig. 0.1 gives a list of chapter dependencies.

The "main track" is listed on the left of the figure: and it represents the minimal dependencies required to continue to the next chapter. Having read the core sections,

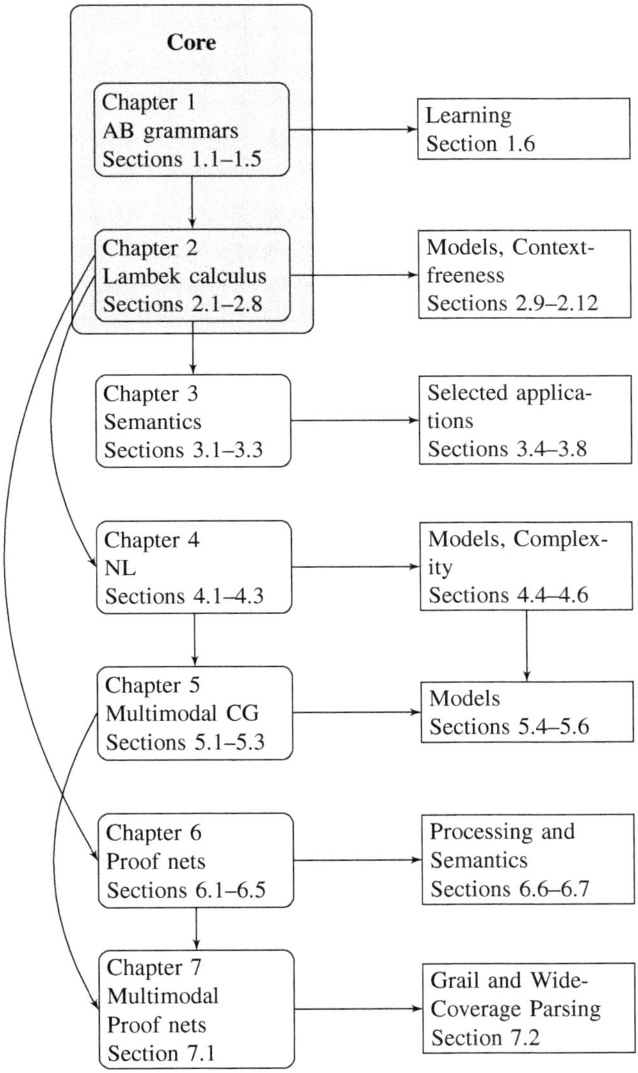

Fig. 0.1. Chapter dependencies

in the gray-shaded box, the reader can continue reading in a number of ways: either continue onward to the chapter on Montague Semantics or jump directly to the non-associative Lambek calculus or to the proof net chapter. It should be noted that this main track is rather restricted and that many fundamental results are found in the "optional" track on the right of the figure — which includes important topics such as model theory, Pentus's context-freeness result for the Lambek calculus and formal learning theory, but which are sections that the reader can skip without it affecting his comprehension of later chapters. The goal of Fig. 0.1 is to allow the eclectic teacher, researcher, or student pick his most efficient path through the material in this book to the topics which interest him most.

Further Reading

For further general references on the logical view of categorial grammars we recommend two chapters of the Handbook of *Logic and Language* (van Benthem and ter Meulen, 1997), namely (Buszkowski, 1997; Moortgat, 1997), and the revised and extended chapter (Moortgat, 2011) in (van Benthem and ter Meulen, 2011) as well as the article on type-logical grammar on the Stanford Encyclopedia of Philosphy website (Moortgat, 2010).

Other recommended recent references, which are in many ways complementary to the current textbook, are Glyn Morrill's textbook (Morrill, 2011), which contains a wealth of material on linguistic applications and on proof nets and processing (as discussed briefly in our Sect. 6.6) and Richard Oehrle's introduction to multimodal categorial grammar (Oehrle, 2011).

Finally, let us say that our logical view of categorial grammar is not the only one, and regarding the combinatorial view of categorial grammars we recommend (Steedman, 1997).

Recent Advances in Categorial Grammars

Though this book discusses only the classical results, categorial grammars in the tradition of the Lambek calculus are still an active area of research. As an indication of some recent developments, we mention only the Lambek–Grishin calculus (Moortgat, 2009; Bernardi and Moortgat, 2010) and the Discontinous Lambek calculus (Morrill, 2011; Morrill et al, 2011). The background provided by the current book should be more than sufficient to allow the interested reader to follow these recent developments.

In the mid-1990s, the introduction of the minimalist program moved mainstream syntax closer to categorial grammars. Working with a formalization of minimalist grammars introduced by Stabler (1997), Lecomte, Retoré and some of their students, Amblard, Anoun, have been able to make this correspondence precise by introducing Categorial Minimalist Grammars, within the parsing-as-deduction paradigm, thus enriching minimalist grammars with a neat computational semantics (Lecomte and Retoré, 1999; Lecomte, 2011).

Let us also mention a different syntactic formalism, the *abstract categorial grammars* of de Groote (2001). They rely on the representation of trees and strings in the simply typed lambda calculus (as opposed to using the structure of the antecedent to represent trees and strings as we do here). Thus, it is closer to the interface between categorial syntax and semantics discussed in our chapter on Montague semantics than to usual categorial grammar with parsing-as-deduction. Abstract categorial grammars define interesting classes of formal languages, with hierarchies that more or less match the ones of formal language theory, and there is a natural correspondence with semantics.

History of the Book

A first version of these notes was written by Christian Retoré, for a lecture on "The Logic of Categorial Grammars" at ESSLLI 2000. The section on categorial grammar acquisition was added for an ACL 2001 tutorial.

Thereafter it was used and improved on various occasions including a lecture at ESSLLI 2003, a crash course at EALING 2003 and 2008, Master lectures in Bordeaux from 2002 to 2011, and in Verona in 2006 and 2010.

Richard Moot extended these early notes, in part for an ESSLLI 2004 course, by adding chapters on the non-associative Lambek calculus, the multimodal Lambek calculus and multimodal proof nets as well as by adding supplementary material and exercises to the other chapters.

Thereafter we produced joint revisions, additions and updates of the book with an independent chapter on Montague semantics.

Acknowledgments

Introductions are always difficult to write and we benefited from the solid advice of Eco (1987) to guide us.

We would like to thank the editorial team at Springer for their support and encouragement during the long gestation period of this book — read Chap. 23 of (Kahneman, 2011) for some thoughtful comments on estimated times until the completion of a book! — and the anonymous Springer referees for their insightful and thorough comments.

We would like to thank our employers, CNRS and Université de Bordeaux gathered in our LaBRI lab, as well as the projects supported by LaBRI, INRIA and the Conseil Régional d'Aquitaine *Grammaire du Français* and *Itipy*. Further financial support was provided by the ANR projects *Prélude* and *Loci*. The current collaboration started with the creation of our research group SIGNES (LABRI-CNRS and INRIA), and special thanks go to their respective directors Serge Dulucq and Claude Puech for their support in creating the team.

Special thanks also go to our proofreaders Ivano Ciardelli and Noémie-Fleur Sandillon-Rezer, who spotted many typos and other mistakes and suggested several improvements and clarifications.

We would also like to thank the following colleagues, course participants and readers for their comments on the lectures and on earlier drafts of the current lecture notes: Anne Abeillé, Vito Michele Abrusci, Maxime Amblard, Houda Anoun, Nicholas Asher, Christian Bassac, Denis Béchet, Nicolas Belgolo, Claire Beyssade, Philippe Blache, Roberto Bonato, Pierre Bourreau, Joan Busquets, Claudia Casadio, Pierre Castéran, Lionel Clément, Francis Corblin, Bruno Courcelle, Dick Crouch, Laurence Danlos, Denis Delfitto, Alexandre Dikovsky, Luca Ducchesi, Gaetano Fiorin, Marie-Renée Fleury, Annie Foret, Christophe Fouqueré,

Nissim Francez, Sean Fulop, Claire Gardent, Christoper Götze, Herman Hendriks, Patrick Henry, Gérard Huet, Dan Klein, Greg Kobele, Joachim Lambek, Anaïs Lefeuvre, Yannick Le Nir, Alda Mari, Jean-Yves Marion, Renaud Marlet, Ralf Matthes, Chiara Melloni, Bruno Mery, Laurent Miclet, Glyn Morrill, Philippe Muller, Reinhard Muskens, David Nicolas, Guy Perrier, Daniele Porello, Laurent Prévot, Myriam Quatrini, Sylvain Salvati, Philippe Schlenker, Géraud Sénizergues, Edward Stabler, Isabelle Tellier, Jacopo Terragrossa, Marc Tommasi, Maria Vender, Willemijn Vermaat, Natalia Vinogradova, Emilie Voisin and Gilles Zémor.

We would also like to thank the people with whom we have worked on material included in or related to this book especially Jean-Yves Girard, Alain Lecomte and Michael Moortgat and also Raffaella Bernardi, Philippe de Groote, François Lamarche, Richard Oehrle, Mario Piazza, Sylvain Pogodalla, and Quintijn Puite.

Last but not least, we have a special thought for one of our commentators who brought his enthusiasm to every ESSLLI: Paul Gochet, who passed away on June 21, 2011.

References

Akmajian, A., Demers, R.A., Farmer, A.K., Harnish, R.M.: Linguistics: An Introduction to Language and Communication, 5th edn. MIT Press (2001)

Bar-Hillel, Y.: Logical syntax and semantics. Language 30(2), 230–237 (1954)

Barnes, J.: Les catégories et les Catégories. In: Bruun, O., Corti, L. (eds.) Les Catégories Et Leur Histoire. Histoire de la philosophie, pp. 11–80. Vrin (2005)

van Benthem, J., ter Meulen, A. (eds.): Handbook of Logic and Language. North-Holland Elsevier, Amsterdam (1997)

van Benthem, J., ter Meulen, A. (eds.): Handbook of Logic and Language, 2nd edn. North-Holland Elsevier, Amsterdam (2011)

Bernardi, R., Moortgat, M.: Continuation semantics for the Lambek-Grishin calculus. Information and Computation 208, 397–416 (2010)

Buszkowski, W.: Mathematical linguistics and proof theory. In: van Benthem and ter Meulen, ch. 12, pp. 683–736 (1997)

Chomsky, N.: Logical syntax and semantics: their linguistic relevance. Language 31(1), 36–45 (1955)

Eco, U.: Come scrivere un'introduzione. L'espresso – La bustina di Minerva Ristampato in Il secondo diario minimo, Bompiani, 1992, pp. 105–106 (1987); English translation: How to write an introduction. In: How to Travel with a Salmon & Other Essays. Harcourt Brace (1994)

Girard, J.Y.: Linear logic: its syntax and semantics. In: Girard, J.Y., Lafont, Y., Regnier, L. (eds.) Advances in Linear Logic. London Mathematical Society Lecture Notes, vol. 222, pp. 1–42. Cambridge University Press (1995)

Girard, J.Y.: The Blind Spot: Lectures on Logic. European Mathematical Society (2011)

Girard, J.Y., Lafont, Y., Taylor, P.: Proofs and Types. Cambridge Tracts in Theoretical Computer Science, vol. 7. Cambridge University Press (1988)

Godel, R. (ed.): Les sources manuscrites du Cours de linguistique générale de F. de Saussure. E. Droz (1957)

de Groote, P.: Abstract categorial grammars. In: Proceedings of the 39th Annual Meeting of the Association for Computational Linguistics, ACL 2001. ACL, Toulouse (2001)

Harrison, M.A.: Introduction to Formal Language Theory. Addison Wesley (1978)

Hopcroft, J.E., Ullman, J.D.: Introduction to Automata Theory, Languages and Computation. Addison-Wesley, Reading (1979)

Kahneman, D.: Thinking, Fast and Slow. Farrar, Straus and Giroux (2011)

Lambek, J.: The mathematics of sentence structure. American Mathematical Monthly, 154–170 (1958)

Lecomte, A.: Meaning, Logic and Ludics. Imperial College Press (2011)

Lecomte, A., Retoré, C.: Towards a minimal logic for minimalist grammars: a transformational use of Lambek calculus. In: Formal Grammar, FG 1999, pp. 83–92. FoLLI (1999)

Moortgat, M.: Categorial type logics. In: van Benthem and ter Meulen, ch. 2, pp. 93–177 (1997)

Moortgat, M.: Symmetric categorial grammar. Journal of Philosophical Logic 38(6), 681–710 (2009)

Moortgat, M.: Typelogical grammar. Stanford Encyclopedia of Philosohpy Website (2010), `http://plato.stanford.edu/entries/typelogical-grammar/`

Moortgat, M.: Categorial type logics. In: van Benthem and ter Meulen, ch. 2, pp. 95–179 (2011)

Morrill, G.: Categorial Grammar: Logical Syntax, Semantics, and Processing. Oxford University Press (2011)

Morrill, G., Valentín, O., Fadda, M.: The displacement calculus. Journal of Logic, Language and Information 20(1), 1–48 (2011)

Oehrle, R.T.: Multi-modal type-logical grammar. In: Borsley, R., Börjars, K. (eds.) Non-transformational Syntax: Formal and Explicit Models of Grammar, ch. 6, pp. 225–267. Wiley-Blackwell (2011)

Pentus, M.: Lambek grammars are context-free. In: Logic in Computer Science. IEEE Computer Society Press (1993)

Stabler, E.P.: Derivational Minimalism. In: Retoré, C. (ed.) LACL 1996. LNCS (LNAI), vol. 1328, pp. 68–95. Springer, Heidelberg (1997)

Steedman, M.: Surface structure and interpretation. Linguistic Inquiry Monographs, vol. 30. MIT Press, Cambridge (1997)

Contents

1

Classical Categorial Grammars: AB Grammars

Summary. This first chapter deals with material from the late fifties and early sixties, but which nevertheless introduces the design of categorial grammars, which are lexcalized grammars, as opposed to the phrase structure grammars like context-free grammars that were introduced afterwards.

Although the success of phrase structure grammars went far beyond that of categorial grammars, their lexicalization was in fact an attractive feature, another one being their connection to logical semantics.

We end with more recent results: a learning algorithm for categorial grammars, which was proved to converge at the end of the nineties. Having a learning algorithm for a class of grammars which can describe (small parts of) natural language is, we think, quite an important feature of categorial grammars. It comes from their lexicalization and logical formulation, which will be further studied in the next chapter.

1.1 Semantic Categories and Ajdukiewicz Fractions

Though many of the ideas behind categorial grammars can be traced to the work of Husserl, Frege and Russell, we begin this introduction to categorial grammars with the work of Ajdukiewiz. For the history of categorial grammars, we refer the reader to (Casadio, 1988; Morrill, 2007). In 1935 Ajdukiewicz defined a calculus of fractions to test the correctness of logical statements (Ajdukiewicz, 1935):

> *The discovery of antinomies, and the method of their resolution have made problems of linguistic syntax the most important problems of logic (provided this word is understood in a sense that also includes meta-theoretical considerations). Among these problems that of syntactic connection is of the greatest importance for logic. It is concerned with the specification of the conditions under which a word pattern constituted of meaningful words, forms an expression which itself has a unified meaning (constituted, to be sure, by the meaning of the single words belonging to it). A word pattern of this kind is called syntactically connected.*

His paper deals with both the formal language of logic and natural language, but is actually more concerned with the language of propositional and predicate logic.

> *If one applies this index symbolism to ordinary language, the semantic categories which we have assumed (in accordance with Leśniewski) will not always suffice, since ordinary languages are richer in semantic categories.*

Each word (or lexical entry) is provided with an index[1] which is a category. Categories are defined inductively as follows:

Basic categories. The two primitive types n (for entities or individuals or first order terms) and s (for propositions or truth values) are categories

Fractions. Whenever N is a category and $D_1, \ldots, D_p$ is a sequence or multiset of categories, then $\frac{N}{D_1 \cdots D_p}$ is itself a category. These complex categories are called functor categories or fractions.

If we formalize the definitions in his article, syntactically connected expressions and their exponents[2] are recursively defined as follows:

- a word or lexical entry is syntactically connected, and its exponent is its index.
- given
  - n syntactically connected expressions $d_1, \ldots, d_n$ of respective exponents $D_1, \ldots, D_n$
  - an expression f of exponent $\frac{N}{D_1 \cdots D_n}$

  the expression $f d_1 \cdots d_n$ (or any permutation of it) is syntactically connected and has exponent N.

This in particular entails that sequences of fractions reduce to a single index using the usual simplifications for fractions. It should be observed that in this "commutative setting" the simplification procedure of the fractions is not that simple if the bracketing corresponding to subexpressions is not given. As Ajdukiewicz is mainly concerned with the language of logic where one can use the Polish notation, word order is not really a problem for him.

1.2 Classical Categorial Grammars or AB Grammars

In 1953, that is a bit before Chomsky introduced his hierarchy of Phrase Structure Grammars (Chomsky, 1955), Bar-Hillel defines bidirectional categorial grammars (Bar-Hillel, 1953), refining Ajdukiewicz types to take constituent order into account. Therefore, his grammars are more adequate for modeling natural language, where

---

[1] Ajdukiewicz uses the word "index" interchangeably with the word "category".

[2] Ajdukiewicz uses the word "exponent" to mean the *result* category of an expression after reduction rules. For example, the exponent of $\frac{s}{n}\ n$ is s, which should be reminiscent of the simplification of a fraction which is multiplied by its denominator in elementary mathematics: $\frac{A}{B} \times B$ simplifying to A.

word order is crucial. We will use Lambek's notation for types (Lambek, 1958), following Bar-Hillel in his later work on categorial grammars.

In the literature, these grammars are called *AB grammars*, *classical categorial grammars* or *basic categorial grammars*.

Types or fractions are defined as follows:

$$\mathsf{L} \quad ::= \quad \mathsf{P} \;\mid\; (\mathsf{L} \setminus \mathsf{L}) \;\mid\; (\mathsf{L} \,/\, \mathsf{L})$$

where P is the set of primitive types, which we will also call atomic types or basic categories, which usually contains S (for sentences) np (for noun phrases) and n (for nouns), and may include pp (for prepositional phrase) inf (for infinitives) etc.

We will often omit the outer brackets when this does not lead to confusion and write $np \setminus S$ instead of $(np \setminus S)$ and $(np \setminus S) / np$ instead of $((np \setminus S) / np)$.

A note on the terminology used throughout this book (and much of the literature on categorial grammars): we often use the terms category, formula and (syntactic) type interchangeably. Since we will discuss *semantic* types only in Chapter 3, unless otherwise indicated, the word 'type' will mean *syntactic* type (or category or formula, that is a member of L).

It is usual to talk about a formula of type $B \setminus A$ or A / B as a *functor*, the formula B as its *argument* and the formula A as its *result*. Speaking colloquially, we will say a formula $B \setminus A$ or A / B (the functor) selects a B (the argument) to form an A (the result). As we will see in Chapter 3, talking about functions and arguments is not just a convenient way to refer to syntactic combinations, the function/argument distinction has direct semantic import. Finally, types of the form $A \setminus A$ or A / A are often called *modifiers* or *A-modifiers*. They take an argument of type A to produce a result of the same type.

The grammar is defined by a lexicon, that is a function Lex which maps words or terminals to finite sets of types (a set of types is needed, since in natural language a single word may admit various constructions: "*eat*" may ask for an object or not, for instance).

An expression, that is a sequence of words or terminals $w_1 \cdots w_n$, is of type u whenever there exists for each w_i a type t_i in $\mathrm{Lex}(w_i)$ such that $t_1 \cdots t_n \longrightarrow u$ with the following reduction patterns:

$$\forall u, v \in \mathsf{L} \qquad \begin{array}{ll} u\,(u \setminus v) \longrightarrow v & \quad (\setminus_e) \\ (v / u)\, u \longrightarrow v & \quad (/_e) \end{array}$$

These rules are called elimination rules, or simplifications, or modus ponens.

An application of an elimination rule is defined as for context-free grammars: inside a sequence of types, we replace the left-hand side of an elimination rule by its right-hand side, in other words, if Γ and Γ' are lists of types then applications of the elimination rules rewrite $\Gamma\ u\ u \setminus v\ \Gamma'$ to $\Gamma\ v\ \Gamma'$ (for $\setminus_e$) and $\Gamma\ v / u\ u\ \Gamma'$ to $\Gamma\ v\ \Gamma'$ (for $/_e$). These rule applications should be compared to the "rewrites immediately" relation for context-free grammars which we will discuss below on page 6.

These rules provide the symbols $\setminus$ and $/$ with an intuitive meaning: an expression y is of type $A \setminus B$ whenever it needs an expression a of type A on its left to obtain an expression ay of type B; symmetrically, an expression z is of type B / A whenever it needs an expression a of type A on its right to obtain an expression za of type B;

The set of sentences or the language generated by the grammar is the set of word sequences of type S.

The derivation tree is simply a binary tree whose leaves are the types t_i and whose nodes are labeled by rules $/_e$ and $\backslash_e$.

Example 1.1 (A Tiny AB Grammar)
Consider the following lexicon:

Word	Type(s)	Translation
cosa	$(S / (S / np))$	*what*
guarda	(S / inf)	*he/she watches*
passare	(inf / np)	*passing by*
il	(np / n)	*the*
treno	n	*train*

The sentence '*guarda passare il treno*' (he/she watches the train passing by) belongs to the generated language:

$$\begin{array}{l} (S / inf) \quad (inf / np) \quad (np / n) \quad n \\ \longrightarrow (S / inf) \quad (inf / np) \quad np \\ \longrightarrow (S / inf) \quad inf \\ \longrightarrow S \end{array} \tag{1.1}$$

The derivation tree for this analysis can be written as:

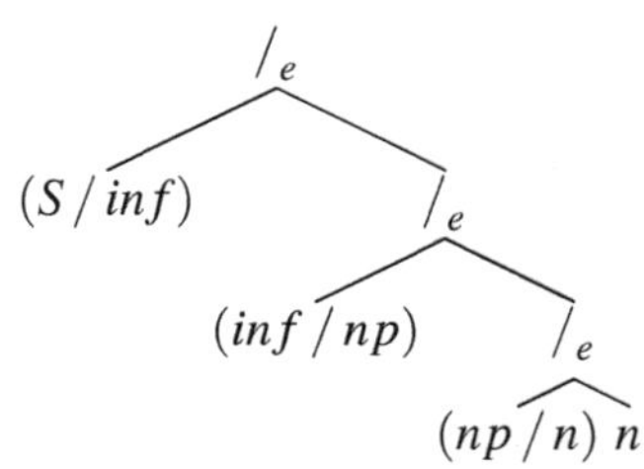

(1.2)

A final way of presenting the same analysis is shown below. Compared to the analysis above, it is written upside-down. It uses explicit Lex rules and gives the result category of each rule application.

$$\dfrac{\dfrac{\text{guarda}}{S / inf}\,\text{Lex} \qquad \dfrac{\dfrac{\text{passare}}{inf / np}\,\text{Lex} \qquad \dfrac{\dfrac{\text{il}}{np / n}\,\text{Lex} \qquad \dfrac{\text{treno}}{n}\,\text{Lex}}{np}\,/_e}{inf}\,/_e}{S}\,/_e \tag{1.3}$$

Though derivation 1.3 appears more detailed, it is actually equivalent to derivation 1.2: the rule name and the daughter categories suffice to deduce the result category in a bottom-up way, from the leaves to the root node. Derivation 1.3 corresponds to the natural deduction derivations we will see in Section 2.2.1. Derivation 1.2 corresponds to the derivation format used for the learning algorithms in Section 1.6. Finally, the flat structure of derivation 1.1 is close to the derivations of context-free grammars, a point we will explore further in the next section.

Returning to the lexicon above, we remark that the sentence '*cosa guarda passare*' (what is he/she looking passing by?) does not belong to the generated language: indeed the sequence

$$(S\,/\,(S\,/\,np))\;(S\,/\,inf)\;(inf\,/\,np)$$

does not contain anything that could be reduced.

It should be observed that AB grammars are lexicalized: that is to say, the grammar consists of the following two components:

1. a universal set of rules which is common to all languages ("universal grammar" in the terminology of (Chomsky, 1995); for AB grammars these are just the two reduction patterns $\backslash_e$ and $/_e$ above)
2. a lexicon, which is responsible for all differences between individual languages.

As such, the universal grammar — and its formal properties — becomes an object of study in itself. This lexicalist view is coherent with many modern linguistic theories, like the minimalist program of Chomsky (Chomsky, 1995) (language variation is only lexical), and with some formalisms for computational linguistics, like (Lexicalized) Tree Adjoining Grammars (Joshi et al, 1975; Joshi and Schabes, 1997) or Head-Driven Phrase Structure Grammars (Pereira and Shieber, 1987; Pollard and Sag, 1994).

Another observation is that the rules are like modus ponens, but in a logic where contraction and weakening are not allowed, and where the order of the hypothesis is taken into account (see Girard, 1995, for a clear and intuitive introduction to the structural rules seen from the perspective of linear logic). The relation between categorial grammars and (linear) logic will be a major theme of this book.

Let us state one of the first results on categorial grammars known as the Gaifman theorem of (Bar-Hillel et al, 1963) which is more or less equivalent to the existence of a Greibach normal form for context-free grammars:

Proposition 1.2. *Every AB grammar is equivalent to an AB grammar containing only types of the form*

$$p \qquad (p\,/\,q) \qquad ((p\,/\,q)\,/\,r)$$

where p, q, r stand for primitive types.

Proof. This theorem is an immediate consequence of Propositions 1.11 and 1.10 to be proved below using the well-known Greibach normal form theorem (Greibach, 1965). □

1.3 AB Grammars and Context-Free Grammars

1.3.1 Context-Free Grammars

Context-Free Grammars (CFGs) were introduced in (Chomsky, 1955) and a good introduction is provided in (Hopcroft and Ullman, 1979); we use the following standard notation:

- M^* stands for the set of finite sequences over the set M.
- M^+ stands for the set of finite non empty sequences over the set M.
- ε stands for the empty sequence of M^*.

Definition 1.3 (Context-free grammar). *A context-free grammar is defined by:*

[Non Terminals] *a set NT of symbols called non terminals, one of them being the start symbol S.*
[Terminals] *set T of symbols, disjoint from NT, called terminals (or words according to the linguistic viewpoint)*
[Production rules] *a finite set of production rules of the form $X \longrightarrow W$ with $X \in NT$ and $W \in (T \cup NT)^*$*

We say a a context-free grammar is lexicalized *if each production rule contains a member of T on its right hand side.*

A sequence $V \in (T \cup NT)^*$ is said to rewrite immediately into a sequence $W \in (T \cup NT)^*$ whenever there exists $W', W'', W''' \in (T \cup NT)^*$ and a non terminal X such that

- $V = W'XW'''$
- $X \longrightarrow W''$ is a production rule.
- $W = W'W''W'''$

The relation $\longrightarrow$ is defined over sequences in $(T \cup NT)^*$ as the transitive closure of "rewrites immediately into". The language generated by a CFG is the smallest subset of T^* containing the sequences into which S rewrites.

Two grammars which generate the same languages are said to be weakly equivalent.

Whenever a non-terminal N rewrites into a sequence of terminals and non terminals $X_1, \ldots, X_n$ it is possible (as linguists often do) to denote the derivation tree by a term T representing this derivation tree as follows:

- a non terminal or a terminal is a derivation tree and its yield is itself.
- if $T_1, \ldots, T_n$ are derivation trees of $X_1, \ldots, X_n$ and if $X \longrightarrow X_1 \cdots X_n$ is a rule of the grammar then $[_X T_1, \ldots, T_n]$ is a derivation tree labeled X and its yield is the concatenation of the yields of $T_1, \ldots, T_n$.

Obviously a sequence of terminals $a_1 \cdots a_n$ is in the language if and only if there exists a derivation tree labeled S the yield of which is $a_1 \cdots a_n$. We denote by ε the empty sequence.

Two grammars which generate the same derivation *trees* are said to be strongly equivalent.

Definition 1.4. *A CFG is said to be ε-free whenever ε does not belong to the generated language.*

It is not difficult to decide whether a CFG is ε-free or not, and if it is not ε-free, the grammar can be written with the production rules of an ε-free CFG, together with the rule: $S \longrightarrow \varepsilon$.

Definition 1.5 (Chomsky normal form). *A CFG is said to be in Chomsky normal from whenever its production rules are of the form $X \longrightarrow Y\,Z$ or of the form $X \longrightarrow a$, with $X,Y,Z \in NT$ and $a \in T$.*

Proposition 1.6. *Any ε-free CFG can be transformed into a weakly equivalent CFG in Chomsky normal form and this transformation can be performed in polynomial time (see Chomsky, 1963; Hopcroft and Ullman, 1979).*

Definition 1.7 (Greibach normal form). *A CFG is said to be in Greibach normal form whenever its production rules are of the form: $X \longrightarrow aX_1 \cdots X_n$ with $a \in T$, $X,X_1,\ldots,X_n \in NT$. It is said to be in strong Greibach normal form whenever $n \leq 2$.*

Proposition 1.8. *Any ε-free CFG can be turned into a CFG in (strong) Greibach normal form, and these transformations can be performed in polynomial time (see Greibach, 1965; Harrison, 1978).*

Transforming a CFG into its Greibach normal form is a way of lexicalizing this grammar, since each rule of a grammar in Greibach normal form contains a terminal. While the derivation trees of a CFG and the ones of its Chomsky normal form are closely related, the derivation trees of the Greibach normal from of a CFG can be very different from the derivation trees of the original CFG: to lexicalize a CFG while preserving the analyses, one has to move to TAGs (Schabes and Waters, 1993; Joshi and Schabes, 1997).

1.3.2 From Context-Free Grammars to AB Grammars

The relationship between CFG and AB grammars was the subject of a detailed investigation in the early sixties (Bar-Hillel et al, 1963).

Proposition 1.9. *Every ε-free Context-Free Grammar in Greibach normal form is equivalent to an AB categorial grammar.*

Proof. Let us consider the following AB grammar:

- Its words are the terminals of the CFG.
- Its primitive types are the non terminals of the CFG.
- Lex(a), the finite set of types associated with a terminal a contains the formulae $((\cdots((X \,/\, X_n) \,/\, X_{n-1}) \,/\, \cdots) \,/\, X_2) \,/\, X_1$ such that there are non terminals $X,X_1,\ldots,X_n$ such that $X \longrightarrow aX_1 \cdots X_n$ is a production rule.

It is then easily observed that the derivation trees of both grammars are isomorphic. □

Proposition 1.10. *Each ε-free context-free grammar is weakly equivalent to an AB grammar containing only types of the form X or X / Y or $(X / Y) / Z$.*

Proof. Here we provide the reader with a simple "modern proof" using the existence of a Greibach normal from: indeed the Gaifman theorem first published in (Bar-Hillel et al, 1963) was proved before the existence of Greibach normal form for CFGs (Greibach, 1965), and these two theorems are actually more or less equivalent.

According to Proposition 1.8, any context-free grammar can be turned into a weakly equivalent CFG in strong Greibach normal form. As can be observed from the construction of an equivalent AB grammar in the previous proof, if the CFG is in strong Greibach normal form that is if rules are of the form: $X \longrightarrow aX_1 \cdots X_n$ with $0 \leq n \leq 2$, then the corresponding AB grammar only uses types of the form X, X / X_1, $(X / X_2) / X_1$. □

1.3.3 From AB Grammars to Context-Free Grammars

Proposition 1.11. *Every AB grammar is strongly equivalent to a CFG in Chomsky normal form.*

Proof. Let G be the CFG defined by:

- Terminals T are the words of the AB grammar.
- Non Terminals NT are all the subtypes of the types appearing in the lexicon of the AB grammar — a type is considered to be a subtype of itself.
- The production rules are of two kinds:
  - $X \longrightarrow a$ whenever $X \in \mathrm{Lex}(a)$
  - $X \longrightarrow (X / Z)\ Z$ and $X \longrightarrow Z\ (Z \backslash X)$ for all $X, Z \in NT$ — keep in mind that from the CFG viewpoint $(Z \backslash X)$ and (X / Z) are both non terminal symbols.

This defines a CFG because the lexicon is finite, so there are only finitely many subtypes of types in the lexicon, hence finitely many production rules. The derivation trees in both formalisms are isomorphic. □

1.4 Parsing AB Grammars

Theorem 1.12. *A sentence of n words can be analyzed using an AB grammar in $O(n^3)$ time using $O(n^2)$ space.*

Proof. (easy exercise) Following the relation between AB grammars and CFGs in Chomsky normal form, it is not difficult to adapt the Cocke Kasami Younger algorithm (see e.g. Sikkel and Nijholt, 1997) to AB grammars. □

1.5 Limitations of AB Grammars

In an AB grammar one is not able to derive $(t\,/\,v)$ from $(t\,/\,u)$ and $(u\,/\,v)$. Consider for instance the Italian sentence '*Cosa guarda passare?*' we've seen in Example 1.1. One is not able to derive it with the simple type assignments given there. We would need transitivity of / to obtain it:

$$(S\,/\,(S\,/\,np)) \quad (S\,/\,inf) \quad (inf\,/\,np) \quad \stackrel{(trans.)}{\longrightarrow} \quad (S\,/\,(S\,/\,np)) \quad (S\,/\,np) \quad \longrightarrow \quad S$$

We would also like to model the behavior of an object relative pronoun like *that/whom*, by providing it with the type $(n\setminus n)\,/\,(S\,/\,np)$ but unfortunately this too requires transitivity — unless a transitive verb also has the type $np\setminus(S\,/\,np)$, but it is rather unusual to analyze English verbs as combining first with their subjects and then with their objects, and, in our view, it is rather unnatural to require the *verb* to have different types when combining with a subject relative pronoun $(n\setminus n)\,/\,(np\setminus S)$ and when combining with an object relative pronoun $(n\setminus n)\,/\,(S\,/\,np)$.

On the mathematical side, one would like to interpret categories by subsets of a free monoid (the intended one being sequences of words), so that the subset of sequences of type S are precisely the correct sentences. This is indeed impossible: one may view the elimination rules as modus ponens, but then what is lacking are introduction rules to get completeness of the calculus with respect to this natural monoidal interpretation. This is solved by the Lambek calculus which we will study in the next chapter.

1.6 Learning AB Grammars

We will end our study of AB grammars with an interesting property: they enjoy good learning algorithms from positive examples, at least when examples are structured. This learning question is important for the following two reasons:

- It models, although very roughly, the process of language acquisition and more precisely of syntax acquisition (Gleitman and Newport, 1995; Pinker, 1995) extensively discussed in generative linguistics; indeed, it is the main justification for the existence of a *universal grammar* see e.g. (Chomsky, 1995).
  - The similarity with natural language acquisition by human beings, is that we only learn from positive examples, and that structure is needed for the language learner.
  - The main difference is that the sequence of languages which converges to the target language is increasing — meaning the learner starts with a set of rules which generate a subset of the target language, seen as a set of sentences, and generalizes this set of rules step by step — while in natural language acquisition the sequence of languages is decreasing — meaning the learner starts with a set of rules which generate a superset of the target language and constrains this set of rules.

- This learning algorithm provides a method for the automated construction of a grammar (that is a lexicon) from a corpus, which also can be viewed as an automated method for completing an existing grammar/lexicon.

1.6.1 Grammatical Inference for Categorial Grammars

Learning (also called grammatical inference) from positive examples is the following problem: define a function Learn from finite sets of positive examples to grammars of a given class $\mathcal{G}$, such that:

- Given a grammar G of the class $\mathcal{G}$ and an enumeration $s_1, s_2, \ldots$ of the sentences G generates, letting $\mathrm{Ex}_i = \{s_1, \ldots, s_i\}$, there exists an N such that for all $n \geq N$ the grammar $\mathrm{Learn}(\mathrm{Ex}_n)$ is constant and exactly generates the sentences produced by G.
- The following is not mandatory, but one usually asks for this extra property: for every set of sentences Ex the grammar $\mathrm{Learn}(\mathrm{Ex})$ generates all the examples in Ex

This definition is the so-called identification in the limit introduced by Gold in 1967 (Gold, 1967). The grammars we are to consider are of course AB grammars, but what will the positive examples be? In our definition the term "sentence" is left vague. Actually we shall use this definition not with mere sequences of words, but we will rather consider the derivation trees produced by the grammar, and so our examples will be derivation trees in which the types of the words are absent: this is not absolutely unrealistic, because the learner of a language has access to some information related to the syntactic structure of the sentences like prosody or semantics; nevertheless it is unrealistic, because the complete syntactic structure is not fully known.

The lexicalization of categorial grammars is extremely helpful for this learning question: indeed we have no rules to learn, but only the types of the words to guess. Observe that we need to bound the number of types per word; otherwise each new occurrence of a word may lead to the introduction of a new type for this word, and this process cannot converge.

As this presentation is just meant to give an idea of learning algorithms, we only present here the simplest case of learning from structures: the algorithm RG of Buszkowski and Penn. The AB grammars considered are rigid, that is to say there is exactly one type per word.

1.6.2 Unification and AB Grammars

The algorithm makes use of type-unification, and this kind of technique is quite common in grammatical inference (see Nicolas, 1999), so let us briefly define it and explain its relation to AB grammars. First, we will consider an extended formula language for AB which has a countable number of type variables, we will use x, y, x_1, x_2, …, y_1, y_2, … to denote type variables. These variables will play much the same role as the atomic formulas, with the exception that we can substitute formulas for type variables. Given a substitution σ', a function from variables to types, we can extend σ' to a substitution σ, a function from types to types as follows (in the definition below, x denotes any type variable in the formula language).

$$\begin{aligned}
\sigma(S) &= S \\
\sigma(x) &= \begin{cases} \sigma'(x) & \text{if } \sigma'(x) \text{ is defined} \\ x & \text{otherwise} \end{cases} \\
\sigma(A \setminus B) &= \sigma(A) \setminus \sigma(B) \\
\sigma(B \,/\, A) &= \sigma(B) \,/\, \sigma(A)
\end{aligned}$$

Given a substitution σ, we can apply it to the lexicon of an AB grammar. If a sentence is generated by an AB grammar defined by a lexicon Lex then it is also generated by the AB grammar defined by the lexicon $\sigma(\text{Lex})$.[3]

A substitution is said to unify a set of types T if for all types A, B in T one has $\sigma(A) = \sigma(B)$. For such kinds of formulae, whenever a unifier exists, there exists a most general unifier (mgu) that is a unifier σ_u such for every unifier τ there exists a substitution σ_τ such that $\tau = \sigma_\tau \circ \sigma_u$.

The relation between two rigid AB grammars with respective lexicons Lex and Lex$'$ defined by *there exists a substitution* σ *such that* $\text{Lex}' = \sigma(\text{Lex})$ defines an order which is a complete lattice, and the supremum of a family corresponds to the least general grammar generating all the trees of all the grammars in the family.

1.6.3 The RG Algorithm

We present here the RG algorithm (learning Rigid Grammars) introduced by W. Buszkowski and G. Penn in (Buszkowski, 1987; Buszkowski and Penn, 1990) and which has been further studied by M. Kanazawa (Kanazawa, 1998).

To illustrate this algorithm, let us take a small set of positive examples:

(1.4) $[_{\setminus_e}[_{/_e}$ a man $]$ swims $]$
(1.5) $[_{\setminus_e}[_{/_e}$ a fish $][_{\setminus_e}$ swims fast $]]$

Typing. As the examples are assumed to be correct sentences, we know the root should be labeled by the type S which is the only type fixed in advance, a constant.

Each time there is a $\setminus_e$ (resp. $/_e$) node labeled y, we know the argument node, the one on the left (resp. on the right) should be x while the function node the one on the right (resp. on the left) should be $x \setminus y$ (resp. $y \,/\, x$)

So by assigning a new variable to each argument node we have typed the whole tree, and so words have been provided with a type (involving the added variables and S).

[3] This is a slight abuse of notation, but its intended meaning should be clear: we apply the substitution σ to the elements of the *range* of the function Lex, so, for any word w, if $\text{Lex}(w)$ is $\{f_1, \ldots f_n\}$ then $\sigma(\text{Lex})(w)$ is $\{\sigma(f_1), \ldots, \sigma(f_n)\}$.

We can do so on our examples; to denote the resulting type, we add it on top of the opening bracket.

(1.6) $[^{S}_{/e}\,[^{x_2}_{\backslash e}$ a:$(x_2\,/\,x_1)$ man:$x_1]$ swims:$(x_2 \backslash S)]$

(1.7) $[^{S}_{\backslash e}\,[^{y_2}_{/e}$ a:$(y_2\,/\,y_3)$ fish:$y_3][^{(y_2\backslash S)}_{\backslash e}$ swims:y_1 fast:$(y_1 \backslash (y_2 \backslash S))]]$

Unification. The previous steps give us several types per word. For instance the examples above yield:

Example 1.13

word	*type*1	*type*2
a:	$x_2\,/\,x_1$	$y_2\,/\,y_3$
fast:		$y_1 \backslash (y_2 \backslash S)$
man:	x_1	
fish:		y_3
swims:	$x_2 \backslash S$	y_1

We now have to unify the set of types associated with a single word, and the output of the algorithm is the grammar/lexicon in which every words gets the single type which unifies the original types, collected from each occurrence of a word in each example. If these sets of types can be unified, then the result of this substitution is a rigid grammar which generates all the examples, and can be shown to be the least general grammar to generate these examples.
In our example, unification succeeds and leads the most general unifier σ_u defined as follows:

Example 1.14

$$\begin{aligned}\sigma_u(x_1) &= z_1\\ \sigma_u(x_2) &= z_2\\ \sigma_u(y_1) &= z_2 \backslash S\\ \sigma_u(y_2) &= z_2\\ \sigma_u(y_3) &= z_1\end{aligned}$$

which yields the rigid grammar/lexicon:

Example 1.15

$$\begin{aligned}\text{a: } & z_2\,/\,z_1\\ \text{fast: } & (z_2 \backslash S) \backslash (z_2 \backslash S)\\ \text{man: } & z_1\\ \text{fish: } & z_1\\ \text{swims: } & z_2 \backslash S\end{aligned}$$

Convergence of the RG Algorithm

This algorithm converges in the sense we defined above, as shown by (Kanazawa, 1998). The technique also applies to learning rigid Lambek grammars from natural deduction trees (Bonato, 2000) and we follow his presentation.

For the proof of convergence, we make use of the following notions and notational conventions:

$G \subset G'$. This reflexive relation between G and G' holds whenever every lexical type assignment $a : T$ in G is in G' as well — in particular when G' is rigid, so is G, and both grammars are identical. Note that this is just the normal subset relation for each of the words in the lexicon G': $\mathrm{Lex}_G(a) \subset \mathrm{Lex}_{G'}(a)$ for every a in the lexicon of G', with $\mathrm{Lex}_G(a)$ non-empty. Keep in mind that in what follows we will also use the subset relation symbol to signify inclusion of the generated *languages*; the intended meaning should always be clear from the context.

size of a grammar. The size of a grammar is simply the sum of the sizes of the occurrences of types in the lexicon, where the size of a type is its number of occurrences of base categories (variables or S).

$G \sqsubset G'$. This reflexive relation between G and G' holds when there exists a substitution σ such that $\sigma(G) \subset G'$ which does not identify different types of a given word, but this is always the case when the grammar is rigid.

FA-structure. An FA-structure is a binary tree whose leaves are labeled with words (terminals) and internal nodes with names of the rules, namely $/_e$ and $\backslash_e$. An analysis in an AB grammar, once the types are erased, is an FA structure, and, conversely, for every type T, every FA structure can be labeled with types in order to obtain an analysis of the sequence of words as having category T — that's what the typing algorithm does, with $T = S$. The positive examples we are using for the RG learning algorithm, see Examples 1.4 and 1.5, are FA-structures.

$FL(G)$. Given a grammar G, $FL(G)$ is the tree language consisting of all the FA-structures with root S derived from G.

$GF(D)$. Given a set of FA-structures D, $GF(D)$ is the lexicon obtained by collecting the types of each word in the various examples of D — as in Example 1.13 above.

$RG(D)$. Given a set of examples D, $RG(D)$ is, whenever it exists, the rigid grammar/lexicon obtained by applying the most general unifier to $GF(D)$ — as in Example 1.15 above.

Proposition 1.16. *Given a grammar G, the number of grammars H such that $H \sqsubset G$ is finite.*

Proof. There are only finitely many grammars which are included in G, since G is a finite set of assignments. Whenever $\sigma(H) = K$ for some substitution σ the size of H is smaller or equal to the size of K, and, up to renaming, there are only finitely many grammars smaller than a given grammar.

By definition, if $H \sqsubset G$ then there exist $K \subset G$ and a substitution σ such that $\sigma(H) = K$. Because there are only finitely many K such that $K \subset G$, and for every K there are only finitely many H for which there could exist a substitution σ with $\sigma(H) = K$ we conclude that there are only finitely many H such that $H \sqsubset G$. □

Proposition 1.17. *If $G \sqsubset G'$ then $FL(G) \subset FL(G')$.*

Proof. $G \sqsubset G'$ means that there exists σ such that $\sigma(G) \subset G'$. Let T be an FA-structure in $FL(G)$, hence T comes from an analysis A of a sequence of words $m_1 \cdots m_n$. If we apply σ to A we obtain an analysis of the same sequence of words in G'. Indeed for a word whose assignment is T in G we have the assignment $\sigma(T)$ which is its assignment in G', and the types obtained inside the tree match the rules since $\sigma(A \setminus B) = \sigma(A) \setminus \sigma(B)$ and $\sigma(A / B) = \sigma(A) / \sigma(B)$ So $\sigma(A)$ is an analysis in G' of $m_1 \cdots m_n$. Hence, by definition, the FA-structure underlying $\sigma(A)$ is in $FL(G')$, and this underlying FA-structure is $\sigma(T)$. □

Proposition 1.18. *If $GF(D) \sqsubset G$ then $D \subset FL(G)$.*

Proof. By construction of $GF(D)$, we have $D \subset FL(GF(D))$. In addition, because of Proposition 1.17, we have $FL(GF(D)) \subset FL(G)$. □

Proposition 1.19. *If $RG(D)$ exists then $D \subset FL(RG(D))$.*

Proof. By definition $RG(D) = \sigma_u(GF(D))$ where σ_u is the most general unifier of all the types of each word. So we have $GF(D) \sqsubset RG(D)$, and applying Proposition 1.18 with $G = RG(D)$ we obtain $D \subset FL(RG(D))$. □

Proposition 1.20. *If $D \subset FL(G)$ then $GF(D) \sqsubset G$.*

Proof. By construction of $GF(D)$, there is exactly one occurrence of a given type variable x in a tree of D typed as done in the example. Now, viewing the same tree as a tree of $FL(G)$ at the place corresponding to x there is a type label, say T. Doing so for every type variable, we can define a substitution by $\sigma(x) = T$ for all type variables x: indeed because x occurs once, such a substitution is well defined. When this substitution is applied to $GF(D)$ it yields a grammar which only contains assignments from G — by applying the substitution to the whole tree, it remains a well-typed tree, and in particular the types on the leaves must coincide. □

Proposition 1.21. *When $D \subset FL(G)$ with G a rigid grammar, the grammar $RG(D)$ exists and $RG(D) \sqsubset G$.*

Proof. By Proposition 1.20 we have $GF(D) \sqsubset G$, so there exists a substitution σ such that $\sigma(GF(D)) \subset G$.

As G is rigid, σ unifies all the types of each word. Hence there exists a unifier of all the types of each word, and $RG(D)$ exists.

$RG(D)$ is defined as the application of most general unifier σ_u to $GF(D)$. By the definition of a most general unifier, which works as usual even though we unify sets of types, there exists a substitution τ such that $\sigma = \tau \circ \sigma_u$.

Hence $\tau(RG(D)) = \tau(\sigma_u(GF(D))) = \sigma(GF(D)) \subset G$;
thus $\tau(RG(D)) \subset G$, hence $RG(D) \sqsubset G$. □

Proposition 1.22. *If $D \subset D' \subset FL(G)$ with G a rigid grammar then $RG(D) \sqsubset RG(D') \sqsubset G$.*

Proof. Because of Proposition 1.21 both $RG(D)$ and $RG(D')$ exist. We have $D \subset D'$ and $D' \subset FL(RG(D'))$, so $D \subset FL(RG(D'))$; hence, by Proposition 1.21 applied to D and $G = RG(D')$ (a rigid grammar) we have $RG(D) \sqsubset RG(D')$. □

Theorem 1.23. *The algorithm RG for learning rigid AB grammars converges in the sense of Gold (see Section 1.6.1).*

Proof. Take $D_i, i \in \omega$ an increasing sequence of sets of examples in $FL(G)$ enumerating $FL(G)$, in other words $\cup_{i\in\omega} D_i = FL(G)$:

$$D_1 \subset D_2 \subset \cdots D_i \subset D_{i+1} \cdots \subset FL(G)$$

Because of Proposition 1.21 for every $i \in \omega$ $RG(D_i)$ exists and because of Proposition 1.22 these grammars define an increasing sequence of grammars w.r.t. $\sqsubset$ which by Proposition 1.21 is bounded by G:

$$RG(D_1) \sqsubset RG(D_2) \sqsubset \cdots RG(D_i) \sqsubset RG(D_{i+1}) \cdots \sqsubset G$$

As they are only finitely many grammars below G w.r.t. $\sqsubset$ (Proposition 1.16) this sequence is stationary after a certain rank, say N, that is, for all $n \geq N$ $RG(D_n) = RG(D_N)$.

We have $FL(RG(D_N)) = FL(G)$:

$FL(RG(D_N)) \supset FL(G)$ Let T be an FA-structure of $FL(G)$. Since $\cup_{i\in\omega} D_i = FL(G)$ there exists a p such that $T \in FL(D_p)$.

- If $p < N$, because $D_p \subset D_N, T \in D_N$, and by Proposition 1.19 $T \in FL(RG(D_N))$.
- If $p \geq N$, we have $RG(D_p) = RG(D_N)$ since the sequence of grammars is stationary after N. By Proposition 1.19 we have $D_p \subset FL(RG(D_p))$ hence $T \in FL(RG(D_N)) = FL(RG(D_p))$.

In all cases, $T \in FL(RG(D_N))$.

$FL(RG(D_N)) \subset FL(G)$ Since $RG(D_N) \sqsubset G$, by Proposition 1.17 we have

$$FL(RG(D_N)) \subset FL(G)$$

□

1.6.4 Other Cases

The learning problem covered by the RG algorithm is very simple and restricted.

Firstly, the class of grammars we are learning is quite limited:

1. They are AB grammars and not richer categorial grammars.
2. They are rigid, that is each word has only a single type of syntactic behavior. This limitation is not too difficult to overcome: different occurrences of the same word corresponding to different syntactic behaviors can be distinguished. This is sound when the occurrences correspond to words which are really different, such as *that* as a demonstrative and *that* as a complementizer, but it is less convincing when the word is the same like the transitive use of *eat* (*I ate an apple.*) and the absolutive use of *eat* (*I already ate*).

Secondly, we are using input structures which are not so easy to obtain, and which are probably too close to the output that we are looking for:

3. Parse structures (or FA structures) are much too precise; instead of having the complete tree structure labeled by rule applications, it would make more sense to have an *unlabeled* tree structure, partial information about the tree structure of the sentence or even just strings as input to the learning algorithms.

The base algorithm that we presented can be adapted in order to go beyond the limitations enumerated above.

1. A first extension is to learn Lambek grammars, which are discussed in the next chapter, from parse structures. Bonato (2000) shows that Lambek grammars are learnable from parse structures. The same learning mechanism works for minimalist grammars when they are viewed as categorial grammars, though with some complications (Bonato and Retoré, 2001). A strong generalization of these results has been proved: reversible regular tree languages (and dependency grammars too) are learnable from positive examples (Besombes and Marion, 2001).
2. An orthogonal extension is to consider k-valued grammars: in this later case, one has to try to unify types in all possible manners in order to have less than k types per word. This has was studied by Kanazawa (Kanazawa, 1998).
3. Regarding the input structures, the simplest generalization is to consider unlabeled trees: then one has to try all possible labeling with $\backslash_e$ and $/_e$. Going even further one can learn from unstructured sentences that are simply sequences of words: once again this is done by considering all possible structures on such sentences. This extension was investigated by Kanazawa (Kanazawa, 1998).

Each of these extensions increases the complexity of the algorithm considerably, as one can imagine, but nevertheless the existence of learning algorithms for categorial grammars is a good property which is shared by few other formalisms for natural language syntax.

1.7 Concluding Remarks

In this chapter, we introduced the core of categorial grammars, AB-grammars, common to the logical and combinatorial approaches. The salient features that distinguish this kind of grammar from phrase structure grammar are the following:

- A finite set of rules acting on categories, which do not depend on the particular language (this one of the reasons for learnability).
- The lexicon, which associates each word with a category describing its syntactic behaviour.
- As opposed to non-terminals, categories themselves have an internal structure which encodes how a word of this category interacts with words of other categories.

These differences explain the learnability in the sense of Gold for various restricted classes of categorial grammars.

The class of languages generated by AB-grammars is the same as the class of languages generated by Lambek grammars, though it is less simple in AB-grammars to give a simple account for syntactic constructions like (peripheral) extraction (as we have seen for our Example 1.1 with *cosa guarda passare*).

In the forthcoming chapters, we therefore develop a connection with logic: categories are formulas and rules are deduction rules, following an idea of Lambek. This also gives a correspondence with semantics, which already works (in a restricted way) for AB-grammars.

Exercises for Chapter 1

Exercise 1.1. Define an AB grammar for $a^n b^n$

Exercise 1.2. Define an AB grammar which generates the language of well-bracketed expressions such as ((())())

Exercise 1.3. Consider the following context-free grammar.

$$S \to \wedge SS \mid \neg S \mid p \mid q \mid r$$

It generates formulae in a minimal logical language containing only the propositions p, q and r, the unary logical symbol $\neg$ and the binary logical symbol $\wedge$. It has the property that it generates expressions in Polish prefix notation. For example, $\neg \wedge \neg p \neg q$ is an expression generated by this grammar corresponding to the more usual infix notation $\neg(\neg p \wedge \neg q)$, which would be generated by the context-free grammar below.

$$S \to (S \wedge S) \mid \neg S \mid p \mid q \mid r$$

An advantage of Polish prefix notation is that it does not require any brackets, though people not used to it tend to find it hard to read.

1. Give an AB lexicon which generates the language of the Polish prefix context-free grammar at the start of this exercise.
2. Give a lexicon which generates the language of the infix notation context-free grammar. *Hint:* assign types to the brackets as well as to the logical symbols.

Exercise 1.4. Consider the following AB lexicon for English:

Word	Type(s)
Amy	np
Ben	np
Chris	np
London	np
some	$(np\,/\,n)$
all	$(np\,/\,n)$
a	$(np\,/\,n)$
students	n
homework	n
dislikes	$((np \backslash S)\,/\,np)$
visited	$((np \backslash S)\,/\,np)$
went	$((np \backslash S)\,/\,pp)$
to	$(pp\,/\,np)$

1. Create, for each of the types in the above lexicon, a derivation which uses this type. Try to find some odd or ungrammatical sentences which are derivable using the above lexicon.

2. Extend the lexicon above in such a way that the following sentences become derivable. Show your lexicon is correct by providing derivations of all sentences.
   a) All intelligent students do their homework.
   b) No lazy students do their homework.
   c) Amy returned from Paris.
   d) Amy returned from Paris yesterday.
   e) Amy just returned from Paris.
   f) Ben considers Chris boring.
   g) Chris dislikes all students who listen to Mozart.
3. In sentences 2c and 2d from the previous exercise, (at least) two type assignments are possible for the constituent "from Paris": one where the constituent is assigned the type *pp* and one where it is assigned the type $(np\backslash s)\backslash(np\backslash s)$. Give a derivation for 2c and 2d corresponding to each of these solutions. How could you argue in favor of one or of the other type assignment?
4. Extend the above lexicon to derive the following sentences.
   a) Ben and Amy visited Paris.
   b) Ben visited and disliked London.
   c) Amy went to England and visited London.
   d) Amy went to Paris and to London.
   e) All students and professors like the new library.
   f) Chris read all new and interesting books.

   What can you say about the different types assigned to "and"?
5. How would you analyze the following sentence?
   a) All students who Amy likes listen to Mozart.

Exercise 1.5. The following exercise is inspired by a remark from Quine (1961). Natural language expressions permit (derivational) ambiguities which are not present in the tiny logical language of the previous exercise.

We have, for example that "Amy and Ben or Chris" is ambiguous between "(Amy and Ben) or Chris" and "Amy and (Ben or Chris)". The words "both" and "either" can help disambiguate between the two readings.

(1.8) Amy and Ben or Chris
(1.9) Both Amy and Ben or Chris
(1.10) Amy and either Ben or Chris

1. Give lexical entries for all words above and verify that there are two ways of deriving Sentence 1.8 above, but only one way to derive Sentence 1.9 and 1.10. *Hint:* assign atomic formulas b and e to "both" and "either" respectively.
2. Which of the two readings for Sentence 1.8 corresponds to Sentence 1.9 and which reading corresponds to Sentence 1.10?
3. Comment on the similarities between the solutions you provided here and your solution to Exercise 1.3.2.

Exercise 1.6. Give the context-free grammar which corresponds the lexicon of Exercise 1.4.

Exercise 1.7. Consider the following context-free grammar, which is already in Greibach normal form.

$$\begin{aligned}
S &\rightarrow aA \\
S &\rightarrow bB \\
S &\rightarrow aSA \\
S &\rightarrow bSB \\
A &\rightarrow a \\
B &\rightarrow b
\end{aligned}$$

1. What is the language generated by this grammar?
2. Transform the grammar into an AB grammar.

References

Ajdukiewicz, K.: Die syntaktische Konnexität. Studia Philosophica 1, 1–27 (1935); English translation in McCall, 207–231 (1967)

Bar-Hillel, Y.: A quasi arithmetical notation for syntactic description. Language 29, 47–58 (1953)

Bar-Hillel, Y., Gaifman, C., Shamir, E.: On categorial and phrase-structure grammars. Bulletin of the Research Council of Israel F(9), 1–16 (1963)

Besombes, J., Marion, J.Y.: Identification of reversible dependency tree languages. In: Popelínský and Nepil, pp. 11–22 (2001)

Bonato, R.: Uno studio sull'apprendibilità delle grammatiche di Lambek rigide — a study on learnability for rigid Lambek grammars. Tesi di Laurea & Mémoire de D.E.A. Università di Verona & Université Rennes 1 (2000)

Bonato, R., Retoré, C.: Learning rigid Lambek grammars and minimalist grammars from structured sentences. In: Popelínský and Nepil, pp. 23–34 (2001)

Buszkowski, W.: Discovery procedures for categorial grammars. In: van Benthem, J., Klein, E. (eds.) Categories, Polymorphism and Unification. Universiteit van Amsterdam (1987)

Buszkowski, W., Penn, G.: Categorial grammars determined from linguistic data by unification. Studia Logica 49, 431–454 (1990)

Casadio, C.: Semantic categories and the development of categorial grammars. In: Oehrle, R.T., Bach, E., Wheeler, D. (eds.) Categorial Grammars and Natural Language Structures, pp. 95–124. Reidel, Dordrecht (1988)

Chomsky, N.: The logical structure of linguistic theory (1955); revised 1956 version published in part by Plenum Press (1975); University of Chicago Press (1985)

Chomsky, N.: Formal properties of grammars. In: Handbook of Mathematical Psychology, vol. 2, pp. 323–418. Wiley, New-York (1963)

Chomsky, N.: The minimalist program. MIT Press, Cambridge (1995)

Girard, J.Y.: Linear logic: its syntax and semantics. In: Girard, J.Y., Lafont, Y., Regnier, L. (eds.) Advances in Linear Logic. London Mathematical Society Lecture Notes, vol. 222, pp. 1–42. Cambridge University Press (1995)

Gleitman, L., Liberman, M. (eds.): An invitation to cognitive sciences, Language, vol. 1. MIT Press (1995)

Gleitman, L., Newport, E.: The invention of language by children: Environmental and biological influences on the acquisition of language. In: Gleitman and Liberman, ch. 1, pp. 1–24 (1995)

Gold, E.M.: Language identification in the limit. Information and Control 10, 447–474 (1967)

Greibach, S.A.: A new normal-form theorem for context-free phrase structure grammars. Journal of the ACM 12(1), 42–52 (1965)

Harrison, M.A.: Introduction to Formal Language Theory. Addison Wesley (1978)

Hopcroft, J.E., Ullman, J.D.: Introduction to Automata Theory, Languages and Computation. Addison-Wesley, Reading (1979)

Joshi, A., Schabes, Y.: Tree adjoining grammars. In: Rozenberg and Salomaa, ch. 2 (1997)

Joshi, A., Levy, L., Takahashi, M.: Tree adjunct grammar. Journal of Computer and System Sciences 10, 136–163 (1975)

Kanazawa, M.: Learnable classes of categorial grammars. Studies in Logic, Language and Information. FoLLI & CSLI, distributed by Cambridge University Press (1998)

Lambek, J.: The mathematics of sentence structure. American Mathematical Monthly, 154–170 (1958)

McCall, S. (ed.): Polish Logic, 1920-1939. Oxford University Press (1967)

Morrill, G.: A chronicle of type logical grammar: 1935-1994. Research on Language and Computation 5(3), 359–386 (2007)
Nicolas, J.: Grammatical inference as unification. Rapport de Recherche RR-3632, INRIA (1999), http://www.inria.fr/
Pereira, F.C.N., Shieber, S.M.: Prolog and Natural-Language Analysis. CSLI Lecture Notes, vol. 10. University of Chicago Press, Chicago (1987)
Pinker, S.: Language acquisition. In: Gleitman and Liberman, ch. 6, pp. 135–182 (1995)
Pollard, C., Sag, I.A.: Head-Driven Phrase Structure Grammar. Center for the Study of Language and Information, Stanford (1994) (distributed by Cambridge University Press)
Popelìnskỳ, L., Nepil, M. (eds.): Proceedings of the third workshop on Learning Language in Logic. LLL 2001, FI MU Report series, FI-MU-RS-2001-08. Faculty of Informatics – Masaryk University, Strabourg (2001)
Quine, W.V.: Logic as a source of syntactical insights. In: Jakobson, R. (ed.) Proceedings of the Symposia in Applied Mathematics, Structure of Language and its Mathematical Aspects, vol. XII, pp. 1–5. American Mathematical Society (1961)
Rozenberg, G., Salomaa, A. (eds.): Handbook of Formal Languages. Springer, Berlin (1997)
Schabes, Y., Waters, R.C.: Lexicalized context-free grammars. In: Proceedings of the 21st Annual Meeting of the Association for Computational Linguistics, Columbus, Ohio, pp. 121–129 (1993)
Sikkel, K., Nijholt, A.: Parsing of context-free languages. In: Rozenberg and Salomaa, ch. 2 (1997)

2

A Logic for Categorial Grammars: Lambek's Syntactic Calculus

Summary. Our second chapter is a rather complete study of the Lambek calculus, which enables a completely logical treatment of categorial grammar.

We first present its syntax in full detail, both with sequent calculus and natural deduction, and explain the relationship between these two presentations. Then we turn our attention to the normal forms for such proofs. Normalization and its dual namely interpolation are not only pleasant mathematical properties; they also are key properties for the correspondence between Lambek grammars and more familiar phrase structure grammars; we give a detailed proof of the theorem of Pentus establishing the weak equivalence between context-free grammars and Lambek grammars.

In addition, we prove completeness for the Lambek calculus with respect to linguistically natural models: in these models categories are interpreted as subsets of a free monoid (eg. as strings of words or lexical items). Providing such a simple and natural interpretation provides another strong justification for the categorial approach.

2.1 Lambek's Syntactic Calculus and Lambek Grammars

We now turn our attention to the Lambek calculus (L) and Lambek grammars (LCG) which were introduced in the seminal paper (Lambek, 1958): we strongly recommend this paper to the reader.

The limitations of AB grammars, and the endless quest for new rules (composition, type raising, Geach laws, etc.) is a way to explain the interest of the Lambek calculus. Another is to place AB-grammar into a richer and more natural mathematical formalism.

A controversial but more interesting justification is the following: syntax is driven by resource consumption, which is neatly handled by resource conscious logics — the Lambek calculus being the first such logic. This viewpoint is not that far from Chomsky's minimalist program (Chomsky, 1995) as discussed in (Retoré and Stabler, 2004).

Lambek (categorial) grammars — or LCGs — are defined in a way very similar to AB grammars. A lexicon Lex provides each word with one or several types, constructed from the usual primitive types $\mathsf{P} = \{S, np, n, pp, \ldots\}$ — sentences, noun phrases, nouns, prepositional phrases... Types are more or less the same as the ones of AB grammars: the only difference is that Lambek types allow for a (non commutative) product or conjunction denoted by $\bullet$:

$$\mathsf{Lp} \quad ::= \quad \mathsf{P} \quad | \quad (\mathsf{Lp} \setminus \mathsf{Lp}) \quad | \quad (\mathsf{Lp} / \mathsf{Lp}) \quad | \quad (\mathsf{Lp} \bullet \mathsf{Lp})$$

When introducing AB grammars, we already explained the intuitive meaning of $A \setminus B$ and B / A: an expression is of type $A \setminus B$ (resp. B / A) when it is looking for an expression of type A on its left (resp. right) to form a compound expression of type B. An expression of type A followed by an expression B is of type $A \bullet B$, and product is related to $\setminus$ and $/$ by the following relations:

$$A \setminus (B \setminus X) = (B \bullet A) \setminus X \qquad\qquad (X / A) / B = X / (B \bullet A)$$

These relations look like currying, but beware of the order, which is required by the behavior of $\setminus$ and $/$: in the left equation both types require a sequence ab on their left, and in the second equation both types require a sequence ba on their right (with a, b of respective types A, B).

Recall that for AB grammars a sequence of words $w_1 \cdots w_n$, is of type u whenever there exists for each w_i a type t_i in $\mathrm{Lex}(w_i)$ such that $t_1 \cdots t_n \longrightarrow u$ with the following reduction patterns:

$$\forall u, v \in \mathsf{Lp} \quad \begin{array}{ll} u\ (u \setminus v) \longrightarrow v & (\setminus_e) \\ (v / u)\ u \longrightarrow v & (/_e) \end{array}$$

Here the logical aspect of these rules — they look like modus ponens — will be emphasized by adding other rules, so that $\setminus$ and $/$ will really be implications (and $\bullet$ will be their associated conjunction). Accordingly $\longrightarrow$ will be written $\vdash$, and our first objective is to define this logical calculus: for the time being we only know the modus ponens of the non commutative implications $\setminus$ and $/$. Therefore we simply replace $\longrightarrow$ with $\vdash$ to obtain the following definition: a sequence of words (or terminals) $w_1 \cdots w_n$ is of type u whenever there exists for each w_i a type t_i in $\mathrm{Lex}(w_i)$ such that $t_1 \cdots t_n \vdash u$, where $\vdash$ is the deductive relation of the Lambek calculus which is defined in the next two sections. The generated language or the set of correct sentences is the set of sequences of type S.

2.2 Natural Deduction for the Lambek Calculus

To the best of our knowledge, natural deduction for Lambek has mainly been studied by van Benthem (van Benthem, 1991), one of the first papers being (van Benthem, 1987).

2.2.1 In Prawitz Style

The simplest way to define product the free Lambek calculus is probably natural deduction in a tree-like setting as shown below (we will have more to say about the requirement that the $\setminus_i$ and $/_i$ rules require at least two free hypotheses in Section 2.5).

this rule requires at least two free hyp.

$$\begin{array}{c} A \text{ leftmost free hyp.} \\ \ldots [A] \ldots\ldots \\ \vdots \\ B \\ \hline A \backslash B \end{array} \backslash_i \text{ binding } A \qquad\qquad \begin{array}{cc} \Delta & \Gamma \\ \vdots & \vdots \\ A & A \backslash B \\ \hline \multicolumn{2}{c}{B} \end{array} \backslash_e$$

this rule requires at least two free hyp.

$$\begin{array}{c} A \text{ rightmost free hyp.} \\ \ldots\ldots [A] \ldots \\ \vdots \\ B \\ \hline B / A \end{array} /_i \text{ binding } A \qquad\qquad \begin{array}{cc} \Gamma & \Delta \\ \vdots & \vdots \\ B / A & A \\ \hline \multicolumn{2}{c}{B} \end{array} /_e$$

These deductions clearly extend the derivation trees of AB grammars. AB simplification or elimination rules are two of the rules of the system, the rules $\backslash_e$ and $/_e$; the other two being the corresponding introduction rules. The fact that these rules are special cases of the rules for intuitionistic logic confirms that the fraction symbols $\backslash$ and $/$ can be viewed as implications.

It should be observed that as opposed to natural deduction for intuitionistic logic, there is no need to specify which hypothesis A is cancelled by an $/_i$ or $\backslash_i$ introduction rule. Indeed in the first case it is the leftmost free hypothesis, and in the second case it is the rightmost free hypothesis. As a consequence the formal structure of a deduction is a plain (binary/unary) tree with leaves labeled with formulae and with nodes labelled by rules: binary nodes are labelled with either $/_e$ or $\backslash_e$ and unary nodes with either $/_i$ or $\backslash_i$. Such a plain tree is enough to reconstruct the deduction, i.e. which hypothesis are free or not and which hypothesis is cancelled by which rule. This remark is the basis of the study of Tiede (2001); we can see the parse structures or proofs of a Lambek grammar as natural deduction trees, and study these trees as tree languages (Gécseg and Steinby, 1997).

Product

Lambek calculus admits a product which is related to the implications by the usual currying rules given above (or, alternatively, by the residuation rules discussed in Section 2.9.1). The product is often skipped in the natural deduction presentation of the Lambek calculus. There is no need to do so, but it is true that these rules are less natural, because of the order on hypotheses: Δ should occur in the place previously occupied by the cancelled (consecutive) A and B hypotheses of the rule.

$$\dfrac{\begin{array}{cc}\Delta & \Gamma\\ \vdots & \vdots\\ A & B\end{array}}{A\bullet B}\,\bullet_i \qquad\qquad \dfrac{\begin{array}{c}\Delta\\ \vdots\\ A\bullet B\end{array} \qquad \begin{array}{c}\textit{no free hyp. between A and B}\\ \ldots[A]_\alpha[B]_\alpha\ldots\\ \vdots\\ C\end{array}}{C}\ \bullet_e(\alpha)\text{ binding } A \text{ and } B$$

The main problem is that in order to apply the product elimination rule there should be no free hypothesis in between the two cancelled assumptions, A and B, and that the order of the premises after the rule is no longer the left-to-right order, but rather has the formulas Δ occurring at the place of the eliminated A and B formulas. Another problem is that, as we shall see, proof-normalization or rather the subformula property is more problematic with the product.

Also observe that there can be several consecutive free A and B hypotheses, so that a labeling (the label α in the rule above) is needed to link specific occurrences of cancelled hypotheses to the instance of the $\bullet_e$ rule: natural deductions are no longer plain trees.

Natural deduction rules of this form were first introduced by Abramsky (1993) for multiplicative linear logic, but in this commutative case the problem of the order of hypotheses disappears.

Example 2.1. As an example in Prawitz Style natural deduction, we give a proof of "Kevin talks to himself" below.

$$\dfrac{\dfrac{\text{Kevin}}{np}\,\text{Lex} \qquad \dfrac{\dfrac{\dfrac{\dfrac{\text{talks}}{(np\backslash S)/pp}\,\text{Lex} \quad \dfrac{\dfrac{\text{to}}{pp/np}\,\text{Lex} \quad [np]}{pp}\,/_e}{np\backslash S}\,/_e}{(np\backslash S)/np}\,/_i \qquad \dfrac{\text{himself}}{((np\backslash S)/np)\backslash(np\backslash S)}\,\text{Lex}}{np\backslash S}\,\backslash_e}{S}\,\backslash_e$$

As already discussed in Example 1.1, the Lex rule used in the proof above is simply an indication that the conclusion formula F of the rule is an element of $\text{Lex}(w)$ for the premise w; that is, F is a formula that the lexicon Lex assigns to the word w.

The type for "himself" is assigned a category which selects a transitive verb to its left to produce an intransitive verb (as we will see in Example 3.2 in the next chapter, there are good semantic reasons for this type assignment).

Seen from the lexical types, the introduction of the np hypothesis is the only step which may not be immediately obvious: it is the type assigned to "himself", which, having a verb phrase $(np\backslash S)$ missing an np as its argument, introduces an np hypothesis. Section 2.6.2 will give a proof search algorithm for (product-free) natural deduction.

Exercises 2.2, 2.6, 2.8 and 2.9 at the end of this chapter will help you get familiar with finding natural deduction proofs for Lambek grammars.

2.2.2 In Gentzen Style

It is sometimes convenient to give a Gentzen style presentation of natural deduction, which specifies at each node what the free hypotheses are; this formulation is possibly clearer, in particular when formulating the rules for the product formulae. Figure 2.1 lists the rules in the calculus.

$$\frac{\Gamma \vdash A \quad \Delta \vdash A \backslash B}{\Gamma, \Delta \vdash B} \backslash_e \qquad \frac{A, \Gamma \vdash C}{\Gamma \vdash A \backslash C} \backslash_i \quad \Gamma \neq \varepsilon$$

$$\frac{\Delta \vdash B / A \quad \Gamma \vdash A}{\Delta, \Gamma \vdash B} /_e \qquad \frac{\Gamma, A \vdash C}{\Gamma \vdash C / A} /_i \quad \Gamma \neq \varepsilon$$

$$\frac{\Delta \vdash A \bullet B \quad \Gamma, A, B, \Gamma' \vdash C}{\Gamma, \Delta, \Gamma' \vdash C} \bullet_e \qquad \frac{\Delta \vdash A \quad \Gamma \vdash B}{\Delta, \Gamma \vdash A \bullet B} \bullet_i$$

$$\frac{}{A \vdash A} \; axiom$$

Fig. 2.1. Gentzen style natural deduction rules for the Lambek calculus

Nevertheless this presentation defines exactly the same logical calculus as the natural deduction rules in tree-like format given above: the proofs of the two systems are isomorphic.

A small note about the notation used in the calculus: the statements of the calculus are expressions of the form $A_1, \ldots, A_n \vdash C$ (sequents), with a comma-separated list of formulae (the *antecedent*, or the hypotheses of the statement) on the left hand side of the turnstile and a single formula on the right hand side (the *succedent* or the conclusion of the statement). Variables $\Gamma, \Delta, \ldots$ range over (possibly empty) lists of formulae which we will call *contexts*, so we can write a sequent as $\Gamma \vdash C$, or a sequent containing a formula A as $\Gamma, A, \Delta \vdash C$.

We will call the sequents above the horizontal line of a rule its *premises* and the single sequent below the horizontal line its conclusion. When given a proof, we will call the conclusion of the last rule the *end-sequent*.

Although we use sequents, this calculus is by no means a sequent calculus: there are no left rules, no cut rule, and the notion of normal proof (for having the subformula property) is completely different — as we will see in Sections 2.6 and 2.7.

Example 2.2 (Our Italian Lexicon Revisited)
Here we take up again our small example of an Italian lexicon:

Word	Type(s)
cosa	$(S\,/\,(S\,/\,np))$
guarda	$(S\,/\,inf)$
passare	$(inf\,/\,np)$
il	$(np\,/\,n)$
treno	n

Remember that the sentence '*Cosa guarda passare*' could not be analyzed in AB grammars, because the transitivity of / was not a rule of AB grammars. Let us show that it can be analyzed with the Lambek calculus (we use Natural Deduction in Gentzen style):

$$\dfrac{(S\,/\,(S\,/\,np)) \vdash (S\,/\,(S\,/\,np)) \qquad \dfrac{\dfrac{(S\,/\,inf) \vdash (S\,/\,inf) \qquad \dfrac{(inf\,/\,np) \vdash (inf\,/\,np) \qquad np \vdash np}{(inf\,/\,np), np \vdash inf}\,/e}{(S\,/\,inf), (inf\,/\,np), np \vdash S}\,/_e}{(S\,/\,inf), (inf\,/\,np) \vdash S\,/\,np}\,/_i}{(S\,/\,(S\,/\,np)), (S\,/\,inf), (inf\,/\,np) \vdash S}\,/_e$$

This example relies on composition for /, which is not provable in AB grammars. Composition is established by first hypothesizing an *np* which is then abstracted by an introduction rule: to make a comparison with Chomsky's theories (Chomsky, 1957, 1995) this hypothetical *np* corresponds to a trace and the introduction rule to movement.

In the Lambek calculus, we can construct sentences with object relatives such as *whom/that* having the type $(n \setminus n)\,/\,(S\,/\,np)$ — which could not be done in AB grammars, as shown in Section 1.5 — and without a need to assign $np \setminus (S\,/\,np)$ to transitive verbs. Indeed, we can derive $np \setminus (S\,/\,np)$ from the "normal" transitive verb type $(np \setminus S)\,/\,np$, since we can rearrange brackets in the Lambek calculus: $(a \setminus b)\,/\,c \vdash a \setminus (b\,/\,c)$, etc. This treatment of peripheral extraction is one of the attractive features of the Lambek calculus. Exercise 2.7 asks you to give proofs of more complicated cases of extraction.

Finally it is easily verified that one has $x \vdash (z\,/\,x) \setminus z$ and $x \vdash z\,/\,(x \setminus z)$ for all categories x and z. As we will see in the next chapter, this is interesting from a semantic viewpoint: an *np* (an individual) can be viewed as a $(S\,/\,np) \setminus S$ or $S\,/\,(np \setminus S)$ (a function form one-place predicates to truth values, that is the set of all the properties of this individual).

2.3 Sequent Calculus

Figure 2.2 shows the rules of the Lambek calculus in Sequent Calculus, as given in the original paper (Lambek, 1958). Although it also handles expressions $A_1, \ldots, A_n \vdash C$,

$$\frac{\Gamma, B, \Gamma' \vdash C \quad \Delta \vdash A}{\Gamma, \Delta, A \backslash B, \Gamma' \vdash C} \backslash_h \qquad \frac{A, \Gamma \vdash C}{\Gamma \vdash A \backslash C} \backslash_i \quad \Gamma \neq \varepsilon$$

$$\frac{\Gamma, B, \Gamma' \vdash C \quad \Delta \vdash A}{\Gamma, B / A, \Delta, \Gamma' \vdash C} /_h \qquad \frac{\Gamma, A \vdash C}{\Gamma \vdash C / A} /_i \quad \Gamma \neq \varepsilon$$

$$\frac{\Gamma, A, B, \Gamma' \vdash C}{\Gamma, A \bullet B, \Gamma' \vdash C} \bullet_h \qquad \frac{\Delta \vdash A \quad \Gamma \vdash B}{\Delta, \Gamma \vdash A \bullet B} \bullet_i$$

$$\frac{\Gamma \vdash A \quad \Delta_1, A, \Delta_2 \vdash B}{\Delta_1, \Gamma, \Delta_2 \vdash B} \mathit{cut} \qquad \frac{}{A \vdash A} \mathit{axiom}$$

Fig. 2.2. Sequent calculus rule for the Lambek calculus

let us insist that it is different from Natural Deduction in sequent style given above: for instance the modus ponens or elimination rules of AB grammars are not rules of this system (they are just derivable) and the notion of a normal proof is rather different (though (Girard et al, 1988) rightly remark there is a “moral equivalence” between the two).

A note on the names of the rules: we will use rule name $\backslash_h$, $/_h$ and $\bullet_h$ (for $\backslash$, $/$ and $\bullet$ used as a *hypothesis*) and $\backslash_i$, $/_i$, $\bullet_i$ (for the *introduction* rules for $\backslash$, $/$ and $\bullet$) instead of the frequently used $L\backslash$, $L/$ and $L\bullet$ (for the connectives on the *left*-hand side of the turnstile) and $R\backslash$, $R/$ and $R\bullet$ (for the connectives on the *right*-hand side of the turnstile). Our notation emphasizes that fact that the $\backslash_i$ rule is shared between the sequent calculus and natural deduction, and that the difference between sequent calculus and natural deduction is whether we add the $\backslash_e$, $/_e$ and $\bullet_e$ rules — as we do for natural deduction — or the $\backslash_h$, $/_h$ and $\bullet_h$ rules — as we do for the sequent calculus.

For a sequent calculus rule (or an instantiation of a rule as we find in an actual proof) it is normal to talk about the *main* formula of the rules for the logical connectives: the main formula is the formula with the connective introduced by the rule as its main connective; so the main formula of the $\bullet_h$ and $\bullet_i$ rule is the formula $A \bullet B$. We will call its direct subformulae, that is the formulae A and B, the *active* formulae of the rule (or the rule application).

Here is an obvious proposition (Exercise 2.3 at the end of this chapter asks you to prove this yourself).

Proposition 2.3. *Every axiom $A \vdash A$ can be derived from axioms $p \vdash p$, with p being a primitive type (and the proof does not use the cut rule).*

Definition 2.4 (Polarity). *The polarity of an occurrence of a propositional variable p in a formula is defined as usual:*

- *p is positive in p*
- *if p is positive in A, then*
  - *p is positive in* $X \bullet A$, $A \bullet X$, $X \backslash A$, A / X
  - *p is negative in* $A \backslash X$, X / A
- *if p is negative in A, then*
  - *p is negative in* $X \bullet A$, $A \bullet X$, $X \backslash A$, A / X
  - *p is positive in* $A \backslash X$, X / A

The polarity of an occurrence of a propositional variable p in a sequent $\Gamma \vdash C$ *is:*

- *if p is in C, the polarity of p in C*
- *if p is in a formula G of* Γ*, the opposite of the polarity of p in G.*

Example 2.5. Polarity is an important notion, which will return in Chapter 6. In a sequent

$$a / b, (a / b) \backslash (d / c) \vdash (d / c)$$

a occurs positively in a / b but negatively in $((a / b) \backslash (d / c)$

If a proof only uses atomic axioms (this is always possible, as said above) ie. $p \vdash p$ with p a primitive type, then one can follow these two occurrences of p, one being negative and the other positive and none of the rules changes the polarity of an occurrence of a primitive type. The two occurrences of p either lead to a cut formula (the formula which disappears from the conclusion of the cut rule) or to the end-sequent. Now observe that the cut rule cancels a formula in positive position (on the right) with the same formula in negative position (on the left), so that the same number of positive and negative occurrences of p disappear. Consequently:

Proposition 2.6. *Each propositional variable has exactly the same number of positive and negative occurrences in a provable sequent.*

Some authors call Proposition 2.6 the *count check* or *count invariant* (van Benthem, 1986; Moortgat, 1988; Roorda, 1991) and it can be used to eliminate sequents which fail to satisfy the condition of Proposition 2.6 at little computational cost.

Example 2.7. Here is an example of a proof in sequent calculus, corresponding to the analysis of '*Cosa guarda passare*' already given in natural deduction format in Example 2.2. It is somewhat less natural, but has other advantages, like an easier subformula property.

$$\dfrac{S \vdash S \qquad \dfrac{\dfrac{\dfrac{S \vdash S \quad inf \vdash inf}{S / inf, inf \vdash S}\,/_h \qquad np \vdash np}{S / inf, inf / np, np \vdash S}\,/_h}{S / inf, inf / np \vdash S / np}\,/_i}{(S / (S / np)), S / inf, inf / np \vdash S}\,/_h$$

2.4 Equivalence of Sequent Calculus and Natural Deduction

As we will see, this equivalence is absolutely clear as far as provability is concerned. In fact there is a correspondence for proofs as well, but it is not a straightforward isomorphism (Girard et al, 1988).

As the introduction rules are common to both formalisms, we just need to mimic elimination rules $\diamond_e$ in sequent calculus and left rules $\diamond_h$ in natural deduction (for $\diamond \in \{\backslash, /, \bullet\}$), and by induction on the height of the proofs the equivalence of both formalisms follows. This section is an easy adaptation of the results in (Girard et al, 1988) for intuitionistic logic.

2.4.1 From Natural Deduction to Sequent Calculus

It is possible to do "better" than the translation we provide here; indeed, when the natural deduction is normal, one can manage to obtain a cut-free proof, and this better translation is implicitly used when one uses proof nets for λ-calculus see e.g. (Girard, 1987; de Groote and Retoré, 1996)

Replace: with:

$$\dfrac{\Delta \vdash A \quad \Gamma \vdash A \backslash B}{\Delta, \Gamma \vdash B}\backslash_e \qquad \dfrac{\Gamma \vdash A \backslash B \quad \dfrac{\Delta \vdash A \quad \dfrac{}{B \vdash B}ax}{\Delta, A \backslash B \vdash B}\backslash_h}{\Delta, \Gamma \vdash B}cut$$

$$\dfrac{\Gamma \vdash B / A \quad \Delta \vdash A}{\Gamma, \Delta \vdash B}/_e \qquad \dfrac{\Gamma \vdash B / A \quad \dfrac{\Delta \vdash A \quad \dfrac{}{B \vdash B}ax}{B / A, \Delta \vdash B}/_h}{\Gamma, \Delta \vdash B}cut$$

$$\dfrac{\Gamma \vdash A \bullet B \quad \Delta, A, B, \Theta \vdash C}{\Delta, \Gamma, \Theta \vdash C}\bullet_e \qquad \dfrac{\Gamma \vdash A \bullet B \quad \dfrac{\Delta, A, B, \Theta \vdash C}{\Delta, A \bullet B, \Theta \vdash C}\bullet_h}{\Delta, \Gamma, \Theta \vdash C}cut$$

2.4.2 From Sequent Calculus to Natural Deduction

By induction on the height of a sequent calculus proof, let us see that it can be turned into a natural deduction. As above, we will not exhibit a translation from cut free proofs to normal deductions, although it is possible.

- If the proof consists in an axiom, its translation is obvious.
- If the proof ends with an introduction rule, $\backslash_i$, $/_i$ or $\bullet_i$ by induction hypothesis we have a deduction of the premise(s) and as these rules also exist in natural deduction and the translation is obvious.
- If the proof ends with an $\backslash_h$ rule:

$$\dfrac{\begin{array}{c}\vdots\ \gamma\\ \Gamma,B,\Gamma'\vdash C\end{array}\quad \begin{array}{c}\vdots\ \delta\\ \Delta\vdash A\end{array}}{\Gamma,\Delta,A\backslash B,\Gamma'\vdash C}\ \backslash_h$$

then by induction hypothesis we have two natural deduction proofs, γ^* of $\Gamma,B,\Gamma'\vdash C$ and δ^* of $\Delta\vdash A$ and a translation of the whole proof is:

$$\begin{array}{ccc} & \begin{array}{c} \dfrac{\begin{array}{c}\Delta\\ \vdots\ \delta^*\\ A\end{array}\quad A\backslash B}{B}\ \backslash_e \end{array} & \\ \Gamma & & \Gamma' \\ & \begin{array}{c}\vdots\ \gamma^*\\ C\end{array} & \end{array}$$

- If the proof ends with $/_h$ we proceed symmetrically.
- If the proof ends with $\bullet_h$:

$$\dfrac{\begin{array}{c}\vdots\ \gamma\\ \Gamma,A,B,\Gamma'\vdash C\end{array}}{\Gamma,A\bullet B,\Gamma'\vdash C}\ \bullet_h$$

by induction hypothesis we have a proof γ^* of $\Gamma,A,B,\Gamma'\vdash C$ and a translation is the following:

$$\dfrac{A\bullet B \qquad \begin{array}{c}\Gamma\ \ A\ B\ \ \Gamma'\\ \vdots\ \gamma^*\\ C\end{array}}{C}\ \bullet_e$$

- If the proof ends with a cut:

$$\dfrac{\begin{array}{c}\vdots\ \gamma\\ \Gamma\vdash X\end{array}\quad \begin{array}{c}\vdots\ \delta\\ \Delta,X,\Delta'\vdash C\end{array}}{C}\ cut$$

by induction hypothesis we have two natural deductions $\gamma^*$of $\Gamma\vdash X$ and δ^* of $\Delta,X,\Delta'\vdash C$ and a translation is:

$$\begin{array}{c}\Gamma\\ \vdots\ \gamma^*\\ \Delta\quad X\quad \Delta'\\ \vdots\ \delta^*\\ C\end{array}$$

2.5 The Empty Sequence

In the formulation of the introduction rules, we have required that the antecedent contains at least two formulae: therefore the antecedent is never empty after an application of an introduction rule. By case inspection we see that this guarantees that the antecedent of a sequent (the sequence on the left of $\vdash$) never is empty in a proof.

This is justified by the intended meaning of the connectives. Indeed by assigning the type $A \setminus B$ to a word or an expression e, we mean that an expression a of type A is *required* before e to obtain an expression ae of type B. This would fail without the "no empty sequence" requirement.

To explain this, let L1 be the calculus L without this restriction. Indeed, assume A is a tautology of L1, i.e. $\vdash_{L1} A$ (*); now let Γ be a sequence of type $A \setminus B$, that is $\Gamma \vdash_{L1} A \setminus B$ (**). Then from (*) and (**) we can infer by $\setminus_e$ the sequent $\Gamma \vdash_{L1} B$ without any sequence preceding Γ. This can actually happen in natural language; indeed some expression, including all modifiers do have such a tautology type, like $X \setminus X$.

For instance, a natural type for English adjectives is n / n and thus *very* gets the type $(n / n) / (n / n)$: when applied to an adjective on its right, one obtains an adjective phrase. Without the exclusion of the empty sequence, one is able to analyze in L1 the expression *"a very book"* as a noun phrase: indeed the adjective following *very* can be provided by the empty sequence, since n / n is derivable in L1. Let us give the proof in L1 using Prawitz-style natural deduction (the rules Lex, with premise w and conclusion formula A, are axioms indicating that A is in $\mathrm{Lex}(w)$):

$$\dfrac{\dfrac{\text{a}}{np\,/\,n}\,\text{Lex} \qquad \dfrac{\dfrac{\dfrac{\text{very}}{(n\,/\,n)\,/\,(n\,/\,n)}\,\text{Lex} \qquad \dfrac{[n]_\alpha}{n\,/\,n}\,/_i{-}\alpha}{n\,/\,n}\,/_e \qquad \dfrac{\text{book}}{n}\,\text{Lex}}{n}\,/_e}{np}\,/_e$$

One may wonder why such a requirement was not needed for AB grammars. As AB grammars contain only elimination rules, no hypotheses are cancelled during a derivation, and since there are hypotheses at the beginning of every sub-analysis (the types of the words in the analyzed sequence) there always is at least one hypothesis.

2.6 Normalization of Natural Deduction

This section is also an easy adaptation of similar results presented in (Girard et al, 1988). Throughout this section, we will mostly be concerned with the product free case, a brief discussion about why normalization is more complicated in the presence of the product is found in Section 2.6.3.

2.6.1 Normalization for the Product-Free Lambek Calculus

A natural deduction is said to be normal whenever it does not contain an introduction rule followed by an elimination rule. There are two such possible configurations:

$$\dfrac{\dfrac{\begin{array}{c}\ldots\ldots[A]_\alpha\ldots\\ \vdots\,\delta'\\ B\end{array}}{B\,/\,A}\,/_i\!-\!\alpha \qquad \begin{array}{c}\Delta\\ \vdots\,\delta\\ A\end{array}}{B}\,/_e \qquad\qquad \dfrac{\begin{array}{c}\Delta\\ \vdots\,\delta\\ A\end{array} \qquad \dfrac{\begin{array}{c}\ldots[A]_\alpha\ldots\ldots\\ \vdots\,\delta'\\ B\end{array}}{A\,\backslash\,B}\,\backslash_i\!-\!\alpha}{B}\,\backslash_e$$

Remember that we call implication formulas $B\,/\,A$ and $A\,\backslash\,B$ *functors* and the formulas A their *arguments*.

Now, whenever such a configuration appears, it can be reduced as follows:

1. find the hypothesis A which has been cancelled in the proof δ' of B under some hypotheses including A
2. replace this hypothesis with the proof δ of A

So the configurations above reduce to:

$$\begin{array}{c}\Delta\\ \vdots\,\delta\\ \ldots A\ldots\\ \vdots\,\delta'\\ B\end{array} \qquad\qquad \begin{array}{c}\Delta\\ \vdots\,\delta\\ \ldots A\ldots\\ \vdots\,\delta'\\ B\end{array}$$

Proposition 2.8. *Natural deduction for L without product enjoys strong normalization, that is there are no infinite reduction sequences.*

Proof. Observe that the size of the proof decreases in each reduction step. □

The proof of strong normalization is so simple because each introduction rule binds exactly one formula A and therefore we never copy nor delete the proof δ, ie. contraction and weakening are not valid in the Lambek calculus. In intuitionistic logic — where the introduction rule for the implication can discharge any number of hypotheses of the formula A — strong normalization is valid as well, but the proof is a bit more delicate (see Girard et al, 1988, Chapters 4 and 6 for details).

Proposition 2.9. *Normalization is a locally confluent process. In other words, if a proof d reduces, in one step, to two different proofs e and f then there exists a proof g such the both e and f reduce, in some number of steps, to g.*

Proof. If a proof d contains two redexes, they correspond to two elimination rules e' and e'' between sub-proofs corresponding to a functor f' applied to an argument a' and to a functor f'' applied to an argument a''. One of the following case applies:

- e'' is in a'
- e'' is in f'

- e' is in a''
- e' is in f''
- e' and e'' can not be compared.

Assume we reduce e'. The redex e'' which is not reduced possesses a unique trace $\bar{e}''$ in the reduced proof d'. Symmetrically if we reduce e'' the redex e' which is not reduced possesses a unique trace $\bar{e}$" in d''. If in d' we reduce $\bar{e}''$ we obtain a proof d''' but if in d'' we reduce $\bar{e}'$ we also obtain d'''. □

We will now show, by an easy induction on the proofs, that whenever a natural deduction is normal (that is without such configuration) each formula is a subformula of a free hypothesis or of the conclusion. More precisely. In order to establish this, let us introduce the notion of principal branch.

Let us call a *principal branch* leading to F a sequence $H_0,\ldots,H_n = F$ of formulae of a natural deduction tree such that:

- H_0 is a free hypothesis
- H_i is the principal premise — the one carrying the eliminated symbol — of an elimination rule whose conclusion is H_{i+1}
- H_n is F

Proposition 2.10. *Let d be a normal natural deduction (without product), then:*

1. *if d ends with an elimination then there is a principal branch leading to its conclusion*
2. *each formula in d is the subformula of a free hypothesis or of the conclusion*

Proof. By induction on d.

[axiom] If d is an axiom, (1) and (2) hold.

[$\backslash_i$ introduction] (1) holds by vacuity. Assume d is made out of d' by the introduction $\backslash_i$ rule: by induction hypothesis each formula in d' is a subformula of A,Γ (the free hypotheses under which B is proved) or a subformula of B; so it is true that each formula in d is a subformula of $\Gamma, A \setminus B$, since A and B are subformulae of $A \setminus B$.

[$\backslash_e$ elimination] Assume d is an elimination rule $\backslash_e$ applied to:

- d' with conclusion A and free hypotheses Γ
- d'' with conclusion $A \setminus B$ and free hypotheses Δ

(1) Since d is normal the last rule of d'' is an elimination: indeed, if it were an introduction rule then it would be a $\backslash_i$ introduction making a redex with the final elimination in d. As d'' ends with an elimination, by induction hypothesis, there is a principal branch leading from H_0 in Δ to $A \setminus B$, so d contains a principal branch leading to its conclusion B.

(2) By induction hypothesis

- all formulae in d' are subformula of A or Γ (the free hypotheses under which A is proved)
- all formulae in d'' are subformulae of $\Delta, A \setminus B$.

Because of the principal branch of d' leading to $A \setminus B$, the conclusion $A \setminus B$ of d' is a subformula of some H_0 in Δ. Thus each formulae in d is a subformula of Γ, Δ hence of Γ, Δ, B

[$/_i$ introduction] as $\setminus_i$ introduction.

[$/_e$ elimination] as $\setminus_e$ elimination □

Here is a proposition of (Cohen, 1967) that we shall use to prove that every context-free grammar is weakly equivalent to a Lambek grammar.

The order $o(A)$ of a formula A is the number of alternating implications, defined formally as follows.

Definition 2.11. *The order of a formula A is defined as follows.*

$$\begin{aligned} o(p) &= 0 \qquad \textit{when } p \textit{ is an atomic type} \\ o(A \setminus B) &= \max(o(A)+1, o(B)) \\ o(B \,/\, A) &= \max(o(A)+1, o(B)) \end{aligned}$$

Thus, the order of $(np \setminus S) \,/\, np$ is 1, but the order of $S \,/\, (np \setminus S)$ is two.

Proposition 2.12. *A provable sequent $A_1, \ldots, A_n \vdash p$ of the product free Lambek calculus with $o(A_i) \leq 1$ and p a primitive type (and therefore of order zero) is provable with $\setminus_e$ and $/_e$ only — in other words AB derivations and L derivations coincide when types are of order at most one.*

Proof. If $A_1, \ldots, A_n \vdash p$ is provable, then it has a normal proof. We claim that this normal proof must contain only $\setminus_e$ and $/_e$. We proceed by contradiction, so we assume that the normal deduction contains an introduction rule, and so there is a lowest introduction rule — one without any introduction rule below.

Let us consider an arbitrary lowest introduction I.

- If the chosen lowest introduction I is an $\setminus_i$ introduction leading from y to $b \setminus y$. This introduction cannot be the last rule, because the conclusion is a primitive type p. So this rule is followed by a an elimination rule E, and there are three possibilities:
  - If $b \setminus y$ is the principal premise of the elimination rule E, then the rule E is an $\setminus_e$ elimination rule other premise b; we then have a redex I, E and this conflicts with the deduction being normal.
  - If $b \setminus y$ is not the principal premise of the elimination rule E, then E is either an $\setminus_e$ elimination rule with principal premise being $(b \setminus y) \setminus z$ or an $/_e$ elimination rule with principal premise $z \,/\, (b \setminus y)$. In both cases the principal premise is of order at least two. This conflicts with d enjoying the subformula property which is forced by d being normal (previous Proposition 2.10).
- If the chosen lowest introduction I is an $/_i$ rule, the argument is symmetrical.

Therefore there is no lowest introduction, hence no introduction at all. □

2.6.2 Decidability of Natural Deduction

We can use the principal branch property and Proposition 2.10 to show that natural deduction in the product free Lambek calculus is decidable (we will prove the classic result of decidability for the Lambek calculus *with* product using cut elimination for the sequent calculus in Section 2.8). In order to do so, we first need to introduce some notation to facilitate talking about left and right arguments of a formula.

Given a formula C, and a sequence of length p of pairs consisting of a letter ε_i (where $\varepsilon_i \in \{l, r\}$) and a formula G_i we denote by

$$C[(\varepsilon_1, G_1), \ldots, (\varepsilon_p, G_p)]$$

the formula defined as follows:

if $p = 0$ $C[] = C$
if $\varepsilon_i = l$ $C[(\varepsilon_1, G_1), \ldots, (\varepsilon_{p-1}, G_{p-1}), (\varepsilon_p, G_p)] = G_p \backslash C[(\varepsilon_1, G_1), \ldots, (\varepsilon_{p-1}, G_{p-1})]$
if $\varepsilon_i = r$ $C[(\varepsilon_1, G_1), \ldots, (\varepsilon_{p-1}, G_{p-1}), (\varepsilon_p, G_p)] = C[(\varepsilon_1, G_1), \ldots, (\varepsilon_{p-1}, G_{p-1})] / G_p$

Figure 2.3 shows an example of a formula with its arguments in list-of-pairs form and its step-by-step conversion to the corresponding formula in standard Lambek calculus notation: note how G_1 is the most deeply embedded argument of C and therefore an element of the first pair on the list.

$$\begin{aligned}
\mathbf{C}[(r, G_1), (l, G_2), (r, G_3), (l, G_4)] &= \\
G_4 \backslash C[(r, G_1), (l, G_2), (r, G_3)] &= \\
G_4 \backslash (C[(r, G_1), (l, G_2)] / G_3) &= \\
G_4 \backslash (G_2 \backslash C[(r, G_1)]) / G_3) &= \\
G_4 \backslash (G_2 \backslash (C[] / G_1)) / G_3) &= \\
G_4 \backslash (G_2 \backslash (C / G_1)) / G_3)
\end{aligned}$$

Fig. 2.3. Example of $C[(\varepsilon_1, G_1), \ldots, (\varepsilon_p, G_p)]$

Similarly, assume we are given a proof d of $\Delta \vdash C[(\varepsilon_1, G_1), (\varepsilon_2, G_2), \ldots, (\varepsilon_p, G_p)]$ — in what follows, d will just be an axiom $C[\ldots] \vdash C[\ldots]$ — and n proofs d_i of $\Gamma_i \vdash G_i$. Let us call $l_1, l_2, ..., l_l$ (resp. $r_1, r_2, ..., r_{p-l}$) the subsequence of indices (hence it's an increasing sequence) such that $\varepsilon_i = l$ (resp. $\varepsilon_i = r$).

We write $d[(\varepsilon_1, d_1), \ldots, (\varepsilon_p, d_p)]$ for the following proof of

$$\Gamma_{l_1}, \Gamma_{l_2}, \ldots, \Gamma_{l_l}, \Delta, \Gamma_{r_{p-l}}, \Gamma_{r_{p-l-1}}, \ldots, \Gamma_{r_1} \vdash C$$

if $p = 0$ $d[] = d$
if $\varepsilon_i = l$ $d[(\varepsilon_1, d_1), \ldots, (\varepsilon_{i-1}, d_{i-1}), (\varepsilon_i, d_i)] = \backslash_e(d_i, d[(\varepsilon_1, d_1), \ldots, (\varepsilon_{i-1}, d_{i-1})])$.
if $\varepsilon_i = r$ $d[(\varepsilon_1, d_1), \ldots, (\varepsilon_{i-1}, d_{i-1}), (\varepsilon_i, d_i)] = /_e(d[(\varepsilon_1, d_1), \ldots, (\varepsilon_{i-1}, d_{i-1})], d_i)$.

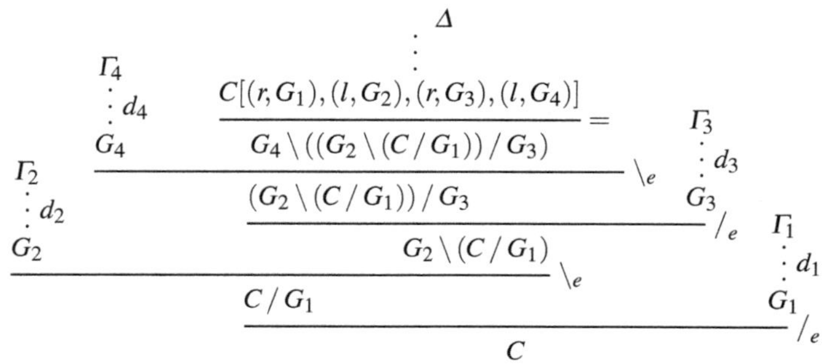

Fig. 2.4. Example: $d[(r,d_1),(l,d_2),(r,d_3),(l,d_4)]$ proving $\Gamma_2,\Gamma_4,\Delta,\Gamma_3,\Gamma_1$

Corollary 2.13. *Whenever a proof d of*

$$H_1,\ldots,H_n \vdash C$$

is normal and the last rule of the proof is an elimination rule, there exists an H_{i_0} *which is equal to* $C[(\varepsilon_1,G_1),\ldots,(\varepsilon_p,G_p)]$ *and subproofs* d_i *of* $\Gamma_i \vdash G_i$ *such that d is* $d[(\varepsilon_1,d_1),\ldots,(\varepsilon_p,d_p)]$.

Proof. Immediate from Proposition 2.10 □

Consequently, with the above notations l_i and r_j, $H_1,\ldots,H_{i_0-1} = \Gamma_{l_1},\Gamma_{l_2},\ldots,\Gamma_{l_l}$ and $H_{i_0+1},\ldots,H_n = \Gamma_{r_{p-l}},\Gamma_{r_{p-l-1}},\ldots,\Gamma_{r_1}$.

Let us prove, by induction on the number of connectives and atoms that $\Gamma \vdash C$ is provable in the Lambek calculus is a decidable question. Because of normalization, if there exists a proof, then there exists a normal proof.

If the (normal) proof ends with an introduction rule, C must be $B\,/\,A$ (or $A \setminus B$) and proving $\Gamma \vdash B\,/\,A$ is equivalent to prove $\Gamma, A \vdash B$ (resp. $A,\Gamma \vdash B$) which is, by induction hypothesis, a decidable question.

Otherwise, the proof ends with an elimination rule. Because of Corollary 2.13 proving $\Gamma \vdash C$ is equivalent to prove a finite number of smaller sequents $\Gamma_i \vdash G_i$ which are obtained as follows: for all hypotheses H_{i_0}in Γ which are of the form $C[(\varepsilon_1,G_1),\ldots,(\varepsilon_p,G_p)]$, with l times $\varepsilon_i = l$ and r times $\varepsilon_i = r$, and for all partitions of the context $H_1,\ldots,H_{i_0-1}$ into l consecutive parts $\Gamma_{l_1},\Gamma_{l_2},\ldots,\Gamma_{l_l}$, and for all partitions of the context $H_{i_0+1},\ldots,H_n$ into r consecutive parts $\Gamma_{r_{p-l}},\Gamma_{r_{p-l-1}},\ldots,\Gamma_{r_1}$ consider the sequent $\Gamma_i \vdash G_i$, which has lesser atoms and connectives and therefore allows us to apply the induction hypothesis.

The above method also provides a decision procedure. The procedure, though simple, is actually rather effective — the only non-deterministic parts are the selection of a formula $C[\ldots]$ corresponding to the conclusion C and partitioning the context for the subproofs. We will improve upon it only in Chapter 6.

The natural deduction decision procedure discussed in this section is also rather close to so-called *normal form* sequent proofs, which are sequent proofs with some procedural restrictions as proposed by König (1989); Hepple (1990); Hendriks (1993) (compare the figure on page 210 of Hendriks, 1993, with our proof search algorithm above), though the correctness of natural deduction proof search is, in our opinion, quite a bit easier to prove using standard proof-theoretic notions such as principal branches. Hepple (1990), in the final section of his article, is the only one to make some brief remarks about a possible connection to natural deduction theorem proving.

2.6.3 Normalization and Lambek Calculus with Product

We have to introduce commutative conversions for the product, otherwise is is possible that a normal proof does not satisfy the subformula property:

$$
\dfrac{\dfrac{\dfrac{\dfrac{\dfrac{A \vdash A \qquad B \vdash B}{A,B \vdash A \bullet B}\bullet_i \qquad D \vdash D}{A,B,D \vdash (A \bullet B) \bullet D}\bullet_i}{A,B \vdash (A \bullet B) \bullet D / D}/_i \qquad A \bullet B \vdash A \bullet B}{A \bullet B \vdash (A \bullet B) \bullet D / D}\bullet_e \qquad D \vdash D}{A \bullet B, D \vdash (A \bullet B) \bullet D}/_e
$$

Let us mention that this can be achieved by adding some "commutative conversions" which basically amount to putting product elimination rules as high as possible (just after the cancelled hypotheses A and B have met), and then rearranging the sub-trees made of product elimination rules with a kind of associativity so that the eliminated product never is the conclusion of another product elimination. Proving this result in full detail is a rather lengthy and technical exercise and is not (in our opinion) very insightful: this kind of result can also be deduced, though indirectly, from the correspondence with the sequent calculus.

2.7 Cut-Elimination for the Sequent Calculus

Cut elimination is the process under which a proof is turned into a proof of the same sequent *without any cut rule* — in other words, the cut rule is redundant.

Cut elimination is one of the fundamental properties of classical and intuitionistic logic, first proved by Gentzen (1934), see e.g. (Girard et al, 1988) for a modern proof and discussion. For L, it was originally proved in (Lambek, 1958).

Cut elimination has an important consequence, which we state before proving cut elimination:

Proposition 2.14. *In a cut-free proof of $A_1, \ldots, A_n \vdash A_{n+1}$ every formula of every sequent is a subformula of some formula A_i ($1 \leq i \leq n+1$).*

Proof. By case inspection it is easily observed that every rule of the sequent calculus except the cut rule satisfies the property that every formula in its premise sequent(s) is a subformula of some formula in its conclusion sequent. □

We give a syntactic proof of cut elimination (while models could be used as well): it is lengthy, tedious and without surprises, but one has to see this kind of proof at least once.

We begin by defining two notions on which we will base the induction. The *degree* of a cut and the *depth* of a cut.

Definition 2.15. *The* degree *of a formula A, written $d(A)$ (or simply d if the formula is clear from the context), is the depth of its subformula tree.*

$$
\begin{aligned}
d(p) &= 1 \qquad \textit{when } p \textit{ is a primitive type} \\
d(A \bullet B) &= \max(d(A), d(B)) + 1 \\
d(A \setminus B) &= \max(d(A), d(B)) + 1 \\
d(B \,/\, A) &= \max(d(A), d(B)) + 1
\end{aligned}
$$

The degree of a cut *is the degree of the cut formula which disappears after application of the rule.*

Definition 2.16. *Let A by a formula which is eliminated by the application of a cut rule c. Let left(A) be the rule which introduces the formula occurrence A on the subproof which ends in the left premise of the cut rule and let right(A) be the rule which introduces the formula occurrence A in the subproof which ends at the right premise of the cut rule. The* depth *of a cut formula A, written r(A) (or r if the cut formula is clear from the context), is the number of rules between left(A) and c plus the number of rules between right(A) and c.*

We show how to remove a single cut of smallest depth, reducing the total number of cut rules by one at each iteration until the proof is cut-free.

We select one of the cut rules with smallest depth and remove it as follows. We proceed by induction on (d,r) with $(d,r) < (d',r')$ if $d < d'$ or $d = d' \wedge r < r'$ where r is the depth of the cut rule, and d the maximal degree of a cut, assumed to be 0 when there is no cut.

$$
\cfrac{\cfrac{\vdots\, \gamma}{\Gamma \vdash X}\, R^a \qquad \cfrac{\vdots\, \delta}{\Delta, X, \Delta' \vdash C}\, R^f}{\Delta, \Gamma, \Delta' \vdash C}\; \textit{cut } d
$$

Notice that because the last rule is a cut of smallest depth, neither R^a nor R^f is a cut rule and neither γ nor δ contain cut rules.

Before exploring all possible values for R^a and R^f, we will first give an overview of the general classes of the reductions; each pair of rules will fall into at least one of the following classes.

1. One of R^a or R^f is an axiom: both the cut and the axiom are suppressed.
2. R^a does not create the cut-formula, — so $R^a \neq \bullet_i, \backslash_i, /_i$. In this case it is possible to apply R^a after the cut. We can apply the induction hypothesis to the proof(s) with R^a and the cut rule reversed — since the depth r of the cut rule is less than the depth of the cut rule before reduction — and turn it into a cut-free proof.
3. If R^f does not create the cut formula, we proceed symmetrically.
4. If both R^a and R^f create the cut formula, then this cut of degree d is replaced by two cuts of strictly smaller degree. Hence, the maximal degree of a cut is strictly smaller (as the last rule was the only cut) and by induction hypothesis we are done.

We only describe the cases for $\backslash$ because the ones for $/$ are strictly symmetrical.

1 R^a or R^f is an axiom The final cut can be suppressed.

$$\frac{X \vdash X \qquad \overset{\vdots\,\delta}{\Gamma, X, \Delta \vdash C}}{\Gamma, X, \Delta \vdash C}\,cut \qquad \text{reduces to} \qquad \overset{\vdots\,\delta}{\Gamma, X, \Delta \vdash C}$$

2 R^a does not create X, the cut formula		
R^a	Before reduction	After reduction
$\bullet_h$	$\dfrac{\dfrac{\overset{\vdots\,\gamma}{\Gamma,A,B,\Gamma' \vdash X}}{\Gamma,A\bullet B,\Gamma' \vdash X}\bullet_h \quad \overset{\vdots\,\delta}{\Delta,X,\Delta' \vdash C}}{\Delta,\Gamma,A\bullet B,\Gamma',\Delta' \vdash C}\,cut\ d$	$\dfrac{\dfrac{\overset{\vdots\,\gamma}{\Gamma,A,B,\Gamma' \vdash X} \quad \overset{\vdots\,\delta}{\Delta,X,\Delta' \vdash C}}{\Delta,\Gamma,A,B,\Gamma',\Delta' \vdash C}\,cut\ d}{\Delta,\Gamma,A\bullet B,\Gamma',\Delta' \vdash C}\bullet_h$
$\backslash_h$	$\dfrac{\dfrac{\overset{\vdots\,\delta}{\Delta,B,\Delta'' \vdash X} \quad \overset{\vdots\,\delta'}{\Delta' \vdash A}}{\Delta,\Delta',A\backslash B,\Delta'' \vdash X}\backslash_h \quad \overset{\vdots\,\gamma}{\Gamma,X,\Gamma' \vdash C}}{\Gamma,\Delta,\Delta',A\backslash B,\Delta'',\Gamma' \vdash C}\,cut\ d$	$\dfrac{\dfrac{\overset{\vdots\,\delta}{\Delta,B,\Delta'' \vdash X} \quad \overset{\vdots\,\gamma}{\Gamma,X,\Gamma' \vdash C}}{\Gamma,\Delta,B,\Delta'',\Gamma' \vdash C}\,cut\ d \quad \overset{\vdots\,\delta'}{\Delta' \vdash A}}{\Gamma,\Delta,\Delta',A\backslash B,\Delta'',\Gamma' \vdash C}\backslash_h$

3 R^f does not create X, the cut formula

R^f	Before reduction	After reduction
$\bullet_h$	$\dfrac{\begin{matrix}\vdots\,\delta\\ \Delta\vdash X\end{matrix}\quad \dfrac{\begin{matrix}\vdots\,\gamma\\ \Gamma,X,\Gamma',A,B,\Gamma''\vdash C\end{matrix}}{\Gamma,X,\Gamma',A\bullet B,\Gamma''\vdash C}\bullet_h}{\Gamma,\Delta,\Gamma',A\bullet B,\Gamma''\vdash C}\,cut\ d$	$\dfrac{\dfrac{\begin{matrix}\vdots\,\delta\\ \Delta\vdash X\end{matrix}\quad\begin{matrix}\vdots\,\gamma\\ \Gamma,X,\Gamma',A,B,\Gamma''\vdash C\end{matrix}}{\Gamma,\Delta,\Gamma',A,B,\Gamma''\vdash C}\,cut\ d}{\Gamma,\Delta,\Gamma',A\bullet B,\Gamma''\vdash C}\bullet_h$
$\bullet_h$	$\dfrac{\begin{matrix}\vdots\,\delta\\ \Delta\vdash X\end{matrix}\quad \dfrac{\begin{matrix}\vdots\,\gamma\\ \Gamma,A,B,\Gamma',X,\Gamma''\vdash C\end{matrix}}{\Gamma,A\bullet B,\Gamma',X,\Gamma''\vdash C}\bullet_h}{\Gamma,A\bullet B,\Gamma',\Delta\Gamma''\vdash C}\,cut\ d$	$\dfrac{\dfrac{\begin{matrix}\vdots\,\delta\\ \Delta\vdash X\end{matrix}\quad\begin{matrix}\vdots\,\gamma\\ \Gamma,X,\Gamma',A,B,\Gamma''\vdash C\end{matrix}}{\Gamma,\Delta,\Gamma',A,B,\Gamma''\vdash C}\,cut\ d}{\Gamma,\Delta,\Gamma',A\bullet B,\Gamma''\vdash C}\bullet_h$
$\backslash_h$	$\dfrac{\begin{matrix}\vdots\,\delta\\ \Delta\vdash X\end{matrix}\quad \dfrac{\begin{matrix}\vdots\,\gamma\\ \Gamma,B,\Gamma'''\vdash C\end{matrix}\quad\begin{matrix}\vdots\,\gamma'\\ \Gamma',X,\Gamma''\vdash A\end{matrix}}{\Gamma,\Gamma',X,\Gamma'',A\backslash B,\Gamma'''\vdash C}\backslash_h}{\Gamma,\Gamma',\Delta,\Gamma'',A\backslash B,\Gamma'''\vdash C}\,cut\ d$	$\dfrac{\begin{matrix}\vdots\,\gamma\\ \Gamma,B,\Gamma'''\vdash C\end{matrix}\quad\dfrac{\begin{matrix}\vdots\,\delta\\ \Delta\vdash X\end{matrix}\quad\begin{matrix}\vdots\,\gamma'\\ \Gamma',X,\Gamma''\vdash A\end{matrix}}{\Gamma',\Delta,\Gamma''\vdash A}\,cutd}{\Gamma,\Gamma',\Delta,\Gamma'',A\backslash B,\Gamma'''\vdash C}\backslash_h$
$\backslash_h$	$\dfrac{\begin{matrix}\vdots\,\delta\\ \Delta\vdash X\end{matrix}\quad \dfrac{\begin{matrix}\vdots\,\gamma\\ \Gamma,B,\Gamma',X,\Gamma''\vdash C\end{matrix}\quad\begin{matrix}\vdots\,\theta\\ \Theta\vdash A\end{matrix}}{\Gamma,\Theta,A\backslash B,\Gamma',X,\Gamma''\vdash C}\backslash_h}{\Gamma,\Theta,A\backslash B,\Gamma',\Delta,\Gamma''\vdash C}\,cut\ d$	$\dfrac{\dfrac{\begin{matrix}\vdots\,\delta\\ \Delta\vdash X\end{matrix}\quad\begin{matrix}\vdots\,\gamma\\ \Gamma,B,\Gamma',X,\Gamma''\vdash C\end{matrix}}{\Gamma,B,\Gamma',\Delta,\Gamma''\vdash C}\,cut\ d\quad\begin{matrix}\vdots\,\theta\\ \Theta\vdash A\end{matrix}}{\Gamma,\Theta,A\backslash B,\Gamma',\Delta,\Gamma''\vdash C}\backslash_h$
$\bullet_i$	$\dfrac{\begin{matrix}\vdots\,\delta\\ \Delta\vdash X\end{matrix}\quad \dfrac{\begin{matrix}\vdots\,\gamma\\ \Gamma,X,\Gamma'\vdash A\end{matrix}\quad\begin{matrix}\vdots\,\theta\\ \Theta\vdash B\end{matrix}}{\Gamma,X,\Gamma',\Theta\vdash A\bullet B}\bullet_i}{\Gamma,\Delta,\Gamma',\Theta\vdash A\bullet B}\,cut\ d$	$\dfrac{\dfrac{\begin{matrix}\vdots\,\delta\\ \Delta\vdash X\end{matrix}\quad\begin{matrix}\vdots\,\gamma\\ \Gamma,X,\Gamma'\vdash A\end{matrix}}{\Gamma,\Delta,\Gamma'\vdash A}\,cut\ d\quad\begin{matrix}\vdots\,\theta\\ \Theta\vdash B\end{matrix}}{\Gamma,\Delta,\Gamma',\Theta\vdash A\bullet B}\bullet_i$
$\bullet_i$	$\dfrac{\begin{matrix}\vdots\,\delta\\ \Delta\vdash X\end{matrix}\quad \dfrac{\begin{matrix}\vdots\,\gamma\\ \Gamma\vdash A\end{matrix}\quad\begin{matrix}\vdots\,\theta\\ \Theta,X,\Theta'\vdash B\end{matrix}}{\Gamma,\Theta,X,\Theta'\vdash A\bullet B}\bullet_i}{\Gamma,\Theta,\Delta,\Theta'\vdash A\bullet B}\,cut\ d$	$\dfrac{\begin{matrix}\vdots\,\gamma\\ \Gamma\vdash A\end{matrix}\quad\dfrac{\begin{matrix}\vdots\,\delta\\ \Delta\vdash X\end{matrix}\quad\begin{matrix}\vdots\,\theta\\ \Theta,X,\Theta'\vdash B\end{matrix}}{\Theta,\Delta,\Theta'\vdash B}\,cut\ d}{\Gamma,\Theta,\Delta,\Theta'\vdash A\bullet B}\bullet_i$
$\backslash_i$	$\dfrac{\begin{matrix}\vdots\,\delta\\ \Delta\vdash X\end{matrix}\quad \dfrac{\begin{matrix}\vdots\,\gamma\\ A,\Gamma,X,\Gamma'\vdash B\end{matrix}}{\Gamma,X,\Gamma'\vdash A\backslash B}\backslash_i}{\Gamma,\Delta,\Gamma'\vdash A\backslash B}\,cut\ d$	$\dfrac{\dfrac{\begin{matrix}\vdots\,\delta\\ \Delta\vdash X\end{matrix}\quad\begin{matrix}\vdots\,\gamma\\ A,\Gamma,X,\Gamma'\vdash B\end{matrix}}{A,\Gamma,\Delta,\Gamma'\vdash B}\,cut\ d}{\Gamma,\Delta,\Gamma'\vdash A\backslash B}\backslash_i$

4 Both R^a and R^f create the cut-formula		
	Before reduction	After reduction
$\bullet$	$\dfrac{\dfrac{\begin{array}{c}\vdots\,\delta\\ \Delta\vdash U\end{array} \quad \begin{array}{c}\vdots\,\theta\\ \Theta\vdash V\end{array}}{\Delta,\Theta\vdash U\bullet V}\bullet_i \quad \dfrac{\begin{array}{c}\vdots\,\gamma\\ \Gamma,U,V,\Gamma'\vdash C\end{array}}{\Gamma,U\bullet V,\Gamma'\vdash C}\bullet_h}{\Gamma,\Delta,\Theta,\Gamma'\vdash C}\,cut\ d$	$\dfrac{\begin{array}{c}\vdots\,\delta\\ \Delta\vdash U\end{array} \quad \dfrac{\begin{array}{c}\vdots\,\theta\\ \Theta\vdash V\end{array} \quad \begin{array}{c}\vdots\,\gamma\\ \Gamma,U,V,\Gamma'\vdash C\end{array}}{\Gamma,U,\Theta,\Gamma'\vdash C}\,cut<d}{\Gamma,\Delta,\Theta,\Gamma'\vdash C}\,cut<d$
$\backslash$	$\dfrac{\dfrac{\begin{array}{c}\vdots\,\delta\\ U,\Delta\vdash V\end{array}}{\Delta\vdash U\backslash V}\backslash_i \quad \dfrac{\begin{array}{c}\vdots\,\gamma\\ \Gamma,V,\Gamma'\vdash C\end{array} \quad \begin{array}{c}\vdots\,\theta\\ \Theta\vdash U\end{array}}{\Gamma,\Theta,U\backslash V,\Gamma'\vdash C}\backslash_h}{\Gamma,\Theta,\Delta,\Gamma'\vdash C}\,cut\ d$	$\dfrac{\dfrac{\begin{array}{c}\vdots\,\theta\\ \Theta\vdash U\end{array} \quad \begin{array}{c}\vdots\,\delta\\ U,\Delta\vdash V\end{array}}{\Theta,\Delta\vdash V}\,cut<d \quad \begin{array}{c}\vdots\,\gamma\\ \Gamma,V,\Gamma'\vdash C\end{array}}{\Gamma,\Theta,\Delta,\Gamma'\vdash C}\,cut<d$

To be fully complete one should check that whenever the original proof contains no sequent with an empty antecedent, so does the cut free proof we inductively defined.

Now let us summarize what we have proved in this section:

Theorem 2.17. *Every proof of a given sequent $\Gamma \vdash C$ can be turned into a cut free proof of the same sequent — all formulae in the cut-free proof being subformulae of the sequent $\Gamma \vdash C$.*

When we compare normalization for natural deduction to cut elimination for the sequent calculus, we saw in Section 2.6.3 that natural deduction required commutative conversions only for the product formulas whereas the "commutative conversions" of the sequent calculus, which are items 2 and 3 of the case analysis for cut elimination above, make up a rather large part of the proof.

2.8 Decidability

One may wonder why we wanted to have normal or cut free proofs since the computational process of cut elimination or normalization is of little interest for categorial grammars.

What is nevertheless very interesting about such a result is that instead of looking for any proof when we want, for instance to parse and analyze a sentence, we can restrict our search space to these canonical proofs, either normal deductions or cut-free proofs. As we have seen, cut elimination (or normalization for natural deduction) entails the subformula property and this makes it quite easy to show that the calculus is decidable:

Proposition 2.18. *There is an algorithm which decides whether a sequent is derivable in L.*

Proof. Assume we want to prove a sequent. Since the cut rule is not needed, we have finitely many rules to try, each of these rules requiring us to prove one or two smaller sequents which, since they contain only subformulae of the sequent we want to prove, are also in finite number. □

Though sequent proof search is decidable, it is rather less efficient than the natural deduction algorithm we presented for the product-free calculus in Section 2.6.2. In particular, there can be many sequent calculus proofs which correspond to a single natural deduction proof and these different sequent calculus proofs differ only for 'bureaucratic' reasons. So though both normal natural deduction proofs and cut-free sequent proofs are canonical proofs in their respective calculi, there are many more of these canonical proofs in the sequent calculus than there are in natural deduction.

This problem of sequent calculus proof search is often called spurious ambiguity. On the other hand, the product formulas are unproblematic for cut-free sequent proof search. We will return to these issues in Chapter 6, were we will introduce proof nets for the Lambek calculus and show how they both handle product formulas unproblematically and make the problem of spurious ambiguity disappear.

2.9 Models for the Lambek Calculus and Completeness

We now turn our attention towards models for the Lambek calculus. As we have seen that as far as provability is concerned, cut-free sequent calculus, sequent calculus and natural deduction are equivalent, we are going to use the most adequate formalism to establish properties of models with respect to the deductive system.

These models have been first investigated in (Buszkowski, 1982) and our presentation follows (Buszkowski, 1997).

As we have said Lambek calculus prohibits the empty sequence, and we will present models for L with this restriction. Let us nevertheless say that all these results can be adapted by adding a unit to residuated semi-groups and to semi-groups — replacing the word "semi-group" with the word "monoid".

2.9.1 Residuated Semi-groups

Let us call a *residuated semi-group*, a structure $(M, \circ, \backslash\backslash, //, \sqsubset)$ where

- M is a set.
- $\circ$ is an associative composition over M — $(M, \circ)$ is a semi-group.
- $\backslash\backslash$ and $//$ are binary composition laws on M.
- $\sqsubset$ is an order on M.

which satisfies the following property:

(RSG) The following order relations are equivalent:

$$a \sqsubset (c \,//\, b)$$
$$(a \circ b) \sqsubset c$$
$$b \sqsubset (a \,\backslash\backslash\, c)$$

Proposition 2.19. *In a residuated semi-group* $(M, \circ, \backslash\backslash, //, \sqsubset)$, *for all* $a, b, x, y \in M$ *one has:*

1. $a \sqsubset b \quad \Rightarrow \quad (a \circ x) \sqsubset (b \circ x)$
2. $a \sqsubset b \quad \Rightarrow \quad (x \circ a) \sqsubset (x \circ b)$
3. $\begin{pmatrix} a \sqsubset b \\ \textit{and} \\ x \sqsubset y \end{pmatrix} \quad \Rightarrow \quad (a \circ x) \sqsubset (b \circ y)$

In other words, a residuated semi-group is in particular an ordered semi-group.

Proof. (1) From $(b \circ x) \sqsubset (b \circ x)$ ($\sqsubset$ is an order) (RSG) yields $b \sqsubset ((b \circ x) // x)$; if we assume $a \sqsubset b$ by transitivity of $\sqsubset$ we have $a \sqsubset ((b \circ x) // x)$ which by (RSG) yields $(a \circ x) \sqsubset (b \circ x)$.

(2) From $(x \circ b) \sqsubset (x \circ b)$ ($\sqsubset$ is an order) (RSG) yields $b \sqsubset (x \backslash\backslash (x \circ b))$; if we assume $a \sqsubset b$ by transitivity of $\sqsubset$ we have $a \sqsubset (x \backslash\backslash (x \circ b))$ which by (RSG) yields $(x \circ a) \sqsubset (x \circ b)$.

(3) The assumption $a \sqsubset b$ yields $(a \circ x) \sqsubset (b \circ x)$ (*) by (1). The assumption $x \sqsubset y$ yields $(b \circ x) \sqsubset (b \circ y)$ (**) by (2). By transitivity of $\sqsubset$, (*) and (**) yields $(a \circ x) \sqsubset (b \circ y)$. □

Given a residuated semi-group, an *interpretation* $[.]$ is a map from primitive types to elements in M, which extends to types and sequences of types in the obvious way:

$$[A, B] = [A] \circ [B] \qquad [A \backslash B] = [A] \backslash\backslash [B]$$
$$[A \bullet B] = [A] \circ [B] \qquad [B / A] = [B] // [A]$$

A sequent $\Gamma \vdash C$ is said to be *valid* in a residuated semi-group whenever $[\Gamma] \sqsubset [C]$.

2.9.2 The Free Group Model

A particular case of residuated semi-group is the free group over primitive types. It will be especially important in Section 2.11. The free group interpretation for L is

- a particular residuated semi-group where
  - $(M, \cdot)$ is the free group over the propositional variables,
  - $a \backslash\backslash b$ is $a^{-1}b$
  - $b // a$ is ba^{-1}
  - $a \sqsubset b$ is $a = b$ (the discrete order)

  One easily observes that the three equalities

  $$ab = c \qquad a = cb^{-1} \qquad b = a^{-1}c$$

  are equivalent — so (RSG) holds.
- a standard interpretation defined by $[p] = p$

Because of the soundness of L w.r.t. residuated semi-groups (next proposition) whenever a sequent $\Gamma \vdash C$ is provable one has $[\Gamma] = [C]$ in the free group. The free group model is of course not complete: indeed it interprets $\vdash$ by a symmetrical relation ($=$) while $\vdash$ is not symmetrical: $n \vdash s / (n \backslash s)$ is provable but not $s / (n \backslash s) \vdash n$.

2.9.3 L Is Sound and Complete w.r.t. Residuated Semi-groups

Proposition 2.20. *A provable sequent is valid in every residuated semi-group, for every interpretation of the primitive types.*

Proof. We proceed by induction on the natural deduction proof.

- If the proof consists in an axiom $X \vdash X$ then the result is true: $[X] \sqsubset [X]$ whatever the semi-group or the interpretation is.
- If the last rule is the introduction rule $\backslash_i$:

$$\frac{A, \Gamma \vdash C}{\Gamma \vdash A \backslash C} \backslash_i \quad \Gamma \neq \varepsilon$$

by induction hypothesis we have $[A] \circ [\Gamma] \sqsubset [C]$, thus, by (RSG) we have $[\Gamma] \sqsubset ([A] \backslash\backslash [C])$, so the sequent $\Gamma \vdash A \backslash C$ is valid as well.

If the last rule is the introduction rule $/_i$ we proceed as for $\backslash_i$.

If the last rule is the elimination rule $\backslash_e$:

$$\frac{\Gamma \vdash A \quad \Delta \vdash A \backslash B}{\Gamma, \Delta \vdash B} \backslash_e$$

then by induction hypothesis we know that $[\Gamma] \sqsubset [A]$, and using Proposition 2.19 we can conclude $[\Gamma] \circ [\Delta] \sqsubset [A] \circ [\Delta]$ (1); we also have $[\Delta] \sqsubset [A] \backslash\backslash [B]$ — hence by (RSG) $([A] \circ [\Delta]) \sqsubset [B]$ (2). Therefore from (1) and (2) we obtain,

$$[\Gamma, \Delta] \quad = \quad [\Gamma] \circ [\Delta] \quad \sqsubset (1) \quad [A] \circ [\Delta] \quad \sqsubset (2) \quad [B]$$

- If the last rule is the elimination rule $/_e$ we proceed as for $/_i$.
- If the last rule is the product elimination rule $\bullet_e$

$$\frac{\Gamma \vdash A \bullet B \quad \Delta, A, B, \Delta' \vdash C}{\Delta, \Gamma, \Delta' \vdash C} \bullet_e$$

By induction hypothesis we know that $[\Gamma] \sqsubset [A \bullet B] = [A] \circ [B]$, and, using Proposition 2.19 we obtain $[\Delta] \circ [\Gamma] \circ [\Delta'] \sqsubset [\Delta] \circ [A] \circ [B] \circ [\Delta']$. We also know that $[\Delta, A, B, \Delta'] = [\Delta] \circ [A] \circ [B] \circ [\Delta'] \sqsubset [C]$. We therefore have

$$[\Delta, \Gamma, \Delta'] = [\Delta] \circ [\Gamma] \circ [\Delta'] \sqsubset [\Delta] \circ [A] \circ [B] \circ [\Delta'] \sqsubset [C]$$

- If the last rule is the product introduction rule $\bullet_i$ by induction hypothesis we know that $[\Delta] \sqsubset [A]$ and that $[\Delta'] \sqsubset [B]$; consequently

$$[\Delta, \Delta'] = [\Delta] \circ [\Delta'] \sqsubset [A] \circ [B] = [A \bullet B]$$

□

Proposition 2.21. *A sequent which is valid in every residuated semi-group is derivable.*

Proof. Let F be the set of formulae and let $M = F / \dashv\vdash$ be the quotient of formulae by the equivalence relation $\dashv\vdash$; this relation $\dashv\vdash$ is defined by $A \dashv\vdash B$ whenever $A \vdash B$ and $B \vdash A$; it is symmetrical, it is reflexive by the axiom rule and the cut rule ensures it is transitive.

Is is easily observed that $\backslash$, $/$, $\bullet$ and $\vdash$ can be defined over equivalence classes, that is: whenever $A \dashv\vdash A'$ and $B \dashv\vdash B'$ one has $(A \diamond B) \dashv\vdash (A' \diamond B')$, for $\diamond \in \{\backslash, /, \bullet\}$. So let us define $\circ, \backslash\backslash, //$ as the corresponding operations over equivalence classes of $\dashv\vdash$: $A^{\dashv\vdash} \circ B^{\dashv\vdash} = (A \bullet B)^{\dashv\vdash}$, $A^{\dashv\vdash} \backslash\backslash B^{\dashv\vdash} = (A \backslash B)^{\dashv\vdash}$ and $B^{\dashv\vdash} // A^{\dashv\vdash} = (B / A)^{\dashv\vdash}$ Finally let $\sqsubset$ be $\vdash$ which can also be defined for equivalence classes: if $A \dashv\vdash A'$ and $B \dashv\vdash B'$ then $A \vdash B$ is equivalent to $A' \vdash B'$.

The property (RSG) is satisfied i.e. $(A^{\dashv\vdash} \circ B^{\dashv\vdash}) \sqsubset C^{\dashv\vdash}$ is equivalent to $A^{\dashv\vdash} \sqsubset (C^{\dashv\vdash} // B^{\dashv\vdash})$ and to $A^{\dashv\vdash} \sqsubset (B^{\dashv\vdash} \backslash\backslash C^{\dashv\vdash})$. Indeed $A \vdash A$ and $B \vdash B$ lead to $A, B \vdash A \bullet B$; thus from $A \bullet B \vdash C$ one obtains $A, B \vdash C$ which yields $A \vdash C / B$ by $/_i$ and $B \vdash A \backslash C$ by $\backslash_i$; from $B \vdash A \backslash C$ (resp. $A \vdash B / C$) using $A \vdash A$ (resp. $B \vdash B$) one obtains $A, B \vdash C$ by $\backslash_e$ (resp. by $/_e$) and $A \bullet B \backslash C$ by $\bullet_h$.

Now let us consider the interpretation $[p] = p^{\dashv\vdash}$ for every primitive type. Then for every formula $[A] = A^{\dashv\vdash}$.

To say that a sequent $H_1, \ldots, H_n \vdash A$ is valid in this model under this interpretation is to say that $[H_1, \ldots, H_n] \sqsubset [A]$. Therefore $H_1 \bullet \cdots \bullet H_n \vdash A$ is provable which entails that $H_1, \ldots, H_n \vdash A$ is provable as well — indeed from $H_1 \bullet \cdots \bullet H_n \vdash A$ one obtains $\vdash H_1 \bullet \cdots \bullet H_n \backslash A$ (*); then by n rules $\bullet_i$ on the axioms $H_i \vdash H_i$ one obtains $H_1, \ldots, H_n \vdash H_1 \bullet \cdots \bullet H_n$ (**) and an application of $\bullet_e$ to (*) and (**) yields $H_1, \ldots, H_n \vdash A$. ☐

2.9.4 L Is Sound and Complete w.r.t. (Free) Semi-group Models

A more interesting class of models is provided by semi-groups. Indeed, the interpretation of a category should be the set of the words and expressions of this category, shouldn't it?

So, given a semi-group $(W, .)$ that is a set W endowed with an associative composition "." one can define a residuated semi-group as follows:

- $M = 2^W$
- $A \circ B = \{ab \mid a \in A \text{ and } b \in B\}$
- $A \backslash\backslash B = \{z \mid \forall a \in A \quad az \in B\}$
- $B // A = \{z \mid \forall a \in A \quad za \in B\}$
- $A \sqsubset B$ whenever $A \subset B$ (as sets).

It is easily seen that this structure really is a residuated semi-group:

- $\circ$ is associative:

$$(A \circ B) \circ C = \{abc \mid a \in A \text{ and } b \in B \text{ and } c \in C\} = A \circ (B \circ C)$$

- $\subset$ is an order on 2^W

  (RSG) The following statements are clearly equivalent:

$$\begin{aligned} (A \circ B) \subset C &: \forall a \in A \ \forall b \in B \quad ab \in C \\ A \subset (C \mathbin{/\!/} B) &: \forall a \in A \quad a \in (C \mathbin{/\!/} B) \\ B \subset (A \mathbin{\backslash\!\backslash} C) &: \forall b \in B \quad b \in (A \mathbin{\backslash\!\backslash} C) \end{aligned}$$

The free semi-group models are particularly interesting, since there are no equations between sequences of words. The following proposition may be understood as stating that L is the logic of free semi-groups:

Proposition 2.22. *Product free L is complete over free semi-group models.*

Proof. Take as semi-group the finite non empty sequences of formulae F^+, endowed with concatenation $(A_1,\ldots,A_n)\cdot(B_1,\ldots,B_p) = A_1,\ldots,A_n,B_1,\ldots,B_p$.

For a primitive type p define $[p]$ by $\{\Gamma \mid \Gamma \vdash p\}$.

Let us first verify that for every formula F, the set of finite sequences $[F]$ defined inductively from the $[p]$'s by the definition of $\backslash\backslash$ and $/\!/$ is precisely $Ctx(F) = \{\Delta \mid \Delta \vdash F\}$. We proceed by induction on F. Is F if some primitive type, it is true by definition. Now assume that $[G] = Ctx(G)$ and $[H] = Ctx(H)$ and let us verify that $[G \backslash H] = Ctx(G \backslash H)$ — the case for $H\,/\,G$ being symmetrical.

$Ctx(G \backslash H) \subset [G \backslash H]$ Let Δ be a sequence such that $\Delta \in Ctx(G \backslash H)$ that is $\Delta \vdash G \backslash H$ (1) and let us see that for every $\Theta \in [G]$ we have $\Theta,\Delta \in [H]$ — which entails $\Delta \in [G \backslash H]$. By induction hypothesis we have $Ctx(G) = [G]$ so $\Theta \vdash G$ (2). From (1) and (2) we obtain $\Theta,\Delta \vdash H$, so $\Theta,\Delta \in Ctx(H)$. Since by induction hypothesis $Ctx(H) = [H]$ we have $\Theta,\Delta \in [H]$. As this holds for every Θ we have $\Delta \in [G \backslash H]$.

$[G \backslash H] \subset Ctx(G \backslash H)$ Let Δ be a sequence such that $\Delta \in [G \backslash H]$. Let us show that $\Delta \vdash G \backslash H$. Since $G \vdash G$ we have $G \in Ctx(G)$ and by induction hypothesis $G \in [G]$. By the definition of $[G \backslash H]$ we thus have $G,\Delta \in [H]$ and, since by induction hypothesis we have $[H] = Ctx(H)$ we obtain $G,\Delta \vdash H$. Now, by the $\backslash_i$ introduction rule we obtain $\Delta \vdash G \backslash H$, that is $\Delta \in Ctx(G \backslash H)$.

If a sequent $A_1,\ldots,A_n \vdash C$ is valid in this model under this interpretation, what does it mean? We have $[A_1] \circ \cdots \circ [A_n] \subset [C]$ and as $A_i \in [A_i]$ we have $A_1,\ldots,A_n \in [C]$ that is $A_1,\ldots,A_n \vdash C$. □

Next follows a very difficult result due to Pentus (Pentus, 1993a), that we state without giving the proof.

Proposition 2.23. *L with product is also complete w.r.t. free semi-groups models.*

2.10 Interpolation

This section presents the interpolation theorem for the Lambek calculus, which appeared in the thesis of Roorda (Roorda, 1991).

Interpolation is somehow the converse of cut elimination. The interest of cut free proofs is that they obey the subformula property. The usual interest of interpolation, say for classical or intuitionistic logic is to be able to factor equal sub-proofs in a given proof. In the Lambek calculus where contraction is prohibited, nothing like this can happen. So the interest is very different, let us explain it in a few words.

Assume we are able to formulate the calculus with a set of axioms, and only the cut rule: viewing $\vdash$ as $\longrightarrow$ (in the opposite direction) the calculus is nothing but a set of context-free production rules — the cut rule is the substitution rule often left implicit in phrase structure grammars.

Indeed a production rule $X \longrightarrow X_1 \cdots X_n$ corresponds to an axiom $X_1, \ldots, X_n \vdash X$ and the cut rule simply states that if we have been able to derive

$$W \longrightarrow V_1 \cdots V_k\, T\, U_1 \cdots U_l$$
$$T \longrightarrow Z_1 \cdots Z_j$$

then we are able to derive

$$W \longrightarrow V_1 \cdots V_k\, Z_1 \cdots Z_j\, U_1 \cdots U_l.$$

Now observe that for a given Lambek grammar because of cut elimination we know that the types appearing in any syntactic analysis are all subformulae of the conclusion sequent: indeed a syntactic analysis is a proof of $t_1, \ldots, t_n \vdash S$ with all t_i in the lexicon. Can we derive any syntactic analysis from a *finite* number of provable sequents by means of the cut rule only? As we shall see in the next section, this is possible and consequently Lambek grammars are weakly equivalent to context-free grammars.

Given a formula or a sequence of formulae Δ and a primitive type p we denote by $\rho_p(\Delta)$ *the number of occurrences of p in* Δ.

Proposition 2.24. *Let $\Gamma, \Delta, \Theta \vdash C$ be a provable sequent in L, with $\Delta \neq \varepsilon$. There exists an interpolant of Δ that is a formula I such that:*

1. $\Delta \vdash I$
2. $\Gamma, I, \Theta \vdash C$
3. $\rho_p(I) \leq \rho_p(\Delta)$ *for every primitive type p*
4. $\rho_p(I) \leq \rho_p(\Gamma, \Theta, C)$ *for every primitive type p*

Proof. We proceed by induction on the size of a cut free proof of $\Gamma, \Delta, \Theta \vdash C$ — there are many cases in this proof, according to the nature of the last rule, and to the respective position of the created formula and Δ.

axiom $X \vdash X$

If the proof is an axiom, then Δ is a formula X and $I = X$ obviously works:

1. $X \vdash X$
2. $X \vdash X$
3. $\rho_p(X) = \rho_p(X)$
4. $\rho_p(X) = \rho_p(\varepsilon, \varepsilon, X)$

$$\frac{\Pi \vdash X \quad \Phi \vdash Y}{\Pi, \Phi \vdash X \bullet Y} \bullet_i$$

- $\Pi = \Pi', \Delta, \Pi''$ — so $\Gamma = \Pi'$ and $\Theta = \Pi'', \Phi$.
  By induction hypothesis we have an interpolant I for Δ in $\Pi', \Delta, \Pi'' \vdash X$, let us see it is an interpolant for Δ in $\Pi', \Delta, \Pi'', \Phi \vdash X \bullet Y$.
  1. We already have $\Delta \vdash I$
  2. From $\Pi', I, \Pi'' \vdash X$ and $\Phi \vdash Y$, we have $\Pi', I, \Pi'', \Phi \vdash X \bullet Y$.
  3. We already have $\rho_p(I) \leq \rho_p(\Delta)$.
  4. From $\rho_p(I) \leq \rho_p(\Pi', \Pi'', X)$ we obtain $\rho_p(I) \leq \rho_p(\Pi', \Pi'', \Phi, X, Y)$.
- $\Phi = \Phi', \Delta, \Phi''$ — so $\Gamma = \Pi, \Phi'$ and $\Theta = \Phi''$.
  Symmetrical to the previous case.
- $\Pi = \Pi', \Delta'$ $\Phi = \Delta'', \Phi''$ and $\Delta = \Delta', \Delta''$ — so $\Gamma = \Pi'$ and $\Theta = \Phi''$.
  By induction hypothesis we have an interpolant I' for Δ' in $\Pi', \Delta' \vdash X$ and an interpolant I'' for Δ'' in $\Delta'', \Phi'' \vdash X$. Then $I = I' \bullet I''$ is an interpolant for $\Delta = \Delta', \Delta''$ in $\Pi', \Delta', \Delta'', \Phi'' \vdash X \bullet Y$.
  1. From $\Delta' \vdash I'$ and $\Delta'' \vdash I''$ we obtain $\Delta', \Delta'' \vdash X \bullet Y$ by $\bullet_i$.
  2. From $\Pi', I' \vdash X$ and $I'', \Phi'' \vdash Y$ we have $\Pi', I', I'', \Phi'' \vdash X \bullet Y$ by $\bullet_i$ and finally $\Pi', I' \bullet I'', \Phi'' \vdash X \bullet Y$ by $\bullet_h$
  3. From $\rho_p(I') \leq \rho_p(\Pi', X)$ and $\rho_p(I'') \leq \rho_p(\Phi'', Y)$ we get $\rho_p(I' \bullet I'') = \rho_p(I') + \rho_p(I'') \leq \rho_p(\Pi', X) + \rho_p(\Phi'', Y) = \rho_p(\Pi', \Phi'', X, Y) = \rho_p(\Pi', \Phi'', X \bullet Y)$.
  4. From $\rho_p(I') \leq \rho_p(\Delta')$ and $\rho_p(I'') \leq \rho_p(\Delta'')$ we get $\rho_p(I' \bullet I'') = \rho_p(I') + \rho_p(I'') \leq \rho_p(\Delta', \Delta'') = \rho_p(\Delta)$.

$$\frac{\Pi, X, Y, \Phi \vdash C}{\Pi, X \bullet Y, \Phi \vdash C} \bullet_h$$

Let Δ' be defined as follows: if Δ contains $X \bullet Y$ then $\Delta' = \Delta[X \bullet Y := X, Y]$, otherwise $\Delta' = \Delta$. Let I be an interpolant for Δ' in $\Pi, X, Y, \Phi \vdash C$. Then I is itself an interpolant for Δ in $\Pi, X \bullet Y, \Phi \vdash C$.

1. From $\Delta' \vdash I$ we have $\Delta \vdash I$ (possibly using $\bullet_h$).
2. From $\Pi, X \bullet Y, \Phi[\Delta' := I] \vdash C$ we get $\Pi, X \bullet Y, \Phi[\Delta := I] \vdash C$.
3. From $\rho_p(\Delta) = \rho_p(\Delta')$ we obtain $\rho_p(I) \leq \rho_p(\Delta)$
4. Since $\rho_p((\Pi, X \bullet Y, \Phi)[\Delta' := \varepsilon], C) = \rho_p((\Pi, X \bullet Y, \Phi)[\Delta := \varepsilon], C)$ we have $\rho_p(I) \leq \rho_p((\Pi, X \bullet Y, \Phi)[\Delta := \varepsilon], C)$.

$$\frac{X, \Gamma, \Delta, \Theta \vdash Y}{\Gamma, \Delta, \Theta \vdash X \backslash Y} \backslash_i$$

By induction hypothesis we have an interpolant I for Δ in $A, \Gamma, \Delta, \Theta \vdash B$. It is an interpolant for Δ in $\Gamma, \Delta, \Theta \vdash X \backslash Y$ as well.

1. We already have $\Delta \vdash I$.
2. From $X, \Gamma, I, \Theta \vdash Y$ we obtain $\Gamma, I, \Theta \vdash X \backslash Y$ by $\backslash_i$.

3. We already have $\rho_p(I) \leq \rho_p(\Delta)$.
4. We have: $\rho_p(I) \leq \rho_p(X,\Gamma,\Theta,Y) = \rho_p(\Gamma,\Theta,X \setminus Y)$.

$$\frac{\Pi \vdash X \quad \Phi,Y,\Psi \vdash C}{\Phi,\Pi,X \setminus Y,\Psi \vdash C} \setminus_h$$

- Δ is included into Π Let I be an interpolant for Δ in the premise containing it. Then I is an interpolant for Δ in $\Phi,\Pi,X \setminus Y,\Psi \vdash C$.
  1. We already have $\Delta \vdash I$
  2. From $\Pi[\Delta := I] \vdash X$ and $\Phi,Y,\Psi \vdash C$, by $\setminus_h$ we obtain $\Phi,\Pi[\Delta := I],X \setminus Y,\Psi \vdash C$
  3. We already have $\rho_p(I) \leq \rho_p(\Delta)$.
  4. From $\rho_p(I) \leq \rho_p(\Pi[\Delta := \varepsilon],X)$ we have $\rho_p(I) \leq \rho_p(\Phi,\Pi[\Delta := \varepsilon],X \setminus Y,\Psi,C)$
- Δ is included in Φ (resp. Ψ) Let I be an interpolant for Δ in the premise containing it. Then I is an interpolant for Δ in $\Phi,\Pi,X \setminus Y,\Psi \vdash C$.
  1. We already have $\Delta \vdash I$
  2. From $\Phi[\Delta := I],Y,\Psi \vdash C$ (resp. $\Phi,Y,\Psi[\Delta := I] \vdash C$) and $\Pi \vdash X$, by $\setminus_h$ we obtain $\Phi[\Delta := I],\Pi,X \setminus Y,\Psi \vdash C$ (resp. $\Phi,\Pi,X \setminus Y,\Psi[\Delta := I] \vdash C$)
  3. We already have $\rho_p(I) \leq \rho_p(\Delta)$.
  4. From $\rho_p(I) \leq \rho_p(\Phi[\Delta := \varepsilon],Y,\Psi,C)$ (resp. $\rho_p(I) \leq \rho_p(\Phi,Y,\Psi[\Delta := \varepsilon],C)$) we have $\rho_p(I) \leq \rho_p(\Phi[\Delta := \varepsilon],\Pi,X \setminus Y,\Psi,C)$ (resp. $\rho_p(I) \leq \rho_p(\Phi,\Pi,X \setminus Y,\Psi[\Delta := \varepsilon],C)$).
- $\Delta = \Delta',\Delta''$ and $\Phi = \Phi',\Delta'$ and $\Pi = \Delta'',\Pi''$.
  Let I' be an interpolant for Δ' in $\Phi',\Delta',Y,\Psi \vdash C$, and let I'' be an interpolant for Δ'' in $\Delta'',\Pi'' \vdash X$. Then $I = I' \bullet I''$ is an interpolant for Δ',Δ'' in $\Phi',\Delta',\Delta'',\Pi''$, $X \setminus Y,\Psi \vdash C$.
  1. From $\Delta' \vdash I'$ and $\Delta'' \vdash I''$ we have $\Delta',\Delta'' \vdash I' \bullet I''$ by $\bullet_i$.
  2. From $I'',\Pi'' \vdash X$ and $\Phi',I',Y,\Psi \vdash C$ we have $\Phi',I',I'',X \setminus Y,\Psi \vdash C$ by $\setminus_h$ and $\Phi',I' \bullet I'',X \setminus Y,\Psi \vdash C$ by $\bullet_i$.
  3. We have $\rho_p(I' \bullet I'') = \rho_p(I') + \rho_p(I'') \leq \rho_p(\Delta') + \rho_p(\Delta'') = \rho_p(\Delta)$.
  4. We have $\rho_p(I' \bullet I'') = \rho_p(I') + \rho_p(I'') \leq \rho_p(\Phi',Y,\Psi,C) + \rho_p(\Pi'',X) = \rho_p(\Phi',\Pi'',X \setminus Y,\Psi,C)$.
- $\Delta = \Phi'',\Pi,X \setminus Y,\Psi'$ with $\Phi = \Phi',\Phi''$ and $\Psi = \Psi',\Psi''$.
  Let I be an interpolant for Φ'',Y,Ψ' in $\Phi',\Phi'',Y,\Psi',\Psi'' \vdash C$. Then I is itself interpolant for $\Phi'',\Pi,X \setminus Y,\Psi'$ in $\Phi',\Phi'',\Pi,X \setminus Y,\Psi',\Phi'' \vdash C$.
  1. From $\Phi'',Y,\Psi' \vdash I$ and $\Pi \vdash X$ we have $\Phi'',\Pi,X \setminus Y,\Psi' \vdash I$ by $\setminus_h$.
  2. We already have $\Phi',I,\Psi'' \vdash C$.
  3. We already have $\rho_p(I) \leq \rho_p(\Phi',\Psi'',C)$
  4. We have $\rho_p(I) \leq \rho_p(\Phi'',Y,\Psi') \leq \rho_p(\Phi'',\Pi,X \setminus Y,\Psi')$.
- $\Delta = \Pi'',X \setminus Y,\Psi'$ with $\Pi = \Pi',\Pi''$ and $\Psi = \Psi',\Psi''$.
  Let I' be and interpolant for Π' in $\Pi',\Pi'' \vdash X$ and let I'' be an interpolant for Y,Ψ' in $\Phi,Y,\Psi',\Psi'' \vdash C$. Then $I' \setminus I''$ is an interpolant for $\Delta = \Pi'',X \setminus Y,\Psi'$ in $\Phi,\Pi',\Pi'',X \setminus Y,\Psi',\Psi'' \vdash C$.

1. From $\Gamma', \Pi'' \vdash X$ and $Y, \Psi' \vdash \Gamma''$ we have $\Gamma', \Pi'', X \setminus Y, \Psi' \vdash \Gamma''$ by $\setminus_h$ and $\Pi'', X \setminus Y, \Psi' \vdash \Gamma' \setminus \Gamma''$ by $\setminus_i$.
2. From $\Phi, \Gamma'', \Psi'' \vdash C$ and $\Pi' \vdash \Gamma'$ we have $\Phi, \Pi', \Gamma' \setminus \Gamma'', \Psi'' \vdash C$.
3. We have $\rho_p(\Gamma' \setminus \Gamma'') \leq \rho_p(\Pi'', X) + \rho_p(Y, \Psi') = \rho_p(\Pi'', X \setminus Y, \Psi')$
4. We have $\rho_p(\Gamma' \setminus \Gamma'') \leq \rho_p(\Pi') + \rho_p(\Phi, \Psi'', C) = \rho_p(\Phi, \Pi', \Psi'', C)$

This ends the proof because $/_i$ and $/_e$ are symmetrical to $\setminus_i$ and $\setminus_e$. □

2.11 Lambek Grammars and Context-Free Grammars

At the beginning of this section we shall see that context-free grammars translate into weakly equivalent Lambek grammars (Cohen, 1967): this is non trivial but unsurprising, and this section is in fact devoted to prove the converse, known as the Chomsky conjecture, stated in 1963 (Chomsky, 1963, p. 413) and proved by Pentus (Pentus, 1993b, 1997): *Languages generated by Lambek grammars are context-free languages*. This result was already suggested in the previous section on interpolation: if we are able to derive all sequents corresponding to syntactic analyses from a *finite* set of sequents by the cut rule only, then Lambek grammars are context-free.

Let us define the *size* $|A|$ of a formula A by its number of primitive types. We are going to show that given an integer m there exists a *finite* set $AX(m)$ of provable sequents such that all provable sequent containing only formulae of size smaller than m are derivable from sequents in $AX(m)$ by means of the cut rule only. This easily entails that Lambek grammars are context-free. Note that the restriction on the size of formulas is essential, since Zielonka (1981, 1989) shows that the Lambek calculus in general (without this size restriction) does *not* permit a formulation consisting of a finite number of sequents AX and the cut rule.

Even though Lambek grammars generate only context-free languages, they have a number of pleasant properties — in addition to their logical foundations — which make them interesting objects of study:

- they are lexicalized,
- they offer a pleasant interface with semantics (as we will see in Chapter 3),
- they permit an elegant treatment of peripheral extraction (as we have seen in Example 2.2, for example). To the best of our knowledge, among the grammar formalisms which generate context-free languages, only the Lambek calculus (and some closely related formalisms) have such a simple and elegant treatment of peripheral extraction.
- finally, let us say that while the derivation trees of a context-free grammars constitute a regular tree language (Thatcher, 1967; Gécseg and Steinby, 1997) the derivation trees (natural deduction trees) of a Lambek grammar can constitute a tree language which is not regular (Tiede, 2001): Kanazawa and Salvati (2009) show that the natural deduction trees of a Lambek grammar correspond to tree languages generated by hyperedge replacement grammars (Engelfriet, 1997; Engelfriet and Maneth, 2000). In other words, if we are interested in *trees* rather than strings, Lambek grammars are more expressive than context-free grammars.

There are basically two ingredients for the Pentus proof that Lambek grammars are context-free. One is interpolation and we already explained its relevance to this question. The other is a property of the free group to be applied to the free group model of Section 2.9.2 on page 45. This property is needed to find, in a sequent where all formulae have sizes lower than m, two (or more) consecutive formulae whose interpolant also has a size less than m — this is of course to be used for the final induction.

We mainly follow (Pentus, 1993b), and borrow a few things from (Pentus, 1997; Buszkowski, 1997).

2.11.1 From Context-Free Grammars to Lambek Grammars

It is natural to think that every AB grammar corresponds to a Lambek grammar, because the Lambek calculus includes the AB elimination rules and is therefore at least as expressive as AB grammars. In fact this result — although not as difficult as the result we will prove in the remainder of this section, where we will show how to translate Lambek grammars to context-free grammars — is not completely trivial: not all theorems of L are also theorems of AB, so we need to be careful about using an AB grammar as a Lambek grammar.

By Proposition 1.2 from Chapter 1, we know that any AB grammar is weakly equivalent to an AB grammar containing only types of order at most 1. Now, by Proposition 2.12 a sequent $A_1,\dots,A_n \vdash S$ with $o(A_i) \leq 1$ is provable with AB elimination rules if and only if it is provable in L. Consequently the language generated by an AB grammar with types of order at most 1 coincides with the language generated by the Lambek grammar with the same lexicon.

Using the weak equivalence between AB grammars and context-free grammars (Propositions 1.11 and 1.10) we have the result of (Cohen, 1967):

Proposition 2.25. *Every ε-free context-free grammar is weakly equivalent to a Lambek grammar.*

2.11.2 A Property of the Free Group

Let w be an element of the free group; then $||w||$ stands for the length of the reduced word corresponding to w — e.g. $||cb^{-1}a^{-1}abc|| = 2$.

This lemma, which is needed for a refinement of interpolation, only concerns the free group. It had actually been proved before in (Nivat, 1971) and was reproved in (Autebert et al, 1984).

Proposition 2.26. *The two following properties of the free group hold:*

1. *Let u,v,w be elements of the free group; if $||u|| < ||uv||$ and $||uv|| \geq ||uvw||$ then $||vw|| \leq \max(||v||,||w||)$.*

2. *Let* u_i $i = 1, \ldots, n+1$ *be elements of the free group with* $u_1 \cdot \cdots \cdot u_{n+1} = 1$. *Then there exists* $k \leq n$ *such that*

$$\|u_k u_{k+1}\| \leq \max(\|u_k\|, \|u_{k+1}\|)$$

Proof. The first part is actually a lemma for the second part.

Proof of 1 We proceed by reductio ad absurdum, so we assume that

a. $\|u\| < \|uv\|$
b. $\|uv\| \geq \|uvw\|$
c. $\|vw\| > \|v\|$
d. $\|vw\| > \|w\|$

There exists three reduced words x, y, z such that

- $u = xy^{-1}$ $\quad v = yz$ $\quad uv = xz$
- xy^{-1} $\quad yz$ $\quad xz$ are reduced.

From (a) we have $\|x\| + \|y\| < \|x\| + \|z\|$ so $\|y\| < \|z\|$ and therefore $\|y\| < \frac{1}{2}\|v\|$ (*).

Similarly there exists three reduced words x', y', z' such that

- $v = x'y'$ $\quad w = y'^{-1}z'$ $\quad vw = x'z'$
- $x'y'$ $\quad y'^{-1}z'$ $\quad x'z'$ are reduced.

From (c) we have $\|y'\| + \|z'\| < \|x'\| + \|z'\|$ so $\|y'\| < \|x'\|$ and therefore $\|y'\| < \frac{1}{2}\|v\|$ (**)

From $v = yz = x'y'$ with $\|y\| < \frac{1}{2}\|v\|$ (*) and $\|y'\| < \frac{1}{2}\|v\|$ (**), there exists a non empty a such that

- $z = ay'$ $\quad x' = ya$ $\quad v = yay'$
- ay' $\quad ya$ $\quad yay'$ are reduced

So we have $uvw = xy^{-1}yay'y'^{-1}z' = xaz'$ — as xa and az' are reduced, xaz' is reduced as well. From (b) we have

$$\|uvw\| = \|xaz'\| \leq \|xay'\| = \|xz\| = \|uv\|$$

and therefore $\|z'\| \leq \|y'\|$.

Since from d we have $\|x'z'\| > \|x'y'\|$ so $\|z'\| > \|y'\|$, there is a contradiction.

Proof of 2 Let k be the first index such that $\|u_1 \cdots u_k\| \geq \|u_1 \cdots u_k u_{k+1}\|$.

If $k = 1$ $\|u_1\| \geq \|u_1 u_2\|$ then $\max(\|u_1\|, \|u_2\|) \geq \|u_1\| \geq \|u_1 u_2\|$.

Otherwise, let

$$u = u_1 \cdots u_{k-1} \qquad v = u_k \qquad w = u_{k+1}$$

we have

$$\|u\| = \|u_1 \cdots u_{k-1}\| < \|uv\| = \|u_1 \cdots u_{k-1} u_k\|$$

and

$$\|uv\| = \|u_1 \cdots u_{k-1} u_k\| \geq \|uvw\| = \|u_1 \cdots u_k u_{k+1}\|$$

so applying the first part (1) of this proposition we obtain

$$\|u_k u_{k+1}\| \leq \max(\|u_k\|, \|u_{k+1}\|)$$

□

2.11.3 Interpolation for Thin Sequents

A sequent $\Gamma \vdash C$ is said to be *thin* whenever it is provable and for all p, $\rho_p(\Gamma, C)$ is at most 2 — where $\rho_p(\Theta)$ is the number of occurrences of a primitive type p in Θ. Notice that by Proposition 2.6 which says that a provable sequent contains as many positive and negative occurrences of a primitive type, $\rho_p(\Gamma, A)$ is either 0 or 2.

Here is a proposition which is very representative of multiplicative calculi, in which formulas can be neither contracted or weakened:

Proposition 2.27. *Each provable sequent may be obtained from a thin sequent by substituting primitive types with primitive types.*

Proof. Given a cut free proof d with only primitive axioms of a sequent $\Gamma \vdash C$, number the axioms and replace each axiom $p \vdash p$ by $p_i \vdash p_i$ where i is the number of the axiom, and also replace all the traces of this occurrence of p in the proof with p_i. Clearly the result is itself a proof of a sequent $\Gamma' \vdash C'$, which contains exactly two or zero occurrences of each primitive type, and which gives back $\Gamma \vdash C$ when each p_i is substituted with p. □

Proposition 2.28. *Let $\Gamma, \Delta, \Theta \vdash C$ be a thin sequent. Then there exists a formula B such that:*

1. *$\Delta \vdash B$ is thin*
2. *$\Gamma, B, \Theta \vdash C$ is thin*
3. *$|B| = \|[\Delta]\|$ — the number of primitive types in B is the size of the interpretation of Δ in the free group (see Section 2.9.2 on page 45).*

Proof. p stands for any primitive type,

Let B be an interpolant of Δ which exists by Theorem 2.24. We then have:

a. $\Delta \vdash B$ is provable
b. $\Gamma, B, \Theta \vdash C$ is provable
c. $\rho_p(B) \leq \min(\rho_p(\Gamma, \Theta, C), \rho_p(\Delta))$

Let Us First Prove 1. As the sequent $\Gamma, \Delta, \Theta \vdash C$ is thin,

$$\rho_p(\Gamma, \Delta, \Theta, C) = \rho_p(\Gamma, \Theta, C) + \rho_p(\Delta)$$

is either 0 or 2; so by c $\rho_p(B)$ is either 0 or 1, and we have

$$\rho_p(\Delta, B) = \rho_p(\Delta) + \rho_p(B) \leq \rho_p(\Gamma, \Delta, \Theta, C) + \rho_p(B) \leq 2 + 1$$

Since $\Delta \vdash B$ is provable (a), $\rho_p(\Delta, B)$ is even, and thus $\rho_p(\Delta, B) \leq 2$. So, being provable, $\Delta \vdash B$ is thin.

Now Let Us Prove 2. Similarly,

$$\rho_p(\Gamma, B, \Theta, C) = \rho_p(\Gamma, \Theta, C) + \rho_p(B) \leq \rho_p(\Gamma, \Delta, \Theta, C) + \rho_p(B) \leq 2 + 1$$

Since $\Gamma, B, \Theta \vdash C$ is provable (b) $\rho_p(\Gamma, B, \Theta, C)$ is even, so $\rho_p(\Gamma, B, \Theta, C) \leq 2$ So, being provable, $\Gamma, B, \Theta \vdash C$ is thin.

Finally Let Us Prove 3

- if p does not occur in Δ then p does neither occur in $[\Delta]$ nor in B, by c.
- if p occurs once in Δ then it occurs once in $[\Delta]$ too — it cannot cancel with another occurrence of p; as $\Delta \vdash B$ is thin it also occurs once in B — it occurs twice in $\Delta \vdash B$ and once in Δ so it occurs once in B.
- if p occurs twice in Δ then it does not occur in Γ, Θ, C; therefore it does not occur in B by (c). The soundness of the interpretation in the free group entails $[\Gamma, \Delta, \Theta] = [C]$ that is $[\Delta] = [\Gamma]^{-1}[C][\Theta]^{-1}$ As p does not occur in Γ, Θ, C, there is no occurrence of p in $[\Gamma]^{-1}[C][\Theta]^{-1}$ and therefore no occurrence of p in $[\Delta]$

So for every primitive type, and whatever its number of occurrences in Δ is, there are exactly as many occurrences of p in B and in $[\Delta]$, so the number of primitive types in B and in $[\Delta]$ are equal: $|B| = \|[\Delta]\|$. □

Proposition 2.29. *Let $A_1, \ldots, A_n \vdash A_{n+1}$ be a thin sequent with $|A_i| \leq m$; then either:*

- *there exists an index k and a type B with $|B| \leq m$ such that the following sequents are thin:*
  $A_1, \ldots, A_{k-1}, B, A_{k+2}, \ldots, A_n \vdash A_{n+1}$
  $A_k, A_{k+1} \vdash B$
- *there exists a type B with $|B| \leq m$ such that the following sequents are thin:*
  $B, A_n \vdash A_{n+1}$
  $A_1, \ldots, A_{n-1} \vdash B$

Proof. Let $u_i = [A_i]$ for $1 \leq i \leq n$ and $u_{n+1} = [C]^{-1}$. Interpreting the provability in the free group we obtain: $u_1 \cdots u_n u_{n+1} = 1$ By Lemma 2.26 there exists an index $k \leq n$ for which $\|u_k u_{k+1}\| \leq \max(\|u_k\|, \|u_{k+1}\|) \leq m$.

- If $k < n$, we apply Proposition 2.28 with
  $\Delta = A_k, A_{k+1}$,
  $\Gamma = A_1, \ldots, A_{k-1}$
  $\Theta = A_{k+2}, \ldots, A_n$.
  So the sequents
  $A_1, \ldots, A_{k-1}, B, A_{k+2}, \ldots, A_n \vdash A_{n+1}$
  $A_k, A_{k+1} \vdash B$
  are thin, and
  $$|B| = \|[A_k, A_{k+1}]\| = \|u_k u_{k+1}\| \leq m$$
- If $k = n$, we apply Proposition 2.28 with
  $\Gamma = \varepsilon$,
  $\Delta = A_1, \ldots, A_{n-1}$
  $\Theta = A_n$.
  So the sequents $A_1, \ldots, A_n \vdash B$ and $B, A_n \vdash B$ are thin.
  Since $[A_1, \ldots, A_{n-1}, A_n] = [C]$
  we have $|B| = \|[A_1, \ldots, A_{n-1}]\| = \|[C][A_n]^{-1}\|$

therefore

$$|B| = \|[C][A_n]^{-1}\| = \|([A_n][C]^{-1})^{-1}\| = \|(u_n u_{n+1})^{-1}\| = \|u_n u_{n+1}\| \leq m \qquad \square$$

2.11.4 From Lambek Grammars to Context-Free Grammars

Proposition 2.30. *If a sequent $A_1, \ldots, A_n \vdash A_{n+1}$ with each $|A_i| \leq m$ is provable in L, then it is provable from provable sequents $U, V \vdash X$ or $U \vdash X$ with $|U|, |V|, |X| \leq m$ by means of the cut rule only.*

Proof. We proceed by induction on n. If $n \leq 2$ then there is nothing to prove. Otherwise, let $A'_1, \ldots, A'_n \vdash A'_{n+1}$ be a corresponding thin sequent obtained as in Proposition 2.27 — using a different primitive type for each axiom in the proof of $A_1, \ldots, A_n \vdash A_{n+1}$. Thus there exists a substitution σ replacing primitive types with primitive types and preserving provability such that $\sigma(A') = A$.

As the substitution replaces primitive types with primitive types, we also have $|A'_i| \leq m$. By Proposition 2.29 there exists a formula B' with $|B'| \leq m$ such that either:

- $A'_1, \ldots, A'_{k-1}, B', A'_{k+2}, \ldots, A'_n \vdash A'_{n+1}$
  $A'_k, A'_{k+2} \vdash B'$

  are thin, and therefore provable.

  Let $B = \sigma(B')$, so B has at most m primitive types as well; applying the substitution we obtain two provable sequents

  $A_1, \ldots, A_{k-1}, B, A_{k+2}, \ldots, A_n \vdash A_{n+1}$
  $A_k, A_{k+1} \vdash B.$

  By induction hypothesis

  $$A_1, \ldots, A_{k-1}, B, A_{k+2}, \ldots, A_n \vdash A_{n+1} \qquad (*)$$

  is provable from provable sequents $U, V \vdash X$ or $U \vdash X$ with $|U|, |V|, |X| \leq m$ by means of the cut rule only.

  Notice that $A_k, A_{k+1} \vdash B \quad (**)$ is of the form $U, V \vdash X$ with $|U|, |V|, |X| \leq m$. A cut rule between the proof of $(*)$ and $(**)$ yields a proof of

  $$A_1, \ldots, A_n \vdash A_{n+1}$$

  from provable sequents $U, V \vdash X$ or $U \vdash X$ with $|U|, |V|, |X| \leq m$ by means of the cut rule only.
- $B', A'_n \vdash A'_{n+1}$ and $A_1, \ldots, A_{n-1} \vdash B$ are thin and therefore provable.
  Let $B = \sigma(B')$, so $|B| \leq m$; applying the substitution we obtain two provable sequents
  $B, A_n \vdash A_{n+1}$
  $A_1, \ldots, A_{n-1} \vdash B.$

By induction hypothesis

$$A_1,\ldots,A_{n-1},B \vdash A_{n+1} \qquad (+)$$

is provable from provable sequents $U,V \vdash X$ or $U \vdash X$ with U,V,X having at most m primitive types by means of the cut rule only.

Notice that $B,A_n \vdash A_{n+1}$ $(++)$ is of the form $U,V \vdash X$ with $|U|,|V|,|X| \leq m$. A cut rule between the proof of $(+)$ and $(++)$ yields a proof of

$$A_1,\ldots,A_n \vdash A_{n+1}$$

from provable sequents $U,V \vdash X$ or $U \vdash X$ with $|U|,|V|,|X| \leq m$ by means of the cut rule only. □

Theorem 2.31. *Let* Lex *be the lexicon of a Lambek grammar G_L, and let and let m the maximal number of primitive types in a formula of the lexicon. Then the language $L(G_L)$ generated by G_L is the same as the language $L(G_C)$ generated by the following context-free grammar G_C:*

- *Terminals: terminals (words) of G_L*
- *Non-Terminals: all formulae A with $|A| \leq m$*
- *Start symbol S, the one of G_L*
- *$X \longrightarrow a$ whenever $X \in \mathrm{Lex}(a)$*
- *$X \longrightarrow A$ whenever $A \vdash X$ is provable in L*
- *$X \longrightarrow A\,B$ whenever $A,B \vdash X$ is provable in L*

Observe that the rules are in finite number, because there are finitely many sequents $U,V \vdash X$ or $U \vdash X$ when U,V,X contains at most m primitive types — hence there are only finitely many provable such sequents.

Proof. Assume $a_1 \cdots a_n \in L(G_C)$. Hence there exist types $X_i \in \mathrm{Lex}(a_i)$ such that $S \longrightarrow X_1 \cdots X_n$. The derivation in the CFG G_C can be turned into a derivation in L using only the cut rule (reversing $\longrightarrow$ into $\vdash$), therefore $a_1 \cdots a_n \in L(G_L)$.

Assume now that $a_1 \cdots a_n \in L(G_L)$. Hence there exist types $X_i \in \mathrm{Lex}(a_i)$ such that $X_1,\ldots,X_n \vdash S$. By Proposition 2.30 such a sequent is provable by means of the sequents corresponding to production rules, and of the cut rule only. By induction on the size of the cut-only proof, it is easily seen that the proof corresponds to a derivation in the CFG G_C. If the proof is reduced to a proper axiom, than this axiom is itself a production rule. If the last rule is a cut, say between $\Gamma,B,\Theta \vdash C$ and $\Delta \vdash B$, then by induction hypothesis we have $B \longrightarrow \Delta$ and $C \longrightarrow \Gamma\,B\,\Theta$ hence $C \longrightarrow \Gamma\,\Delta\,\Theta$. Thus, if $a_1 \cdots a_n \in L(G_L)$, we have $S \longrightarrow X_1 \cdots X_n$ with $X_i \in \mathrm{Lex}(A_i)$; as $X_i \in \mathrm{Lex}(a_i)$ we have $S \longrightarrow a_1 \cdots a_n$. □

2.12 Concluding Remarks

This concludes our chapter on the Lambek calculus. We have given proofs of many of the classic results (cut elimination, soundness and completeness, equivalence of

Lambek grammars and context-free grammars). In the following chapters, we discuss some variants of the Lambek calculus.

Since complexity is not one of the major themes of the current book, we have decided not to include a recent proof of NP-completeness of the Lambek calculus (Pentus, 2006) in this chapter, although Section 4.6.1 gives a very brief overview of the known complexity results for the Lambek calculus and some of it variants.

In the next chapter we will see another aspect of the Lambek calculus, which is its direct link with natural language semantics in the style of Montague.

Exercises for Chapter 2

Exercise 2.1. Give sequent calculus derivations for each of the following sequents.

1. $(A \backslash B) / A \vdash A \backslash (B / A)$
2. $A / B \vdash (A / C) / (B / C)$
3. $(A \bullet B) \backslash C \vdash B \backslash (A \backslash C)$
4. $B \backslash (A \backslash C) \vdash (A \bullet B) \backslash C$
5. $(B / A) \backslash B \vdash ((A \backslash B) / A) \backslash (A \backslash B)$

Exercise 2.2. Give natural deduction derivations for each of the sequents of the previous exercise.

Exercise 2.3. Prove Proposition 2.3 on page 29.

Exercise 2.4. Definition 2.11 on page 36 defines the order of formulae. Calculate the order of the following formulae.

1. np / n
2. $((np \backslash S) / pp) / np$
3. $(S / np) \backslash S$
4. $((np \backslash S) / np) \backslash (np \backslash S)$
5. $((n / n) / (n / n)) / ((n / n) / (n / n))$

Exercise 2.5. Prove Proposition 2.6 on page 30.

Exercise 2.6. Using the following lexicon, find two different normal derivations in natural deduction for "Someone loves everyone" and "Someone is_missing"

Word	Type(s)
someone	$(S / (np \backslash S))$
everyone	$((S / np) \backslash S)$
loves	$((np \backslash S) / np)$
is_missing	$((S / (np \backslash S)) \backslash S)$
gave	$(((np \backslash S) / pp) / np)$
a_book	np
to	pp / np

Exercise 2.7. Following Carpenter (1996, Section 6.3), look at the following lexicon.

Word	Type(s)
kid	n
who	$((n \backslash n) / (S / np))$
Kelly	np
Terry	np
Robin	np
likes	$((np \backslash S) / np)$
believes	$((np \backslash S) / S)$
knows	$((np \backslash S) / S)$

Give natural deduction derivations showing that the lexicon above allows us to show that all of the following expressions are of type n.

(2.1) kid who Kelly likes.
(2.2) kid who Kelly believes Terry likes.
(2.3) kid who Kelly believes Terry knows Robin likes.

Exercise 2.8. Extend the lexicon from Exercise 2.6 with types for "every" and "a", as well as for "child" and "toy" in such a way that "every child loves a toy" obtains exactly two natural deduction derivations.

Exercise 2.9. Using the lexicon from Exercise 2.6, give two natural deduction derivations for "Someone gave a_book to everyone" .

Exercise 2.10. Section 2.6.2 gives a decision procedure for natural deduction without product. Use this decision procedure to find all proofs for the sequent of Example 2.1 on page 26. How many proofs are there? What is their relation to the natural deduction proof shown in Example 2.1?

References

Abramsky, S.: Computational interpretations of linear logic. Theoretical Computer Science 111, 3–57 (1993)

Autebert, J.M., Boasson, L., Sénizergues, G.: Langages de parenthèses, langages n.t.s. et homomorphismes inverses. RAIRO Informatique Théorique 18(4), 327–344 (1984)

van Benthem, J.: Categorial grammar. In: Essays in Logical Semantics, ch. 7, pp. 123–150. Reidel, Dordrecht (1986)

van Benthem, J.: Categorial grammars and lambda calculus. In: Skordev, D. (ed.) Mathematical Logic and its Applications. Plenum Press (1987)

van Benthem, J.: Language in Action: Categories, Lambdas and Dynamic Logic. Sudies in logic and the foundation of mathematics, vol. 130. North-Holland, Amsterdam (1991)

Buszkowski, W.: Compatibility of a categorial grammar with an asssociated category system. Zeitschrift für Matematische Logik und Grundlagen der Mathematik 28, 229–238 (1982)

Buszkowski, W.: Mathematical linguistics and proof theory. In: van Benthem, J., ter Meulen, A. (eds.) Handbook of Logic and Language, ch. 12, pp. 683–736. North-Holland Elsevier, Amsterdam (1997)

Carpenter, B.: Lectures on Type-Logical Semantics. MIT Press, Cambridge (1996)

Chomsky, N.: Syntactic structures. Janua linguarum. Mouton, The Hague (1957)

Chomsky, N.: Formal properties of grammars. In: Handbook of Mathematical Psychology, vol. 2, pp. 323–418. Wiley, New-York (1963)

Chomsky, N.: The minimalist program. MIT Press, Cambridge (1995)

Cohen, J.M.: The equivalence of two concepts of categorial grammars. Information and Control 10, 475–484 (1967)

Engelfriet, J.: Context-free graph grammars. In: Rosenberg, G., Salomaa, A. (eds.) Handbook of Formal Languages 3: Beyond Words, pp. 125–213. Springer, New York (1997)

Engelfriet, J., Maneth, S.: Tree Languages Generated by Context-Free Graph Grammars. In: Ehrig, H., Engels, G., Kreowski, H.-J., Rozenberg, G. (eds.) TAGT 1998. LNCS, vol. 1764, pp. 15–29. Springer, Heidelberg (2000)

Gécseg, F., Steinby, M.: Tree languages. In: Rozenberg, G., Salomaa, A. (eds.) Handbook of Formal Languages, vol. 3, ch. 1. Springer, Berlin (1997)

Gentzen, G.: Untersuchungen über das logische Schließen. Mathematische Zeitschrift 39, 176–210, 405–431 (1934)

Girard, J.Y.: Linear logic. Theoretical Computer Science 50(1), 1–102 (1987)

Girard, J.Y., Lafont, Y., Taylor, P.: Proofs and Types. Cambridge Tracts in Theoretical Computer Science, vol. 7. Cambridge University Press (1988)

de Groote, P., Retoré, C.: Semantic readings of proof nets. In: Kruijff, G.J., Morrill, G., Oehrle, D. (eds.) Formal Grammar, pp. 57–70. FoLLI, Prague (1996)

Hendriks, H.: Studied flexibility: Categories and types in syntax and semantics. PhD thesis, University of Amsterdam, ILLC Dissertation Series (1993)

Hepple, M.: Normal form theorem proving for the Lambek calculus. In: Proceedings of COLING 1990, Helsinki, pp. 173–178 (1990)

Kanazawa, M., Salvati, S.: On the derivations of lambek grammars (2009) (unpublished manuscript)

König, E.: Parsing as natural deduction. In: Proceedings of the Annual Meeting of the Association for Computational Linguistics, pp. 272–297 (1989)

Lambek, J.: The mathematics of sentence structure. American Mathematical Monthly, 154–170 (1958)

Moortgat, M.: Categorial Investigations: Logical and Linguistic Aspects of The Lambek Calculus. Foris, Dordrecht (1988)

Nivat, M.: Congruences de thue et t-langages. Studia Scientiarum Mathematicarum Hungarica 6, 243–249 (1971)

Pentus, M.: Lambek calculus is L-complete. Tech. Rep. LP-93-14, Institute for Logic, Language and Computation, Universiteit van Amsterdam (1993a)

Pentus, M.: Lambek grammars are context-free. In: Logic in Computer Science. IEEE Computer Society Press (1993b)

Pentus, M.: Product-free Lambek calculus and context-free grammars. Journal of Symbolic Logic 62(2), 648–660 (1997)

Pentus, M.: Lambek calculus is NP-complete. Theoretical Computer Science 357(1), 186–201 (2006)

Retoré, C., Stabler, E.: Generative grammar in resource logics. Research on Language and Computation 2(1), 3–25 (2004)

Roorda, D.: Resource logic: proof theoretical investigations. PhD thesis, FWI, Universiteit van Amsterdam (1991)

Thatcher, J.W.: Characterizing derivation trees of context free grammars through a generalization of finite automata theory. Journal of Computer and System Sciences 1, 317–322 (1967)

Tiede, H.-J.: Lambek Calculus Proofs and Tree Automata. In: Moortgat, M. (ed.) LACL 1998. LNCS (LNAI), vol. 2014, pp. 251–265. Springer, Heidelberg (2001)

Zielonka, W.: Axiomatizability of the Adjukiewicz-Lambek calculus by means of cancellation schemes. Zeitschrift für Matematische Logik und Grundlagen der Mathematik 27(13-14), 215–224 (1981)

Zielonka, W.: A simple and general method of solving the finite axiomatizability problems for Lambek's syntactic calculi. Studia Logica 48(1), 35–39 (1989)

3

Lambek Calculus and Montague Grammar

Summary. This chapter discusses one of the important advantages of using (Lambek) categorial grammars: the straightforward correspondence between Lambek calculus proofs and derivations in Montague-style semantics, which extends straightforwardly to modern theories like DRT. In order to keep the exposition simple, we will only briefly discuss the intensional operators of Montague.

3.1 Introduction

The Lambek calculus is a lexicalized formalism: that is, the Lambek calculus consists of a universal set of deduction rules — the logical properties of which we have studied in detail in the previous chapter — and to obtain a Lambek *grammar* we only add a lexicon, a function Lex which assigns a finite set of types to each word.

Now we will turn our attention to its relation to Montague semantics, introduced in (Montague, 1970a,b, 1973) which is a very important feature of categorial grammars.

We do not intend to give a lecture on Montague semantics, which is a large research topic in itself — the reader interested in this topic is referred to (Dowty et al, 1981; Gamut, 1991; Partee and Hendriks, 2011) for general introductions, and to (Carpenter, 1996) for many more applications of the Lambek calculus and Montague grammar to different semantic phenomena — but only to illustrate Montague semantics viewed from the perspective of the Lambek calculus. Montague semantics is also a controversial view of semantics. Indeed it has nothing fancy to say about mental representation or the organization of concepts as for instance in (Jackendoff, 1995) or (Pustejovsky, 1995) (though we believe that Montague grammar is at least compatible with a more comprehensive cognitive theory of meaning): the semantics of a sentence is given by formulae of (higher-order) predicate calculus, possibly of intensional logic, and the elementary expression are interpreted by logical constants: the word "Paul" is interpreted by the logical constant *Paul* and the word "car" by the logical constant *car* (or equivalently, as is more usual in Montague semantics, $\lambda x^e.car(x)$). Nevertheless it enables a neat and computational treatment of (co)reference and of quantifiers and this is an important step towards modeling meaning computationally.[1] Interpreting sentences in a logical syntax also allows us to model entailment: a set of sentences $s_1, \dots, s_n$ *entails* another sentence s if and only is the truth of all of the sentences $s_1, \dots, s_n$ implies the truth of s. If all of the sentences are interpreted as logical formulae, then this semantic notion of entailment corresponds to the logical notion of entailment, that is to $s_1, \dots, s_n \vdash s$ where '$\vdash$' is the logical entailment relation of the logic we use for our interpretation.[2]

Although Montague himself thought that one should forget about whatever lies between the sentence and its interpretation in possible world semantics, we shall choose a more intermediate level, which could be called the "syntax of semantics", as the endpoint of our semantic interpretation: the logical formula itself. We agree with Montague and others that the intermediate steps, being unobservable, are difficult to study (or even to substantiate). The logical form is already less observable than the sentence, though tools like entailment allow us to reason about properties these logical forms must have.

3.2 Logic and Lambda Calculus

We will first give an extremely brief introduction to the typed lambda calculus, then see how to model logical formulae in typed lambda calculus.

3.2.1 Typed Lambda Calculus and Intuitionistic Propositional Calculus

We do not pretend to include here a presentation of the typed lambda calculus, many of them exist (Krivine, 1990; Seldin and Hindley, 1980; Girard et al, 1988) but we provide a brief reminder of the minimal background necessary to follow a presentation of Montague semantics. There are at least two unrelated ways to consider the relation between the typed lambda calculus and logic:

[Church]. Typed lambda terms as logical formulae disregarding their truth and provability. Provided the type systems includes a type **e** for entities (also called individuals) and one, **t**, for truth values (or propositions), constants for logical connectives and constants for the predicates, functions and constants, every closed formula correspond to a normal lambda term of type **t**, and conversely, every closed normal lambda term (with constants as indicated) of type **t** corresponds to a formula (this will be made precise in next section).

[1] We consider these phenomena part of semantics. However, in generative grammar, they are considered part of syntax. For a good introduction to semantics from the point of view of generative grammar, see (Heim and Kratzer, 1997).

[2] There are, of course, many caveats to using logical entailment to model semantic entailment: semantic entailment depends, in many cases, on complex world knowledge and there has been much debate about the difference between the logical implication "$A \Rightarrow B$" and the construction "if A then B" in natural language.

[Curry-Howard]. Typed lambda terms as proofs in intuitionistic logic. Any proof in the pure implicative propositional logic of C under the assumptions $H_1, \ldots, H_n$ with propositional variables in P can be identified with a lambda term (not necessarily normal) of type C with free variables of type $H_1, \ldots, H_n$.

When studied from the context of a non proof-theoretic syntax, Montague semantics is only concerned with the first point of view, since it computes logical formulae. Here we shall also consider the second viewpoint, because the starting point for our calculations are the syntactic analyses in type logical grammars (Lambek grammars, multimodal categorial grammars, etc.), which are proofs, and which we convert into lambda terms that are proofs as well, though they are proofs which represent logical formulae.

Historically, the first viewpoint was introduced by Church in order to give a proper account of substitution in Hilbert-style deductive systems. The second one, which appears later, is the viewpoint responsible for typed functional programming languages such as ML, CaML, Haskell, etc. One may wonder about the connection between the two viewpoints, if any. It is actually a tiny one: what is a formula a proof of? A formula is a proof of its own correctness, and there is no relation with a proof of the formula itself! We are to see this tiny connection at work.

Definition 3.1 (Types). *From a (finite) set of basic types P, also called atomic types, we define the set of types as follows.*

$$\mathscr{T} \quad ::= \quad P \quad | \quad (\mathscr{T} \to \mathscr{T})$$

Types which are not basic types are called compound *types.*

For instance, if $a, b \in P$ then $a \to ((a \to b) \to b)$ is a type. Note that some authors write $\langle a, \langle\langle a, b\rangle, b\rangle\rangle$; using $\langle X, Y\rangle$ when we write $X \to Y$.

To represent formulae we require that P contains the type $\mathbf{t}$ for truth values and the type $\mathbf{e}$ for individuals (in case one wishes to represent a multisorted logic with sorts $e_1, \ldots, e_n$ these sorts should be in P).

Any type $T \in \mathscr{T}$, be it atomic or compound, is given a countable set of variables of this type. Each of these variables, written as $x_i : T$ or x_i^T, is a term of type T with free variable x_i. Types may be provided with a finite or countable set of *constants* of this type. They behave much like free variables, with the exception that they cannot be bound. A term t of type T is written $t : T$ or t^T. We will switch between these two notations for terms and variables freely throughout this chapter.

- Variables: if x is a variable of type T written $x : T$, then x is a term of type T with one free occurrence of the variable x. The only subterm of x is x.
- Constant: if k is a constant of type T written $k : T$, then k is a term of type T without free variables. The only subterm of k is k itself.
- Abstraction: if t is term of type T with the free variables V, and if x is a variable of type U then $w = \lambda x^U.t$ is a term of type $U \to T$ with free variables $F \setminus \{x\}$ the free occurrences of $x : U$ in t are bound in w by the initial λx^U. Occurrences of other free variables remain free. The subterms of w are w and the subterms of t.

- Application: if t is a term of type $U \to V$ with free variables F and if u is a term of type U with free variables G then $w = (t(u))$ it a term of type V with free variables $F \cup G$. Any free occurrence of a free variable in t or in u is free in $(t(u))$, but observe there might be free variables common to t and u. The subterms of w are w itself, the subterms of t and the subterms of u.

An occurrence of a variable x which is not a free occurrence of this variable x is a *bound* variable and therefore it is associated with its λx binder.

We use implicit right bracketing for types: $a{\to}b{\to}c = a{\to}(b{\to}c)$, which goes with implicit left bracketing for λ-terms: $w\ v\ u = (w\ v)\ u$.

Let $w = t[u]$ be a term with the subterm $u = \lambda x\, t'$, and let z be a variable not occurring in w. Then $w' = t[\lambda z\, t'[x := z]]$ is a term which is said to be *alpha equivalent* to w. All operations defined on lambda terms are defined up to alpha equivalence, that is up to the renaming of free variables.[3]

The notation $t[x := u]$, where t is a term, x a variable and u a lambda term of the same type as x, represents the term obtained from t by replacing all free occurrences of x by the lambda term u. Note that we may need to substitute u into a term t' alpha equivalent to t to make sure all free variables of u are free variables of $t[x := u]$. For example, $(\lambda y.f(x,y))[x := y]$ is *not* equal to $\lambda y.f(y,y)$, where the free occurrence y is accidentally bound, but — after alpha conversion to $(\lambda z.f(x,z))[x := y]$ — to $\lambda z.f(y,z)$.

A one-step beta reduction is defined by: $(\lambda x\ t)\ u \overset{\beta}{\leadsto}_1 t[x := u]$ and it preserves the typing. Such a step of beta reduction may also take place in a subterm of a term. Finally beta reduction can be iterated: beta reduction, written $\overset{\beta}{\leadsto}$ is the reflexive, transitive closure of the one-step beta reduction.

One should know the following results:

- Beta reduction is confluent, one also says Church-Rosser, that is to say if a term t beta reduces to two terms t' and t'' then there exists a term w such that t' and t'' both beta reduce to w.
- There is no infinite path of beta reduction in the typed lambda calculus (strong normalization).
- As a consequence of the first two points: every typed lambda term has a unique normal form, which is reached no matter how the beta reductions are performed.

One may also consider eta expansion, which is defined as follows, for a term t without free occurrence of x:

$$t^{A\to B} \overset{\eta}{\leadsto} \lambda x^A\ (t(x))$$

It will be useful to consider β-normal η-long forms Huet (1976) which are defined using the auxiliary notion of atomic forms:

- Variables are atomic forms.
- $(M\ N)$ is an atomic form when $M : A \to B$ is an atomic form and $N : A$ is in β-normal η-long form.

[3] Basically, alpha equivalence says that $x \mapsto 2x$ is the same function as $z \mapsto 2z$.

- Atomic terms whose type is a base type are β-normal η-long forms.
- $\lambda x^A\ t$ is a β-normal η-long form whenever t is.

Every term as a unique β-normal η-long form, which is obtained by beta reduction and then some eta expansion steps. Roughly speaking every term of type $A \to B$ without arguments is of the form $\lambda x^A\ t$ with $t : B$ (the beta-normal eta-long form requires replacing an $f : A \to B$ without argument with $(\lambda x^A\ (f\ x)) : A \to B$.

Observe that a λ-term consists of λ-abstractions, $\lambda x_1 \cdots \lambda x_n$, $1 \leq i \leq n$ (possibly none when $i = 0$) and then one subterm f which is successively applied to several arguments, λ-terms $t_1, \ldots t_p$ (possibly none when $p = 0$). If the lambda term is normal, f cannot be a λ-abstraction. If it were, $f = \lambda x_0\ u$, because the term is normal, then there would be no application of f (p=0) but in this case the λy should be considered as λx_{n+1}. Hence f is either a constant or a variable.

3.2.2 First Order Logic, Mono and Multisorted

First order logic has a single sort of individuals denoted here, as in Montague's work by $\mathbf{e}$ (entities) — Church call this type ι. Here we shall consider several sorts $\mathbf{e}_i$ with $1 \leq i \leq N$ the standard first-order case being $N = 1$.

Let us define functional types as $\mathbf{e}_{q1} \to (\mathbf{e}_{q2} \to (.... \to \mathbf{e}_{qs_q} \to \mathbf{e}_{q0}))$ (in case $s_q = 0$ this type is $\mathbf{e}_{q0}$)) and relational types as $\mathbf{e}_{q1} \to (\mathbf{e}_{q2} \to (.... \to \mathbf{e}_{qs_q} \to \mathbf{t}))$ (in case $s_q = 0$, this type is $\mathbf{t}$) .

To define a multisorted language one needs several sets of symbols which are assumed to be at most countable and possibly empty:

- Constants: for each sort $\alpha = \mathbf{e}_{q0}$ we have a set of constants a_n^α of sort α.
- Variables: for each sort $\alpha = \mathbf{e}_{q0}$ we have a set of variables x_n^α of sort α.
- Function symbols: for every functional type $\phi = \mathbf{e}_{q1} \to (\mathbf{e}_{q2} \to (.... \to \mathbf{e}_{qs_q} \to \mathbf{e}_{q0}))$ we have a countable set of function symbols f_q^ϕ of this type ϕ.[4]
- Relational symbols: for every relational type $\psi = \mathbf{e}_{q1} \to (\mathbf{e}_{q2} \to (.... \to \mathbf{e}_{qs_q} \to \mathbf{t}))$ we have a countable set of relational symbols R_k^ψ of this type ψ. Propositional constants, if any, are viewed as relational symbols with $s_q = 0$ i.e. with $\psi = \mathbf{t}$.

We first need to define *terms*:

- A constant a_n^α is a term of sort α.
- A variable x_n^α is a term of sort α.
- If f_q is function symbol of type $\phi = \mathbf{e}_{q1} \to (\mathbf{e}_{q2} \to (.... \to \mathbf{e}_{qs_q} \to \mathbf{e}_{q0}))$ and if t_i for $i = 1, \ldots, s_q$ are terms of respective sorts $\mathbf{e}_{q_i}$ then $f_q(t_1, \ldots, t_{s_q})$ is a term of sort $\mathbf{e}_{q0}$.
- Nothing else is a term. Terms without variables are called closed terms; every occurrence of a variable in a term is a free occurrence.

[4] Constants can be viewed as function symbols with $s_q = 0$ i.e. with $\phi = \mathbf{e}_{q0}$ for some q_0, but as we shall often use languages with constants and without proper symbol we do not use this generalization.

Then atomic formulae are defined as follows: if R is a relational symbol of type $\psi = \mathbf{e}_{q1} \rightarrow (\mathbf{e}_{q2} \rightarrow (.... \rightarrow \mathbf{e}_{qs_q} \rightarrow \mathbf{t}))$ and if t_i for $1 \leq i \leq s_q$ terms of respective sorts $\mathbf{e}_{q_i}$, then the following is an *atomic formula*:

$$R(t_1, \dots, t_{s_q})$$

The free variables in an atomic formula are the free variables in the terms.

Next we can define *formulae*

- An atomic formula is a formula.
- If F is a formula, so is $\neg F$.
- If F and G are two formulae so are $F \wedge G$, $F \vee G$, $F \Rightarrow G$.
- If F is a formula, and if x is a variable of some sort $\mathbf{e}_i$ then $\forall x : \mathbf{e}_i.F$ and $\exists x : \mathbf{e}_i.F$ are formulae.
- Nothing else is a formula.

Models for multisorted logic

The intended models for many-sorted logic with N sorts $(\mathbf{e}_i)_{1 \leq i \leq N}$ are defined by a partition of a set $\bar{\mathbf{e}}$ into N classes $\bar{\mathbf{e}}_i$. In order to define the model, we shall need an interpretation I as well as an assignment A which maps every free variable into an object of $\bar{\mathbf{e}}_i$.

An interpretation is fully defined by

- an element of $\bar{\mathbf{e}}_i$ for every constant of sort $\mathbf{e}_i$.
- a function $\bar{f}_q$ from $\bar{\mathbf{e}}_{q1} \times \dots \times \bar{\mathbf{e}}_{qs_q} \mapsto \bar{\mathbf{e}}_{q0}$ for each function symbol f_q of type $\phi = \mathbf{e}_{q1} \rightarrow (\mathbf{e}_{q2} \rightarrow (.... \rightarrow \mathbf{e}_{qs_q} \rightarrow \mathbf{e}_{q0}))$, that is a function from s_q elements of the model (each of the appropriate sort $\bar{\mathbf{e}}_{q_i}$) to an element $\bar{\mathbf{e}}_{q0}$.
- a subset $\bar{R}$ of $\bar{\mathbf{e}}_{q1} \times \dots \times \bar{\mathbf{e}}_{qs_q}$ for each relational symbol R of type $\psi = \mathbf{e}_{q1} \rightarrow (\mathbf{e}_{q2} \rightarrow (.... \rightarrow \mathbf{e}_{qs_q} \rightarrow \mathbf{t}))$, that is a set of tuples with s_q elements of the model (each of sort $\bar{\mathbf{e}}_{q_i}$) for which the relation R is true.

An assignment A is a map from variables to elements in the corresponding set: $A(x) \in \bar{\mathbf{e}}_i$ whenever x is of sort $\mathbf{e}_i$.

An interpretation together with an assignment maps terms of sort $\mathbf{e}_i$ to the set $\bar{\mathbf{e}}_i$ as follows: constants of sort $\mathbf{e}_i$ are mapped to elements in $\bar{\mathbf{e}}_i$ by the interpretation, variables of sort $\mathbf{e}_i$ are mapped to elements in $\bar{\mathbf{e}}_i$ by the assignment, and, after that, if f_q is function symbol of type $\phi = \mathbf{e}_{q1} \rightarrow (\mathbf{e}_{q2} \rightarrow (.... \rightarrow \mathbf{e}_{qs_q} \rightarrow \mathbf{e}_{q0}))$ and if t_i for $i = 1, \dots, s_q$ are terms of respective sorts $\mathbf{e}_{q_i}$ which the interpretation and the assignment map to $\bar{\mathbf{e}}_{q_i}$ then the interpretation of $f_q(t_1, \dots, t_{s_q})$ which is in $\bar{\mathbf{e}}_{q0}$ is simply $\bar{f}_q(\bar{\mathbf{e}}_1, \dots, \bar{\mathbf{e}}_{s_q})$.

Given an interpretation I and an assignment A, formulae are interpreted as usual taking sorts into account:

- An atomic formula $R(t_1, \dots, t_{s_q})$ with R a relational symbol of type $\psi = \mathbf{e}_{q1} \rightarrow (\mathbf{e}_{q2} \rightarrow (.... \rightarrow \mathbf{e}_{qs_q} \rightarrow \mathbf{t}))$ and t_i for $1 \leq i \leq s_q$ terms of respective sorts $\mathbf{e}_{q_i}$, is true whenever $(\bar{\mathbf{e}}_{q1} \times \bar{\mathbf{e}}_{q2} \times \dots \times \bar{\mathbf{e}}_{qs_q})$ is in $\bar{R}$.

- If F_1 and F_2 are two formulae and $\diamond$ is a connective, the truth of the formula $F_1 \diamond F_2$ and $\neg F_1$ is given by the usual truth tables.
- If F is a formula, and if F is true for the interpretation I and for some assignment A' which coincides with A except, possibly, for the variable x of sort $\mathbf{e}_i$ (i.e. $A'(x)$ may differ from $A(x)$) then $\exists x F$ is true with interpretation I and assignment A.
- If F is a formula, and if F is true for the interpretation I and for all assignments A' which coincides with A except, possibly, for the variable x of sort $\mathbf{e}_i$ (i.e. $A'(x) \neq A(x)$) then $\forall x.F$ is true with interpretation I and assignment A.

We will not have a lot more to say about these models in this book, but we hope that it helps to give a clearer picture about multisorted logic.

3.2.3 Second Order and Higher Order Logic

Multisorted second order logic is a rather simple extension of the logic above: for each type $\psi = \mathbf{e}_{q1} \to (\mathbf{e}_{q2} \to (.... \to \mathbf{e}_{q_{sq}} \to \mathbf{t}))$ in addition to the relational symbols one has relational *variables*. Terms are defined as above and are called first order terms. Atomic formulae are defined as above, except that R can also be a relational variable. Hence the main difference is that one may quantify, existentially and universally, over relational variables: given a formula F and a relational variable X of type $\psi = \mathbf{e}_{q_1} \to (\mathbf{e}_{q_2} \to (.... \to \mathbf{e}_{q_{sq}} \to \mathbf{t}))$ the expressions $\forall X.F$ and $\exists X.F$ are formulae. The cleanest way to treat quantifiers over relational variables of different types is to treat them as different symbols.

Higher order logic is more complex to define in a standard logical syntax than in the lambda calculus. Usually, one skips function symbols and higher order function symbols, since they are not really necessary: they are particular cases of predicates. However it is non standard to have heterogenous higher order variables, ranging over objects with distinct status and quantification over those: e.g. it would be weird, although not impossible, to quantify over variables of relation between relations and individuals.

Intended models of such logic, which are usually called principal models, are as expected: a relational variable X of type $\psi = \mathbf{e}_{q_1} \to (\mathbf{e}_{q_2} \to (.... \to \mathbf{e}_{q_{sq}} \to \mathbf{t}))$ is interpreted as any subset of $\bar{\mathbf{e}}_{q_1} \times \bar{\mathbf{e}}_{q2} \times \ldots \times \bar{\mathbf{e}}_{q_{sq}}$. The formula $\forall X.F$ (resp. $\exists X.F$) is true whenever it is true for any (resp. some) interpretation of the X as a subset of f $\bar{\mathbf{e}}_{q1} \times \bar{\mathbf{e}}_{q2} \times \ldots \times \bar{\mathbf{e}}_{q_{sq}}$.

Nevertheless, we need to remember here that there is no completeness result with respect to such principal models, even for the simplest case, second order logic. Completeness can be stated as: every non-contradictory set of formulae admits a model. Compactness says that when any finite subset of a set of formulae admits a model, so does the whole set. Clearly, completeness implies compactness. Indeed, if a set of formulae does not admit a model then it proves "false". But a proof is finite, hence if there is a proof of "false" from the set of formulae T then a finite number of the formulae of T is enough to prove "false". Hence if no finite subset of T contains a contradiction, then neither does T, and therefore T has a model. Now consider the following closed formulae F_n $(n \geq 1) : \exists x_1 \cdots \exists x_n \wedge_{i=1}^{n} \wedge_{j=1}^{i} x_i \neq x_j$ and

$F_0 : \forall X(\_,\_)$ *X is an injective application* $\Rightarrow$ *X is surjective*[5] then clearly every finite subset S of $F_i, i \in N$ has a principal model (assuming N is the largest integer such that $F_N \in S$, take a set $\{a_i \mid i = 1, \dots, N\}$), but there cannot be a set which is finite and with at least n elements for all n.[6]

For second order logic, the reader is referred to van Dalen (1983).

3.2.4 Lambda Terms and Logical Formulae

Assume that the base types are $\mathbf{e}_i$ $1 \leq i \leq N$ and $\mathbf{t}$ and that the only constants are

- The logical constants:
  - $\neg$ of type $\mathbf{t} \to \mathbf{t}$
  - $\Rightarrow, \wedge, \vee$ of type $\mathbf{t} \to (\mathbf{t} \to \mathbf{t})$
  - for every with i $1 \leq i \leq n$, two constants $\forall_i$ and $\exists_i$ of type $(\mathbf{e}_i \to \mathbf{t}) \to \mathbf{t}$
- The language constants:
  - R_q of type $\mathbf{e}_{q1} \to (\mathbf{e}_{q2} \to (\dots. \to \mathbf{e}_{q s_q} \to \mathbf{t}))$
  - f_q of type $\mathbf{e}_{q1} \to (\mathbf{e}_{q2} \to (\dots. \to \mathbf{e}_{q s_q} \to \mathbf{e}_{q0}))$
- For second order logic, we need second order variables (relational variables) and quantifiers:
  - X_q of type $(\mathbf{e}_{q1} \to (\mathbf{e}_{q2} \to (\dots. \to \mathbf{e}_{q s_q} \to \mathbf{t}))) = \tau_q$
  - $\forall^2_{\tau_q} : ((\tau_q \to \mathbf{t}) \to \mathbf{t})$ and $\exists^2_{t_q} : ((\tau_q \to \mathbf{t}) \to \mathbf{t})$

Note that the logical connectives and quantifiers are treated simply as constants of the lambda calculus, so a first-order logic formula

$$\forall x.(f(x) \Rightarrow g(x)) \tag{3.1}$$

will be written as.

$$\forall(\lambda x((\Rightarrow \ f(x))\ g(x))) \tag{3.2}$$

Gamut (1991, Section 4.4.3) calls treating quantifiers the way we do in this chapter (exemplified by lambda-term 3.2 above) the *categorematic* way of introducing quantifiers (as opposed to the syncategorematic shown as formula 3.1 above). Introducing lambda-terms categorematically has the advantage that we do not need to modify our notions of free and bound variables, since only the λ operator binds variables. So instead of binding a variable x in a formula (as a syncategorematic quantifier does) we have a categorematic quantifier which takes a term of type $\mathbf{e}_i \to \mathbf{t}$ as an argument, which, because it is eta-expanded, is of the form $\lambda x^{\mathbf{e}_i}.y^{\mathbf{t}}$ for some y,

[5] This can be formulated as follows:

- X defines an application: $(\forall x.\exists y.X(x,y)) \wedge (\forall x.\forall y.\forall y'.\,(X(x,y) \wedge X(x,y')) \Rightarrow y = y')$
- X is injective $\forall x.\forall x'.\forall y.\,(X(x,y) \wedge X(x',y)) \Rightarrow x = x'$
- X is surjective: $(\forall y.\exists x.X(x,y))$

[6] Completeness and a compactness can be obtained but with non principal models in which n-ary relations are allowed to vary among sub-boolean algebras of powersets.

with the lambda operator taking care of binding the variable x in y. However, in the interest of readability, we will often write $\forall x.y$ instead of $\forall(\lambda x\ y)$ and $x \Rightarrow y$ instead of $((\Rightarrow x)\ y)$.

We can show by induction that every eta-expanded normal λ-term of type $\mathbf{e}_i$ with free variables x_k of type $\mathbf{e}_{i_k}$ corresponds to a term whose free variables are the x_k of sort $\mathbf{e}_{i_k}$.

Because the λ-term is normal and of type $\mathbf{e}_i$ it does not start with a λ, it is either

- a variable x of type $\mathbf{e}_i$ corresponding to the term $x : \mathbf{e}_i$ of multisorted logic (the free variable of the λ-term corresponds to the variable of the term),
- a constant f_q (in this case $s_q = 0$) of type $\mathbf{e}_i$ corresponding to the term $f : \mathbf{e}_i$ of multisorted logic (no free variable in the λ-term and none in the logical term),
- some function f_q applied to s_q arguments t_i of type $\mathbf{e}_{q_i}$ (because of the result type $\mathbf{e}_i$, it cannot be another kind of constant since they cannot yield $\mathbf{e}_i$ when applied to some arguments, and f must be provided with all its arguments to obtain this $\mathbf{e}_i$). The arguments t_i are themselves λ-terms of type $\mathbf{e}_{q_i}$,and they correspond, by induction hypothesis, to logical terms $t_i^{\triangledown}$ of sort $\mathbf{e}_{q_i}$, with the same variables as the free variables of the λ-term. The logical term, in this case is $f_q(t_1^{\triangledown},\ldots,t_{s_q}^{\triangledown})$ and its free variables are the same as those of the λ-term.

Now let us prove that every λ-term of type $\mathbf{t}$ in beta normal eta long form corresponds to a formula, with the same free variables as the formula.

Because the λ-term is normal and of type $\mathbf{t}$, we note that it does not start with a λ and that it cannot be of the form f_q, because this would imply a result type $\mathbf{e}_i$. We proceed by case analysis on the form of the λ-term.

- the λ-term is some R_q or X_q applied to s_q arguments t_i of type $\mathbf{e}_{q_i}$. The arguments t_i are λ-terms of type $\mathbf{e}_{q_i}$,and they correspond, by induction hypothesis, to logical terms $t_i^{\triangledown}$ of sort $\mathbf{e}_{q_i}$, with the same variables as the free variables of the λ-term . The logical formula, in this case is $R_q(t_1^{\triangledown},\ldots,t_{s_q}^{\triangledown})$ and its free variables are the same as those of the λ-term .
- a binary logical constant $*$ applied to two λ-terms (otherwise the result would not be of type $\mathbf{t}$) t_1 and t_2 of type $\mathbf{t}$: by induction hypothesis each of these corresponds to a formula whose free variables are those of the λ-term. The corresponding formula is $t_1^{\triangledown} * t_2^{\triangledown}$.
- a unary logical constant $\neg$ applied to a λ-term (otherwise the result would not be of type $\mathbf{t}$) t_1 of type $\mathbf{t}$: by induction hypothesis each of these corresponds to a formula whose free variables are those of the λ-term. The corresponding formula is $\neg t_1^{\triangledown}$.
- $\forall_i$ (resp. $\exists_i$) of type $(\mathbf{e}_i \to \mathbf{t}) \to \mathbf{t}$ applied to a term u of type $\mathbf{e}_i \to \mathbf{t}$. Because the λ-term is β-normal η-long, $u = \lambda x : \mathbf{e}_i\ v$ with $v : \mathbf{t}$ and v having an extra free variable, namely $x : \mathbf{e}_i$. The free variable of $\forall_i(\lambda x : \mathbf{e}_i\ v)$ say $y_1,\ldots,y_l$ are the ones of $v : \mathbf{t}$ minus $x : \mathbf{e}_i$. By induction hypothesis, $v : \mathbf{t}$ corresponds to a formula $v^{\triangledown}$ with free variables $x : \mathbf{e}_i$ and $y_1,\ldots,y_l$. The formula corresponding to the complete λ-term is simply $\forall x : \mathbf{e}_i v^{\triangledown}$ (resp. $\exists x : \mathbf{e}_i v^{\triangledown}$), its free variable are he ones of the λ-term , namely $y_1,\ldots,y_l$.

- If there are second order variables and quantifiers, it can be one of them, applied to a term. The quantifier $\forall^2_{t_q} : ((\tau_q \to \mathbf{t}) \to \mathbf{t})$ (resp. $\exists^2_{\tau_q} : ((\tau_q \to \mathbf{t}) \to \mathbf{t})$ is applied to a term of type $(\tau_q \to \mathbf{t})$. Because term are in $\beta\eta$ normal form, the term maybe assumed to be $\lambda X^{\tau_q}.u$ with $u : \mathbf{t}$. By induction hypothesis, u corresponds to a second order formula $u^\triangledown$, as one more free predicate variable X and the corresponding formula is $\forall X : \tau_q\ u^\triangledown$ (resp. $\exists X : \tau_q\ u^\triangledown$).

3.3 From Categorial Analysis to Montague Semantic Analysis

Let us come back to the relation between Montague semantics and categorial grammars. This relation, which was emphasized in particular by van Benthem (van Benthem, 1991) (see also Hendriks, 1993, for a systematic study of this relation), is a consequence of the following fact: simply typed λ-terms, which represent formulae of predicate calculus and neatly handle substitution, are very close to proofs in the Lambek calculus, which are the syntactic analyses of Lambek grammars. Indeed, via the Curry-Howard isomorphism (see e.g. Girard et al, 1988) simply typed λ-terms are proofs in intuitionistic logic, which extends the Lambek calculus. Indeed, reading $a \backslash b$ and $b \,/\, a$ as $a{\to}b$ (intuitionistic implication) each rule of the Lambek calculus is a rule of intuitionistic logic.[7] Assume our Lambek grammar uses the primitive types: *np*, *n*, *S*. First let us define a morphism from syntactic types to semantic types: these semantic types are formulae are define from two types *e* (entities) and *t* (truth values or propositions) with the intuitionistic implication $\to$ as their only connective:

$$types \quad ::= \quad e \quad | \quad t \quad | \quad (types {\to} types)$$

Thus a common noun like *chair* or an intransitive verb like *sleep* have the type $e{\to}t$ (the set of entities which are chairs or who sleep) a transitive verb like *takes* is a two place predicate of type $e{\to}(e{\to}t)$ (the pairs of entities such that the first one takes the second one) etc.

Thus we can define a morphism from syntactic types to semantic types.

(Syntactic type)* = Semantic type	
$S^* = t$	a sentence is a proposition
$np^* = e$	a noun phrase is an entity
$n^* = e{\to}t$	a noun is a subset of the set of entities
$(a \backslash b)^* = (b \,/\, a)^* = a^* \to b^*$	extends $(\_)^*$ to all syntactic types

[7] We will ignore the product formula $a \bullet b$ in this chapter, though this is just to keep the exposition of the lambda-calculus simple. The Curry-Howard isomorphism for the implication/conjunction fragment of intuitionistic logic is discussed in detail in (Girard et al, 1988) and carries over to the Lambek calculus without problems (see Abramsky, 1993; Moortgat, 1997, for two different solutions).

The lexicon associates to each syntactic type $t_k \in \text{Lex}(m)$ of a word m a λ-term τ_k *whose type is precisely* t_k^*, the semantic counterpart of the syntactic type t_k.

As discussed in Section 3.2.4, we need constants for the usual logical operations like quantification, conjunction etc.:

Constant	Type
$\exists$	$(e \to t) \to t$
$\forall$	$(e \to t) \to t$
$\wedge$	$t \to (t \to t)$
$\vee$	$t \to (t \to t)$
$\Rightarrow$	$t \to (t \to t)$

and proper constants for the denotation of the words in the lexicon:

likes	$\lambda x \lambda y$ (likes x) y	$x : e,\ y : e$, likes : $e \to (e \to t)$
`<< likes >> is a two-place predicate`		
Pierre	λP (P Pierre)	$P : e \to t$, Pierre : e
`<< Pierre >> is viewed as the set of properties which hold for the entity Pierre`		

These constants can include intensionality operators ˆ and ˇ (Gamut, 1991) which we will discuss briefly in Section 3.7.

Given

- *a syntactic analysis of* $m_1, \ldots, m_n$ in the Lambek calculus, that is a proof $\mathscr{D}$ of $t_1, \ldots, t_m \vdash S$ and
- *the semantics of each word* $m_1, \ldots, m_n$, that are λ-terms $\tau_i : t_i^*$,

we obtain the semantics of the sentence by the following algorithm:

1. Replace every syntactic type in $\mathscr{D}$ with its semantic counterpart; since intuitionistic logic extends the Lambek calculus the result $\mathscr{D}^*$ of this operation is a proof in intuitionistic logic of $t_1^*, \ldots, t_n^* \vdash t = S^*$.
2. Via the Curry-Howard isomorphism, this proof in intuitionistic logic can be viewed as a simply typed λ-term $\mathscr{D}_\lambda^*$ which contains one free variable x_i of type t_i^* per word m_i.
3. Replace in $\mathscr{D}_\lambda^*$. each variable x_i by the λ-term τ_i — whose type is also type t_i^*, so this is a correct substitution.
4. Reduce the resulting λ-term: this provides the semantics of the sentence (another syntactic analysis of the same sentence can lead to a different semantics).

We use natural deduction, because natural deduction is closer to λ-terms, but computing λ-terms from sequent calculus proofs is possible too, though this essentially corresponds to translating the sequent proof into natural deduction, as we have done in Section 2.4.2. Even though we remarked in Section 2.2.1 that it was unnecessary in the Lambek calculus to mark the hypothesis of the $/_i$ and $\backslash_i$ rules, we will coindex hypotheses and introduction rules throughout this chapter: intuitionistic proofs require us to mark which hypotheses correspond to which introduction rules, and since we translate Lambek proofs to intuitionistic proofs, indicating explicitly in *both* logics which hypotheses are withdrawn will make the translation from Lambek calculus proofs to intuitionistic proofs more transparent. For the same reason, we will use the $\rightarrow_e$ rule both with the principal branch on the left (for the translation of a $/_e$ rule) and with the principal branch on the right (for the translation of a $\backslash_e$ rule): the context of the rule application always allows us to determine the principal branch of the rule unambiguously.

3.4 Some Typical Examples

Example 3.2. Consider the following lexicon:

word	***syntactic type*** u ***semantic type*** u^* ***semantics:*** λ***-term of type*** u^* x^v ***means that the variable or constant*** x ***is of type*** v
some	$(S/(np\backslash S))/n$ $(e\rightarrow t)\rightarrow((e\rightarrow t)\rightarrow t)$ $\lambda P^{e\rightarrow t}\,\lambda Q^{e\rightarrow t}\,(\exists^{(e\rightarrow t)\rightarrow t}\,(\lambda x^e(\wedge^{t\rightarrow(t\rightarrow t)}(P\,x)(Q\,x))))$
statements	n $e\rightarrow t$ $\lambda x^e(\mathtt{statement}^{e\rightarrow t}\,x)$
speak_about	$(np\backslash S)/np$ $e\rightarrow(e\rightarrow t)$ $\lambda y^e\,\lambda x^e\,((\mathtt{speak\_about}^{e\rightarrow(e\rightarrow t)}\,x)\,y)$
themselves	$((np\backslash S)/np)\backslash(np\backslash S)$ $(e\rightarrow(e\rightarrow t))\rightarrow(e\rightarrow t)$ $\lambda P^{e\rightarrow(e\rightarrow t)}\,\lambda x^e\,((P\,x)\,x)$

Let us first show that "*Some statements speak about themselves*" belongs to the language generated by this lexicon. So let us prove (in natural deduction) the following:

$$(S/(np\backslash S))/n\,,\ n\,,\ (np\backslash S)/np\,,\ ((np\backslash S)/np)\backslash(np\backslash S)\vdash S$$

using the abbreviations *So* (some) *Sta* (statements) *SpA* (speak about) *Refl* (themselves) for the syntactic types.

$$\dfrac{\dfrac{So\vdash(S/(np\backslash S))/n\quad Sta\vdash n}{So,Sta\vdash(S/(np\backslash S))}/_e\qquad\dfrac{SpA\vdash(np\backslash S)/np\quad Refl\vdash((np\backslash S)/np)\backslash(np\backslash S)}{SpA,Refl\vdash(np\backslash S)}\backslash_e}{So,Sta,SpA,Refl\vdash S}/_e$$

Using the homomorphism from syntactic types to semantic types we obtain the following intuitionistic deduction, where So^*, Sta^*, SpA^*, $Refl^*$ are abbreviations for the semantic types respectively associated with the syntactic types: So, Sta, SpA, $Refl$.

$$\dfrac{\dfrac{So^* \vdash (e\to t)\to(e\to t)\to t \quad Sta^* \vdash e\to t}{So^*, Sta^* \vdash (e\to t)\to t}\to_e \quad \dfrac{SpA^* \vdash e\to e\to t \quad Refl^* \vdash (e\to e\to t)\to e\to t}{SpA^*, Refl^* \vdash e\to t}\to_e}{So^*, Sta^*, SpA^*, Refl^* \vdash t}\to_e$$

The λ-term representing this deduction simply is

$$((\mathit{some\ statements})\ (\mathit{themselves\ speak\_about}))\ \text{of type}\ t$$

where *some*, *statements*, *themselves*, *speak_about* are variables with respective types So^*, Sta^*, $Refl^*$, SpA^*. Let us replace these variables with the semantic λ-terms (of the same type) which are given in the lexicon. We obtain the following λ-term of type t (written on two lines) that we reduce:

$$\Big(\big(\lambda P^{e\to t}\ \lambda Q^{e\to t}\ (\exists^{(e\to t)\to t}\ (\lambda x^e(\wedge(P\,x)(Q\,x))))\big)\big(\lambda x^e(\mathtt{statement}^{e\to t}\ x)\big)\Big)$$
$$\Big(\big(\lambda P^{e\to(e\to t)}\ \lambda x^e\ ((P\,x)x)\big)\big(\lambda y^e\ \lambda x^e\ ((\mathtt{speak\_about}^{e\to(e\to t)}\ x)y)\big)\Big)$$

$$\downarrow\beta$$

$$\big(\lambda Q^{e\to t}\ (\exists^{(e\to t)\to t}\ (\lambda x^e(\wedge^{t\to(t\to t)}(\mathtt{statement}^{e\to t}\ x)(Q\,x))))\big)$$
$$\big(\lambda x^e\ ((\mathtt{speak\_about}^{e\to(e\to t)}\ x)x)\big)$$

$$\downarrow\beta$$

$$\big(\exists^{(e\to t)\to t}\ (\lambda x^e(\wedge(\mathtt{statement}^{e\to t}\ x)((\mathtt{speak\_about}^{e\to(e\to t)}\ x)x)))\big)$$

This term represent the following formula of predicate calculus (in a more pleasant format):

$$\exists x : e\ (\mathtt{statement}(x) \wedge \mathtt{speak\_about}(x,x))$$

This is the semantics of the analyzed sentence.

Example 3.3 (Sorts and Adverbs). The following example illustrates the use of a multisorted logic with sorts h (human beings) v (events) — and possibly others — and the possible treatment of higher order predicate by reification.

(3.3) Michele works hard.

When properties likes "works" can be the argument of a higher order predicate like the adverb "hard" it is common to introduce an event variable to designate "work" to which we apply the adverb "hard", thereby avoiding the need for to analyze "hard" as a property of a property. In this case, the event variable should be bound, hence, one should translate the syntactic type S by $v \to \mathbf{t}$, rather than by $\mathbf{t}$, and at the end of the computation an existential quantifier over events should be applied to event variables.

word	***syntactic type*** u ***semantic type*** u^* ***semantics: λ-term of type*** u^* x^v ***means that the variable or constant*** x ***is of type*** v
Michele	np h $\mathtt{M}^h$
works	$np \setminus S$ $h \to (v \to t)$ $\lambda u^h\, (\mathtt{works}^{h \to (v \to t)}\; u)$
hard	$(np \setminus S) \setminus (np \setminus S)$ $(h \to (v \to t)) \to (h \to (v \to t))$ $\lambda P^{h \to (v \to t)}\, \lambda x^h\, \lambda e^v (\wedge((P\, e)x)(\mathtt{hard}^{v \to t} e))$

Here is the syntactic analysis of the sentence:

$$\dfrac{np \qquad \dfrac{(np \setminus S) \quad (np \setminus S) \setminus (np \setminus S)}{(np \setminus S)} \setminus_e}{S} \setminus_e$$

This Lambek proof can be turned into an intuitionistic proof:

$$\dfrac{h \qquad \dfrac{(h \to (v \to S)) \quad (h \to (v \to S)) \to (h \to (v \to S))}{(h \to (v \to S))} \to_e}{(v \to S)} \to_e$$

The lambda term issued from the syntactic analysis is $((H\,W)\,M)$, and after inserting the lambda terms provided by the lexicon it becomes

$$\left(\left(\lambda P^{h \to (v \to t)}\, \lambda x^h\, \lambda e^v (\wedge((P\, e)\, x)(\mathtt{hard}^{v \to t} e))\right)\left(\lambda u^h \lambda i^v\, ((\mathtt{works}^{h \to (v \to t)}\; u)\; i)\right)\right) \mathtt{M}^h$$

$$\left(\lambda x^h\, \lambda e^v ((\wedge(((\left(\lambda u^h \lambda i^v\, ((\mathtt{works}^{h \to (v \to t)}\; u)\; i)\right) x))\; e)(\mathtt{hard}^{v \to t} e)))\right) \mathtt{M}^h$$

$$\left(\lambda x^h\, \lambda e^v ((\wedge(((\mathtt{works}^{h \to (v \to t)} x))\; e))(\mathtt{hard}^{v \to t} e))\right) \mathtt{M}^h$$

$$\lambda e^v ((\wedge(((\mathtt{works}^{h \to (v \to t)} \mathtt{M}^h))\; e))(\mathtt{hard}^{v \to t} e))$$

In this reified vision, in order to obtain a logical formula, it is common to consider, once the sentence or discourse has been analyzed, the existential closure of the result. Hence the semantic representation becomes

$$\exists^{(v\to\mathbf{t})\to\mathbf{t}}\left(\lambda e^{v}((\wedge(((\mathtt{works}^{h\to(v\to t)}\mathtt{M}^{h}))\ e))(\mathtt{hard}^{v\to t} e))\right)$$

which, in standard notation is

$$\exists e : \mathit{event}\ \mathtt{works}(e,\mathtt{M}^{h}) \wedge \mathtt{hard}(e)$$

There is a slightly different analysis — without reification but using a higher order predicate — of the same sentence. The lexicon is as follows:

word	***syntactic type*** u ***semantic type*** u^* ***semantics: λ-term of type*** u^* x^{v} ***means that the variable or constant*** x ***is of type*** v
Michele	np h $\mathtt{M}^{h}$
works	$np \setminus S$ $h \to t$ $\lambda u^{h}\ (\mathtt{works}^{h\to\mathbf{t}}\ u)$
hard	$(np \setminus S) \setminus (np \setminus S)$ $(h\to\mathbf{t})\to(h\to\mathbf{t}))$ $\lambda P^{h\to\mathbf{t}}\ (\mathtt{hard}^{(h\to\mathbf{t})\to(h\to\mathbf{t})} P)$

The syntactic analysis is just the same, yielding a term $((H\ W)\ M)$. If one inserts the new semantic lambda terms, one obtains:

$$\left(\lambda P^{h\to\mathbf{t}}\ (\mathtt{hard}^{(h\to\mathbf{t})\to(h\to\mathbf{t})} P)\right)(\lambda u^{h}\ (\mathtt{works}^{h\to\mathbf{t}}\ u))\ \mathtt{M}^{h}$$

$$(\mathtt{hard}^{(h\to\mathbf{t})\to(h\to\mathbf{t})}(\lambda u.(\mathtt{works}^{h\to\mathbf{t}}\ u))\ \mathtt{M}^{h}$$

which can be written in higher order logic as:

$$(\mathtt{hard}(\mathtt{works}))(\mathtt{M})$$

Example 3.4 (Relative pronouns and relative clauses)
The semantic effect of the pronoun is to introduce a conjunction between the relative clause of type $\mathbf{e} \to \mathbf{t}$ and the common noun it is attached to $\mathbf{e} \to \mathbf{t}$. We will use French because it makes a neater distinction between the relative pronoun acting as a subject (*qui*) and relative pronouns acting as an object (*que*).

(3.4) *Un enfant qui courait est_tombé.*
A child who ran fell.

'A child who ran fell.'

Such an example should be structured as follows:

$\exists\, x\, ((child(x) \wedge past\_run(x)) \wedge (past\_fall(x))$

The existential quantifier is correctly applied to the conjunction of two predicates: the restriction *((child(x)* $\wedge$ *past_run(x))* and the predicate *fall(x)*.

The lexicon should look as follows.

word	***syntactic type*** u ***semantic type*** u^* ***semantics: λ-term of type*** u^* x^v ***means that the variable or constant*** x ***is of type*** v
courait	$np \backslash S$ $\mathbf{e} \to \mathbf{t}$ $\lambda x^{\mathbf{e}}(past\_run(x))$
est_tombé	$np \backslash S$ $\mathbf{e} \to \mathbf{t}$ $\lambda x^{\mathbf{e}}(past\_fall(x))$
enfant	n $\mathbf{e} \to \mathbf{t}$ $\lambda x^{\mathbf{e}}\ (child(x))$
qui	$(n \backslash n) / (np \backslash S)$ $(\mathbf{e} \to \mathbf{t}) \to (\mathbf{e} \to \mathbf{t}) \to (\mathbf{e} \to \mathbf{t})$ $\lambda P^{\mathbf{e}\to\mathbf{t}} \lambda Q^{\mathbf{e}\to\mathbf{t}} (\lambda x^{\mathbf{e}} \wedge (Px)(Qx))$
un	$(S / (np \backslash S)) / n$ $(\mathbf{e} \to \mathbf{t}) \to (\mathbf{e} \to \mathbf{t}) \to \mathbf{t}$ $\lambda P^{\mathbf{e}\to\mathbf{t}} \lambda Q^{\mathbf{e}\to\mathbf{t}} \exists (\lambda x^{\mathbf{e}} \wedge (Px)(Qx))$

The syntactic analysis is

$$\dfrac{\dfrac{(S/(np\backslash S))/n \qquad \dfrac{n \qquad \dfrac{(n\backslash n)/(np\backslash S) \qquad np\backslash S}{n\backslash n}/_e}{n}\backslash_e}{(S/(np\backslash S))}/_e \qquad np\backslash S}{S}\backslash_e$$

The corresponding proof with semantic types instead is:

$$\dfrac{\dfrac{(\mathbf{e}\to\mathbf{t})\to((\mathbf{e}\to\mathbf{t})\to\mathbf{t}) \qquad \dfrac{(\mathbf{e}\to\mathbf{t}) \qquad \dfrac{(\mathbf{e}\to\mathbf{t})\to(\mathbf{e}\to\mathbf{t})\to(\mathbf{e}\to\mathbf{t}) \qquad \mathbf{e}\to\mathbf{t}}{(\mathbf{e}\to\mathbf{t})\to(\mathbf{e}\to\mathbf{t})}\to_e}{(\mathbf{e}\to\mathbf{t})}\to_e}{((\mathbf{e}\to\mathbf{t})\to\mathbf{t})}\to_e \qquad \mathbf{e}\to\mathbf{t}}{\mathbf{t}}\to_e$$

The corresponding lambda term is *un ((qui courait) enfant) est_tombé* when one inserts the lexical lambda terms it yields:

$$((\lambda P^{\mathbf{e}\to\mathbf{t}}\lambda Q^{\mathbf{e}\to\mathbf{t}}\exists(\lambda x{:}\mathbf{e} \wedge (Px)(Qx)) \\ ((\lambda P^{\mathbf{e}\to\mathbf{t}}\lambda Q^{\mathbf{e}\to\mathbf{t}}(\lambda x^{\mathbf{e}} \wedge (P\,x)\ (Q\,x))\lambda x^{\mathbf{e}}\ (past\_run(x)))\lambda x^{\mathbf{e}}\ (child(x)))) \\ \lambda x^{\mathbf{e}}\ (past\_fall(x)))$$

This reduces to:

$$(\lambda P^{\mathbf{e}\to\mathbf{t}}\lambda Q^{\mathbf{e}\to\mathbf{t}}\exists(\lambda x^{\mathbf{e}} \wedge (P\,x)\ (Q\,x))(\lambda x^{\mathbf{e}}(\wedge(past\_run(x))child(x)))(\lambda x^{\mathbf{e}}\ (past\_fall(x))))$$

an finally to:

$$\exists(\lambda x^{\mathbf{e}}(\wedge(\wedge(past\_run(x)\ child(x))\ (past\_fall(x))))$$

In other words:

$$\exists x(past\_run(x) \wedge child(x) \wedge past\_fall(x))$$

Example 3.5 (Quantifier scope). A classical example is scope ambiguity in a sentence like

(3.5) Every child ate a pizza.

word	***syntactic type*** u ***semantic type*** u^* ***semantics: λ-term of type*** u^* x^v ***means that the variable or constant*** x ***is of type*** v
every	$(S\,/\,(np\,\backslash\,S))\,/\,n$ (subject) $((S\,/\,np)\,\backslash\,S)\,/\,n$ (object) $(e\to t)\to((e\to t)\to t)$ $\lambda P^{e\to t}\ \lambda Q^{e\to t}\ (\forall^{(e\to t)\to t}\ (\lambda x^e(\Rightarrow^{t\to(t\to t)}\ (P\,x)(Q\,x))))$
a	$((S\,/\,np)\,\backslash\,S)\,/\,n$ (object) $(S\,/\,(np\,\backslash\,S))\,/\,n$ (subject) $(e\to t)\to((e\to t)\to t)$ $\lambda P^{e\to t}\ \lambda Q^{e\to t}\ (\exists^{(e\to t)\to t}\ (\lambda x^e(\wedge^{t\to(t\to t)}(P\,x)(Q\,x))))$
child	n $e\to t$ $\lambda x^e(\mathtt{child}^{e\to t}\ x)$
pizza	n $e\to t$ $\lambda x^e(\mathtt{pizza}^{e\to t}\ x)$
ate	$(np\,\backslash\,S)\,/\,np$ $e\to(e\to t)$ $\lambda y^e\ \lambda x^e\ ((\mathtt{ate}^{e\to(e\to t)}\ x)y)$

There are two possible syntactic analyses that will lead to the two different readings. Let us completely compute the first analysis and the first semantic representation.

One of the two possible syntactic analyses is:

∃∀

$$\dfrac{\dfrac{\dfrac{\dfrac{(S/(np\backslash S))/n \quad n}{(S/(np\backslash S))}/_e \quad \dfrac{(np\backslash S)/np \quad [np]^1}{(np\backslash S)}/_e}{S}/_e}{S/np}/_i(1) \quad \dfrac{((S/np)\backslash S)/n \quad n}{(S/np)\backslash S}\backslash_e}{S}\backslash_e$$

The corresponding semantic analysis is the following:

∃∀

$$\dfrac{\dfrac{\dfrac{\dfrac{\overset{every}{(\mathbf{e}\to\mathbf{t})\to(\mathbf{e}\to\mathbf{t})\to\mathbf{t}} \quad \overset{child}{(\mathbf{e}\to\mathbf{t})}}{(\mathbf{e}\to\mathbf{t})\to\mathbf{t}}\to_e \quad \dfrac{\overset{ate}{\mathbf{e}\to\mathbf{e}\to\mathbf{t}} \quad \overset{o}{[\mathbf{e}]^1}}{\mathbf{e}\to\mathbf{t}}\to_e}{\mathbf{t}}\to_e}{\mathbf{e}\to\mathbf{t}}\to_i(1) \quad \dfrac{\overset{a}{(\mathbf{e}\to\mathbf{t})\to(\mathbf{e}\to\mathbf{t})\to\mathbf{t}} \quad \overset{pizza}{(\mathbf{e}\to\mathbf{t})}}{(\mathbf{e}\to\mathbf{t})\to\mathbf{t}}\to_e}{\mathbf{t}}\to_e$$

The corresponding lambda terms is:

$\exists\forall = (a\ pizza)(\lambda o^{\mathbf{e}}(every\ child)(ate\ o))$

where words have to be replaced with the corresponding lambda terms.

Let us first compute the following lambda terms

$(a\ pizza)$
$= (\lambda P^{e\to t}\,\lambda Q^{e\to t}\,(\exists^{(e\to t)\to t}\,(\lambda x^e(\wedge^{t\to(t\to t)}(P\,x)(Q\,x)))))(\lambda z^e(\mathtt{pizza}^{e\to t}\,z))$
$= (\lambda Q^{e\to t}\,(\exists^{(e\to t)\to t}\,(\lambda x^e(\wedge^{t\to(t\to t)}((\lambda z^e(\mathtt{pizza}^{e\to t}\,z))\,x)(Q\,x)))))$
$= (\lambda Q^{e\to t}\,(\exists^{(e\to t)\to t}\,(\lambda x^e(\wedge^{t\to(t\to t)}((\mathtt{pizza}^{e\to t}\,x)))(Q\,x))))$

$(every\ child)$
$= (\lambda P^{e\to t}\,\lambda Q^{e\to t}\,(\forall^{(e\to t)\to t}\,(\lambda x^e(\Rightarrow^{t\to(t\to t)}\,(P\,x)(Q\,x)))))(\lambda u^e(\mathtt{child}^{e\to t}\,u))$
$= (\lambda Q^{e\to t}\,(\forall^{(e\to t)\to t}\,(\lambda x^e(\Rightarrow^{t\to(t\to t)}\,((\lambda u^e(\mathtt{child}^{e\to t}\,u))\,x)(Q\,x)))))$
$= (\lambda Q^{e\to t}\,(\forall^{(e\to t)\to t}\,(\lambda x^e(\Rightarrow^{t\to(t\to t)}\,(\mathtt{child}^{e\to t}\,x)(Q\,x)))))$

$(every\ child)(ate\ o) =$
$(\lambda Q^{e\to t}\,(\forall^{(e\to t)\to t}\,(\lambda w^e(\Rightarrow^{t\to(t\to t)}(\mathtt{child}^{e\to t}\,w)(Q\,w)))))((\lambda y^e\,\lambda x^e\,((\mathtt{ate}^{e\to(e\to t)}\,x)\,y))\,o)$
$= (\lambda Q^{e\to t}\,(\forall^{(e\to t)\to t}\,(\lambda w^e(\Rightarrow^{t\to(t\to t)}\,(\mathtt{child}^{e\to t}\,w)(Q\,w)))))(\lambda x^e\,((\mathtt{ate}^{e\to(e\to t)}\,x)\,o))$
$= \forall^{(e\to t)\to t}\,(\lambda w^e(\Rightarrow^{t\to(t\to t)}\,(\mathtt{child}^{e\to t}\,w)((\lambda x^e\,((\mathtt{ate}^{e\to(e\to t)}\,x)\,o))\,w)))$
$= \forall^{(e\to t)\to t}\,(\lambda w^e(\Rightarrow^{t\to(t\to t)}\,(\mathtt{child}^{e\to t}\,w)(((\mathtt{ate}^{e\to(e\to t)}\,w)\,o))))$

$(a\ pizza)(\lambda o\ (every\ child)(ate\ o))$
$= (\lambda Q^{e\to t}\,(\exists^{(e\to t)\to t}\,(\lambda x^e(\wedge^{t\to(t\to t)}((\mathtt{pizza}^{e\to t}\,x)))(Q\,x))))$
$\quad (\lambda o\forall^{(e\to t)\to t}\,(\lambda w^e(\Rightarrow^{t\to(t\to t)}\,(\mathtt{child}^{e\to t}\,w)(((\mathtt{ate}^{e\to(e\to t)}\,w)\,o)))))$
$= (\exists^{(e\to t)\to t}\,(\lambda x^e(\wedge^{t\to(t\to t)}((\mathtt{pizza}^{e\to t}\,x)))$
$\quad ((\lambda o\forall^{(e\to t)\to t}\,(\lambda w^e(\Rightarrow^{t\to(t\to t)}\,(\mathtt{child}^{e\to t}\,w)(((\mathtt{ate}^{e\to(e\to t)}\,w)\,o))))\,x)))$

$= (\exists^{(e\to t)\to t}\ (\lambda x^e (\wedge^{t\to(t\to t)}((\texttt{pizza}^{e\to t}\ x)))$
$\quad (\forall^{(e\to t)\to t}\ (\lambda w^e (\Rightarrow^{t\to(t\to t)}\ (\texttt{child}^{e\to t}\ w)((\texttt{ate}^{e\to(e\to t)}\ w)\ x))))))$

that one usually writes:

$$\exists x.\ pizza(x) \wedge \forall w.\ (child(w) \Rightarrow ate(w,x))$$

The other possible syntactic analysis is:
$\forall\exists$

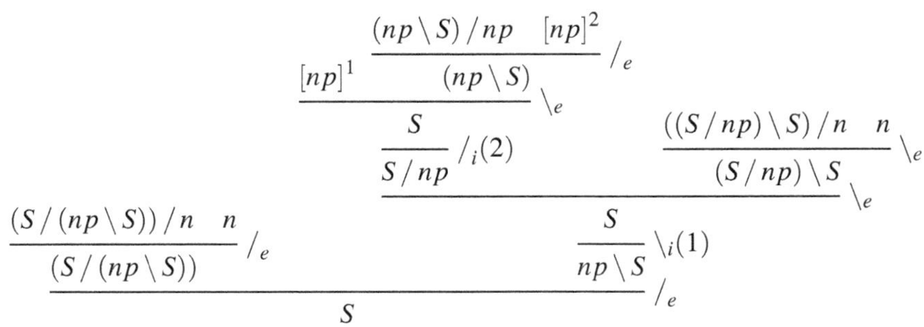

It corresponds to the following analysis:
$\forall\exists$

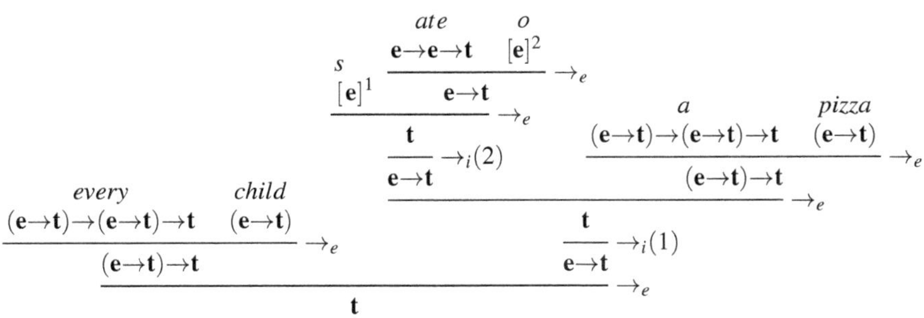

This proof corresponds to the lambda term

$$\forall\exists = (every\ child)(\lambda s.\ (a\ pizza)(\lambda o\ ((ate\ o)\ s)))$$

where the words are to be replaced with the lambda terms from the lexicon and that we are going to reduce.

Let us compute the following lambda terms, using the terms we computed before for $(a\ pizza)$ and $(every\ child)$.

$\quad (a\ pizza)(\lambda o\ ((ate\ o)\ s))$
$= (\lambda Q^{e\to t}\ (\exists^{(e\to t)\to t}\ (\lambda x^e (\wedge^{t\to(t\to t)}((\texttt{pizza}^{e\to t}\ x)))(Q\ x))))$
$(\lambda o\ (((\lambda y^e\ \lambda x^e\ ((\texttt{ate}^{e\to(e\to t)}\ x)\ y))\ o)\ s)))$
$= (\lambda Q^{e\to t}\ (\exists^{(e\to t)\to t}\ (\lambda x^e (\wedge^{t\to(t\to t)}((\texttt{pizza}^{e\to t}\ x)))(Q\ x))))(\lambda o\ ((\texttt{ate}^{e\to(e\to t)}\ s)\ o))$
$= (\exists^{(e\to t)\to t}\ (\lambda x^e (\wedge^{t\to(t\to t)}((\texttt{pizza}^{e\to t}\ x)))((\lambda o\ ((\texttt{ate}^{e\to(e\to t)}\ s)\ o))\ x)))$
$= (\exists^{(e\to t)\to t}\ (\lambda x^e (\wedge^{t\to(t\to t)}((\texttt{pizza}^{e\to t}\ x)))((\texttt{ate}^{e\to(e\to t)}\ s)\ x)))$

$$\begin{array}{l}\forall\exists = (every\ child)(\lambda s.\ (a\ pizza)(\lambda o\ ((ate\ o)\ s)))\\ = (\lambda Q^{e\to t}\ (\forall^{(e\to t)\to t}\ (\lambda u^e(\Rightarrow^{t\to(t\to t)}\ (\mathtt{child}^{e\to t}\ u)(Q\ u)))))\\ (\lambda s.\ (\exists^{(e\to t)\to t}\ (\lambda x^e(\wedge^{t\to(t\to t)}((\mathtt{pizza}^{e\to t}\ x)))((\mathtt{ate}^{e\to(e\to t)}\ s)\ x))))\\ = (\forall^{(e\to t)\to t}\ (\lambda u^e(\Rightarrow^{t\to(t\to t)}\ (\mathtt{child}^{e\to t}\ u)\\ ((\lambda s.\ (\exists^{(e\to t)\to t}\ (\lambda x^e(\wedge^{t\to(t\to t)}((\mathtt{pizza}^{e\to t}\ x)))((\mathtt{ate}^{e\to(e\to t)}\ s)\ x))))\ u)))))\end{array}$$

$$\begin{array}{l}= (\forall^{(e\to t)\to t}\ (\lambda u^e(\Rightarrow^{t\to(t\to t)}\ (\mathtt{child}^{e\to t}\ u)\\ (\exists^{(e\to t)\to t}\ (\lambda x^e.\ (\wedge^{t\to(t\to t)}((\mathtt{pizza}^{e\to t}\ x)))((\mathtt{ate}^{e\to(e\to t)}\ u)\ x))))))\end{array}$$

which is usually written as:

$$\forall u.\ child(u) \Rightarrow \exists.x\ pizza(x) \wedge ate(u,x)$$

3.5 Determiners, Quantifiers and Type Raising

So far we did not interpret the definite article. To be simple, one can use Hilbert's $\tau : (\mathbf{e} \to \mathbf{t}) \to \mathbf{e}$ which, given a predicate $P^{\mathbf{e}\to\mathbf{t}}$ picks up an element satisfying P. As a definite article needs to be first introduced it would be better to replace logic in the lambda calculus by Discourse Representation Theory (DRT) in the lambda calculus. DRT correctly handles many puzzles concerning discourse referents. We discuss this compositional version of DRT, as well as several motivating examples, in Section 3.6.

Hitherto we stuck to a strict correspondence (often called homomorphism) between syntactic categories and semantic types. As we have seen, this complicates the syntactic categories for quantifiers and determiners somewhat, since they depend on the syntactic position. Why should the syntactic category of *someone* be different in *Someone forgot his wallet.* and in *I met someone*? Here, we will discuss alternative semantic solutions, which also have their own drawbacks.

Systematic Type Raising

The first alternative is to translate np into $(\mathbf{e} \to \mathbf{t}) \to \mathbf{t}$ rather than into $\mathbf{e}$. Thus the semantics of "*Peter*" is $\lambda P^{\mathbf{e}\to\mathbf{t}}.\ P(peter)$ The advantage of this solution is that that quantifiers have the same syntactic category no matter where they appear. Determiners will therefore get a lifted type: since "*the cat*", of syntactic type np should be, of semantic type $(\mathbf{e} \to \mathbf{t}) \to \mathbf{t}$, the determiner "*the*", of syntactic type $np\,/\,n$, should be of type $(\mathbf{e} \to \mathbf{t}) \to ((\mathbf{e} \to \mathbf{t}) \to \mathbf{t})$ and the term should be $\lambda P^{\mathbf{e}\to\mathbf{t}}\lambda Q^{\mathbf{e}\to\mathbf{t}}Q(\tau(P))$. As a consequence, the semantic types of verbs become more complicated: "*sleeps*", of syntactic type $np \setminus S$, should be of semantic type $((\mathbf{e} \to \mathbf{t}) \to \mathbf{t}) \to \mathbf{t}$ with term $\lambda R^{(\mathbf{e}\to\mathbf{t})\to\mathbf{t}}R(\lambda x^{\mathbf{e}} sleeps(x))$, so we simplify the syntactic types but complicate the semantic types and, consequently, the lambda terms.

A problem of this approach is that we no longer generate two distinct readings for "every child ate a pizza": in Example 3.5 we obtained two proofs for this sentence, each of which corresponded to a different semantic reading. The use of systematic type raising would require us to find another way to generate the two readings;

one possibility would be to assign to different lambda-terms to the verb, but the treatment of quantifier scope as verbal polysemy seems hardly appropriate.

Substituting terms for terms rather than terms for variables

The second alternative, which also provides the same syntactic category for quantifiers wherever they appear in the sentence, is to change the way we substitute the lexical lambda terms: as we have used it before, a Lambek calculus derivation corresponds to a lambda-term with a free variable w for each of the hypotheses and we replace these free variables by lexical lambda terms of the same type. We can make this replacement operation a bit more complicated by allowing the replacement of any lambda term of the correct type provided that its only free variable is the lexical variable w. Let us take a simple example:

word	***syntactic type*** u ***semantic type*** u^* ***semantics: λ-term of type*** u^* x^v ***means that the variable or constant*** x ***is of type*** v
everyone	np $((e\to t)\to t)$ $\lambda Q^{e\to t}\ (\forall^{(e\to t)\to t}\ (\lambda x^e (Q\, x)))$
sleeps	$(np\setminus S)$ $(e\to t)$ $\lambda x^e\ (\mathtt{sleep}^{e\to t}\ x)$

(3.6) Everyone sleeps

Note the type clash between the syntactic type np, corresponding to the semantic type e, and the semantic type $((e\to t)\to t)$ which is assigned to it in the lexicon. This sentence can be analyzed by a non normal proof in the Lambek calculus, with a subproof of conclusion $(S\,/\,(np\setminus S))$ and a single free variable (or hypothesis) np (observe that it is not an analysis in AB grammars because of the introduction rule).

$$\dfrac{\dfrac{\dfrac{np \qquad [np\setminus S]_1}{S}\setminus_e}{S\,/\,(np\setminus S)}/_i \qquad np\setminus S}{S}/_e$$

The semantic translation of this proof is:

$$\dfrac{\dfrac{\dfrac{\mathbf{e} \qquad [\mathbf{e}\to\mathbf{t}]_1}{\mathbf{t}}\to_e}{(\mathbf{e}\to\mathbf{t})\to\mathbf{t}}\to_i \qquad \mathbf{e}\to\mathbf{t}}{\mathbf{t}}\to_e$$

The corresponding lambda term is $(\lambda P^{\mathbf{e}\to\mathbf{t}}\, P(\texttt{everyone}))^{(\mathbf{e}\to\mathbf{t})\to\mathbf{t}}\texttt{sleep}$ and the first part only has everyone as a free variable. With the relaxed rule, one can replace the whole subterm by the semantic lambda term associated with *everyone*,

$$\lambda Q^{e\to t}\ (\forall^{(e\to t)\to t}\ (\lambda x^{e}(Q\, x)))$$

thus obtaining as the semantics of the sentence the expression:

$$(\lambda Q^{e\to t}\ (\forall^{(e\to t)\to t}\ (\lambda x^{e}(Q\, x))))(\lambda x^{e}\ (\texttt{sleep}^{e\to t}\ x))$$

which reduces to.

$$(\forall^{(e\to t)\to t}\ (\lambda x^{e}(\texttt{sleep}\, x))))$$

Because this analysis forces us to consider non normal proofs with a subproof of the semantic type, it is not very satisfactory.

To have a single syntactic category for quantifiers wherever they are located, and to handle discourses referents properly, one may go for categorial minimalist grammars, with lambda-mu DRT for computing the semantics (Amblard et al, 2010), but this is goes beyond the scope (!) of the present book.

We will give another solution for unifying the two type assignments in Section 5.1.1.

3.6 Lambek Calculus and Discourse Representation Theory

Though the link between the Lambek calculus and Montague grammar is traditional and well-known, a lot of modern research has been done in what is called *dynamic* semantics. Dynamic semantics takes into account the *context* of an utterance and models how this utterance changes the context. One of the typical applications of this point of view is the semantics of anaphors: anaphors, like "he" or "she" must — in a coherent discourse — refer to a previously introduced individual, as indicated by the following example discourse.

(3.7) The president of the board/A friend from Amsterdam/Thomas arrived yesterday.
(3.8) I'll meet him for lunch today.

Here the first sentence is interpreted as adding a new element to the context[8], corresponding to "the president of the board", "a friend from Amsterdam" or "Thomas". In each of these cases it is possible for the second sentence to refer back to this previously introduced entity by means of the pronoun "him".

On the other hand, it is well-known that we cannot just refer back to *any* noun phrase or quantifier of the preceding discourse, as indicated by the following discourse ("#" denotes the sentence is incoherent in the given discourse).

[8] We stay, for the moment, deliberately vague as to the nature of the context and the elements in it. We will give a more concrete interpretation in what follows.

(3.9) Four of my friends/nobody/everyone arrived yesterday.
(3.10) # I'll meet him for lunch today.

Here none of the expressions "four of my friends", "nobody" or "everyone" can be referred to by the pronoun "him" and a theory of anaphoric reference needs to explain this. Though the contrast between plural and singular explains at least part of the difference, it does not explain all of it: neither "him" nor "them" can refer to "nobody".

As another set of examples (after van Eijck and Kamp, 2011), look at the following contrast.

(3.11) Someone didn't smile.
(3.12) He had a terrible headache.

(3.13) Not everyone smiled.
(3.14) # He had a terrible headache.

Though from a truth-conditional point of view "Someone didn't smile" and "Not everyone smiled" are logically equivalent, they are clearly different in terms of the anaphoric possibilities they allow, as shown by the second sentence of each of the two minimal discourses above.

As one final example, let's return to Example 3.5, repeated as the first sentence below.

(3.15) Every child ate a pizza.
(3.16) It had lots of ham and mozzarella.

In this case, continuing the discourse with the second sentence forces us to give the first sentence the interpretation where there is a single pizza which was eaten by all the children.

One very successful theory of dynamic semantics which gives an account of all these and many more phenomena is Discourse Representation Theory (Kamp and Reyle, 1993). The basic objects of Discourse Representation Theory (DRT) are Discourse Representation Structures. Dynamic semantics and Discourse Representation Theory are both fields with a long history and a large volume of research and this short section cannot really do more than give a flavor of some of the very basic ideas. For a more detailed treatment of these subjects, the reader is referred to (Kamp and Reyle, 1993; Muskens et al, 2011; van Eijck and Kamp, 2011; Kamp et al, 2011).

Figure 3.1 shows two examples Discourse Representation Structures (DRSs).

It shows the two readings of "Every child ate a pizza" which we computed for Example 3.5 in DRS form. Each DRS is drawn as a box, divided into two parts by a horizontal line. Above the line, there is a list of variables which are called the *discourse referents*. The set of discourse referents is sometimes called the *universe* of the DRS. Below the line are the *conditions* of the DRS: these can be predicates, like *pizza*(x) but also complex conditions, which recursively contain other DRSs. In the example there is an implication $\Rightarrow$ between two DRSs, with the obvious

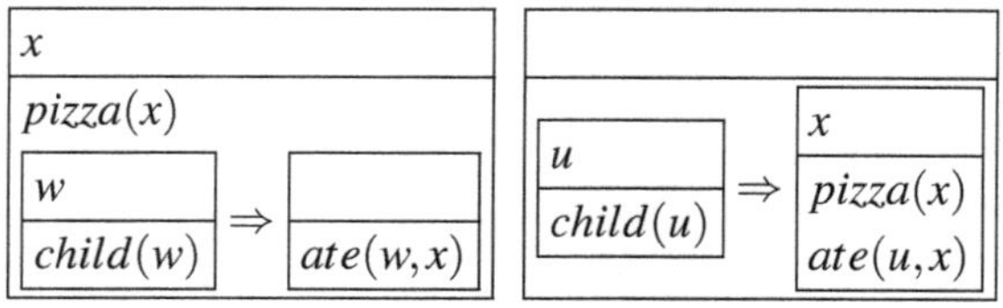

Fig. 3.1. Two example Discourse Representation Structures corresponding to the two readings of Example 3.5

intended meaning that *if* the DRS on the left of the arrow evaluates to true *then* the DRS on the right of the arrow evaluates to true. The intuitive meaning of a DRS is that the discourse referents are existentially quantified variables (and, by duality, the discourse referents on the left hand side of an implication are universally quantified).

Definition 3.6 (DRS). *A discourse representation structure is a pair $\langle V, C\rangle$ such that V is a set of variables and C is a set of conditions.*

A condition is either an atomic condition, a predicate $p(x_0,\ldots,x_n)$ where all the x_i are discourse referents (we treat $x = y$, often called anaphoric link, as a two-place predicate which is interpreted as the identity relation) or a complex condition: if K_1 and K_2 are DRSs then $K_1 \Rightarrow K_2$, $K_1 \vee K_2$ and $\neg K_1$ are DRS conditions.

One of the crucial notions is the notion of *accessibility* of discourse referents: if we add a sentence containing an anaphoric pronoun, such as "he", "she" or "it" to a DRS, we need to link it to an accessible discourse referent. We say a discourse referent x is accessible from DRS K if:

1. x is a member of the universe of K,
2. K is a DRS which occurs as a condition $\neg K$, $K \vee K_1$, $K_1 \vee K$, $K \Rightarrow K_1$ or $K_1 \Rightarrow K$ in a DRS K_p and x is accessible from K_p,
3. K is a DRS which occurs as a condition $K_1 \Rightarrow K$ and x is accessible from K_1.

So, intuitively, from a DRS K we can "see" all discourse referents which occur in a DRS which contains K or which occur to the left of an implication from K. If x_i is an accessible discourse referent occurring in an atomic condition $p(x_0,\ldots,x_n)$ we will call x_i *bound*[9]; if not, we will call x_i *free*. A DRS without free variables is called a *proper* DRS.

Given this definition of accessibility, we see that for the example DRSs of Figure 3.1, the outermost level of the leftmost DRS has only the discourse referent x which is accessible in the "outer" DRS, whereas w is accessible for the DRS to the left of the implication and both w and x are accessible DRS to the right of the implication. The rightmost DRS has no discourse referents accessible at the outer level, u is accessible in the DRS left of the arrow and both u and x are accessible right of the arrow.

[9] It is possible to distinguish between two types of bound occurrences (see van Eijck and Kamp, 2011, for further details and discussion), however, in order to keep the current discussion simple, we will not do so here.

As a slight abuse of notation, we will write $x = ?$ for an unresolved anaphoric link, indicating that we need to substitute an accessible discourse referent for '?' before we can interpret the DRS. In other words, the actual resolution of anaphora, though subject to syntactic constraints, is handled in the semantics. The alternative is to handle binding in the syntax, by coindexing an anaphor and its antecedent, then ensuring in the translation that expressions labeled by the same index are translated by the same variable (see Muskens, 1996, for example). We will have a bit more to say on this topic when we talk about the semantics assigned to "he".

A useful operation on two DRSs K_1 and K_2 is the *merge* operation. The idea behind the merge operation is that we add the information of K_2 to the information of K_1 to obtain a new DRS K. We can see K_1 as the DRS corresponding to the preceding discourse, K_2 as the DRS corresponding to the sentences which have just been uttered and their merge K as the DRS corresponding to the resulting discourse. Defining merge is easy when the universes of K_1 and K_2 are disjoint and none of the conditions of either DRS contains free variables. However, this seems much too restrictive. In order to give an appropriate account of binding phenomena, we allow the discourse referents of K_1 to bind the variables of the conditions in K_2, but not vice versa and we rename the variables in the universe of K_2 in order to avoid naming conflicts with variables of the universe K_1, as indicated by the following definition (this is not the only possibility; see van Eijck and Kamp, 2011, for a discussion of different strategies for defining merge).

Definition 3.7 (Merge). *Let K_1 and K_2 be two Discourse Representation Structures with variables $x_1,\dots,x_n$ and $y_1,\dots,y_n$ respectively. We define the* merge *of K_1 and K_2, written $K_1 \oplus K_2$ as follows*

- *we replace all free variables x_i of K_1 which are identical to one of the y_i of K_2 by unused variables v_i, obtaining a new DRS K_1',*
- *we replace all bound variables y_i of K_2 which are identical to one of the x_i of K_1 by unused variables z_i, obtaining a new DRS K_2',*
- *$K_1 \oplus K_2$ is defined as the union of the discourse referents of K_1' with the discourse referents of K_2' and the union of the conditions of K_1' with the conditions of K_2'.*

Note that the merge of two DRSs is always defined and unique up to variable renaming, but asymmetric: discourse referents of K_1 can bind variable occurrences of conditions of K_2 (and this is, of course, the essence of anaphoric reference) but discourse referents of K_2 can never bind conditions of K_1; free variables of K_1 will be free variables of $K_1 \oplus K_2$ by definition.

The relation between Discourse Representation Structures and first-order logic is rather transparent, as shown by the following translations.

Definition 3.8. *We define two translations: one from first-order logic — where we restrict the translation to formulae which use a different variable for each quantifier and to formulae for which no free variable uses the same variable name as a quantified variable — to Discourse Representation Structures, shown in Table 3.1 and one from Discourse Representation Structures to first-order logic, shown in*

Table 3.2. Since Discourse Representation Structures and conditions are defined by mutual recursion, the translation from DRSs and conditions is defined by two mutually recursive function d (from DRSs to formulas of first-order logic) and c (from DRS conditions to formulas of first-order logic) as well.

Table 3.1. Translating first-order formulae into Discourse Representation Structures

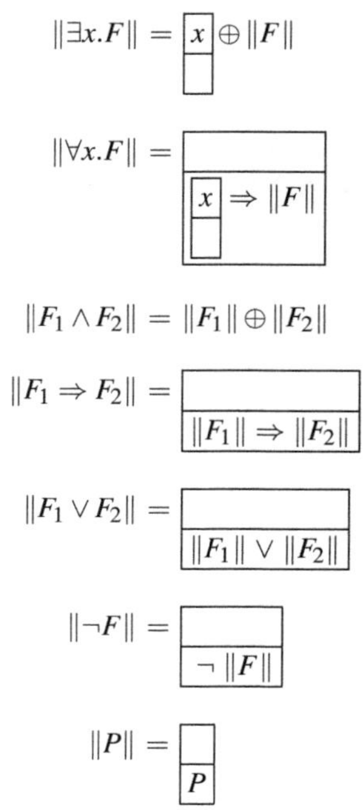

Table 3.2. Translation functions d from Discourse Representations structures to first-order formulae and c from DRS conditions to first-order formulae

$$d(\langle [x_0,\ldots,x_n],[C_1,\ldots,C_m]\rangle) = \exists x_0,\ldots,x_n.c(C_1)\wedge\ldots c(C_m)$$

$$\begin{aligned}
c(p(x_0,\ldots,x_n)) &= p(x_0,\ldots,x_n)\\
c(D_1 \vee D_2) &= d(D_1)\vee d(D_2)\\
c(\langle [x_0,\ldots,x_n],[C_1,\ldots,C_m]\rangle \Rightarrow D) &= \forall x_0,\ldots,x_n(c(C_1),\ldots,c(C_m)\Rightarrow d(D))\\
c(\neg D) &= \neg d(D)
\end{aligned}$$

Muskens (1994, 1996) shows that DRT semantics is fully compatible with both Montague semantics and the Lambek calculus, and that the combination of both approaches into a unified theory based on the simply typed lambda calculus allows us to use the strong points of both formalisms. For example, as pointed out by Muskens (1996), the combined calculus handles the interaction of anaphors with coordination rather easily. In addition, this combination of Montague grammar and DRT, being based on the simply type lambda calculus, can easily handle DRT semantics within the Lambek calculus (as shown by Muskens, 1994).

The essential idea is very simple: we replace all terms of type t by DRSs, which are of type $i \to (i \to t)$, where i is an information state: a function which assigns each discourse referent in the universe of the DRS an element of the domain." by "a function which assigns each discourse referent in the universe of the DRS an element of the domain (there is an additional subtlety: Muskens has a type distinction between discourse referents and entities; however, for ease of exposition we will only use the type entity in our discussion). Given that the discourse we have processed before the current sentence may have already assigned some values to discourse referents, what a sentence does is *update* this information — possibly adding new discourse referents and discarding old ones. The interpretation of the conditions of the DRS must therefore be a term of type $i \to t$, a function which given an information state i — that is, an assignment to the variables in the universe of the DRS — returns true or false; when occurring inside a DRS this term is therefore applied to the current context i to give a term of type t. So the intuition behind the semantics of a DRS in this calculus is that a DRS is interpreted as a relation between input context c_i and an output context c_o, such that all the variables $x_0, \dots, x_n$ are added to c_o (there are different opinions about what to do if any of the x_i is already assigned a value in the assignment c_i, we will only assume here that in the context which follows x_i is treated just like a free variable) and that all the conditions $C_0, \dots, C_n$ are true for the assignment c_o. In case there is no assignment c_o satisfying all C_i then the DRS evaluates to false, otherwise it evaluates to true. If we make the intuitions of this paragraph formally precise, we can — as shown in Muskens (1996, Section III.2) — treat the DRSs as *abbreviations* for terms which manipulate contexts directly. In what follows, we follow the simpler strategy where boxes are simply objects of type $i \to (i \to t)$ and conditions are terms of type t — ie. the application of a term of type $i \to t$ applied to the current context c. This choice means that the accessibility relation and context updates are handled at the level of the DRSs and not at the level of the lambda terms. Though the solution of coding the accessibility in the lambda terms is rather neat, the current solution produces lambda-terms which convert to Discourse Representation Structures and therefore permits a more transparent comparison with DRS semantics.

In what follows, to simplify the notation of types, we will often abbreviate the type $i \to (i \to t)$ as σ. The new constants of DRS semantics have the following types.

Constant	**Type**	**Simplified type**
$\neg$	$(i{\to}i{\to}t){\to}t$	$\sigma{\to}t$
$\Rightarrow, \vee$	$(i{\to}i{\to}t){\to}(i{\to}i{\to}t){\to}t$	$\sigma{\to}\sigma{\to}t$
$\oplus$	$(i{\to}i{\to}t){\to}(i{\to}i{\to}t){\to}(i{\to}i{\to}t)$	$\sigma{\to}\sigma{\to}\sigma$

Example 3.9. Translating the lexicon of Example 3.5 to a DRS lexicon gives us the following.

word	***syntactic type*** u ***semantic type*** u^* ***semantics:*** λ***-term of type*** u^* x^v ***means that the variable or constant*** x ***is of type*** v
every	$(S/(np\backslash S))/n$ (subject) $((S/np)\backslash S)/n$ (object) $(e\to\sigma)\to((e\to\sigma)\to\sigma)$ $\lambda P^{e\to\sigma}\lambda Q^{e\to\sigma}\begin{array}{\|c\|}\hline \\ \hline (((\Rightarrow^{(\sigma\to\sigma\to t)}\ ((\oplus^{\sigma\to\sigma\to\sigma}\begin{array}{\|c\|}\hline x\\ \hline \\ \hline\end{array})\ (P\,x)))\ (Q\,x))\\ \hline\end{array}$
a	$((S/np)\backslash S)/n$ (object) $(S/(np\backslash S))/n$ (subject) $(e\to\sigma)\to((e\to\sigma)\to\sigma)$ $\lambda P^{e\to\sigma}\ \lambda Q^{e\to\sigma}\ ((\oplus^{\sigma\to\sigma\to\sigma}\ ((\oplus^{\sigma\to\sigma\to\sigma}\begin{array}{\|c\|}\hline x\\ \hline \\ \hline\end{array})\ (P\,x)))\ (Q\,x))$
child	n $e\to\sigma$ $\lambda x^e\begin{array}{\|c\|}\hline \\ \hline child^{e\to t}(x)\\ \hline\end{array}$
pizza	n $e\to\sigma$ $\lambda x^e\begin{array}{\|c\|}\hline \\ \hline pizza^{e\to t}(x)\\ \hline\end{array}$
ate	$(np\backslash S)/np$ $e\to(e\to\sigma)$ $\lambda y^e\ \lambda x^e\begin{array}{\|c\|}\hline \\ \hline ate^{e\to e\to t}(x,y)\\ \hline\end{array}$
Leon	$S/(np\backslash S)$ (subject) $(S/np)\backslash S$ (object) $(e\to\sigma)\to\sigma$ $\lambda P^{e\to\sigma}((\oplus^{\sigma\to\sigma\to\sigma}\begin{array}{\|l\|}\hline x\\ \hline Leon(x)\\ \hline\end{array})(P\,x))$
he	$S/(np\backslash S)$ $(e\to\sigma)\to\sigma$ $\lambda P^{e\to\sigma}((\oplus^{\sigma\to\sigma\to\sigma}\begin{array}{\|l\|}\hline x\\ \hline x=?\\ \hline\end{array})(P\,x))$

Apart from the fact that we have replaced type t by type σ, the types of the lambda-terms have remained them same. Words like "pizza" and "child" are now functions from entities to DRSs with the condition $pizza(x)$ (resp. $child(x)$). The interesting cases are the quantifiers "a" and "every": "every" is a function from two properties P and Q — that is, two functions from entities to DRSs — to a DRS. The resulting DRS will have a single complex condition of the form $K_1 \Rightarrow K_2$, where the left hand side K_1 will be the merge of a DRS containing only the unused discourse referent x with the DRS (Px) and the right hand side K_2 will be the DRS (Qx). The DRS introducing the unused discourse referent x will therefore bind the occurrences of x in both K_1 and K_2. The existential quantifier "a" will simply merge a DRS containing an unused discourse referent x with the DRSs obtained by applying properties P and Q to x.

While the above lexicon corresponds rather closely for quantifiers and common nouns, we need to change the way we analyze noun phrases a bit. Before, we assigned proper names the syntactic type np, corresponding to the semantic type e. However, in the standard DRS account of names, we want a name *Name* to introduce both a discourse referent x corresponding to this name into the universe of the DRS[10] *and* an atomic condition of the form $Name(x)$. When we lift the np type to $S/(np\backslash S)$, the semantic type becomes $(e\rightarrow\sigma)\rightarrow\sigma$ and we can introduce a discourse referent x, with the condition that $Name(x)$, to the universe as shown above.

As for the pronoun "he", it introduces a new discourse referent x, but introduces an anaphoric link as well, requiring this discourse referent to be bound to one of the accessible discourse referents — which we can decide only in the context of a larger DRS, of course. As we said at the beginning of the chapter, many authors consider finding the reference for x to be a question of *syntax*, typically noted by coindexing the pronoun "he" with a noun phrase elsewhere in the discourse. Here, we will treat the resolution of the anaphor to be a *semantic* problem (though subject to various well-known syntactic constraints). The lexical entry, with the condition $x = ?$ notes this unresolved anaphor, indicating it needs to be resolved in order for the final DRS in which it occurs to make sense.

3.7 A Word about Intensional Logic

Let us briefly explain how the method we give for computing semantic representations straightforwardly extends to the intensionality operators popularized in semantics by Montague and his followers.

It is quite common to consider Kripke models which can be referred to within the syntax itself with type $\mathtt{s}$ for worlds and with the $\hat{}$ and $\check{}$ operators. Possible worlds are just common classical models, with an accessibility relation between them, usually a partial order. Phenomena whose truth in a given world depend on truth in other worlds are said to be *intensional*. For instance, to say that A is possible (resp. necessary) at world w means that for some (resp. every) world w' accessible from w we have that A is true in w'.

One is thus able to model modalities and to deal with opaque contexts like belief verbs. For instance, if "Max believes that Chomsky is a computer scientist" is

[10] Actually, names should be added to the *outermost* DRS. However, will ignore this complication here.

true, than we cannot infer that Chomsky is a computer scientist. It actually has little to do with the truth of the believed sentence. A common interpretation is that in any possible world where the beliefs of "Max" are true, "Chomsky" happens to be a computer scientist. In intensional logic, we therefore distinguish between the truth value of a proposition and the proposition itself: we will interpret a proposition as a set of worlds, a term of type $\mathtt{s} \rightarrow \mathbf{t}$: the set of worlds in which the proposition is true. Its truth value is therefore this term of type $\mathtt{s} \rightarrow \mathbf{t}$ applied to the current world w.

In order to provide an account of intensional phenomena, the syntax of the lambda calculus is commonly enriched with two operators $^\wedge : \mathtt{s} \rightarrow a$ and $^\vee : (\mathtt{s} \rightarrow a) \rightarrow a$ where $\mathtt{s}$ is the type for the (indices of the) possible worlds with $^{\vee\wedge}u$ reducing to u.

Gallin (1975) proposes to translate the intensional operators in a two sorted logic with two types $\mathtt{s}$ and $\mathbf{e}$ (in addition to $\mathbf{t}$), with the particularity that there is only a single variable of type $\mathtt{s}$. It is possible to see $^\wedge t^a$, of type $\mathtt{s} \rightarrow a$, as an abbreviation of $\lambda w^{\mathtt{s}} t^a$ and $^\vee t^{\mathtt{s}\rightarrow a}$, of type a, as an abbreviation of $(t^{\mathtt{s}\rightarrow a}\ w^{\mathtt{s}})$ (see also Gamut, 1991; Morrill, 1994). According to Gallin's translation, $^{\vee\wedge}u$ becomes $(\lambda w^{\mathtt{s}}\ u)w$ which reduces to u by beta reduction. Its inverse $^{\wedge\vee}u$ corresponds to $\lambda w^{\mathtt{s}}(uw)$, which reduces to u only if there are no free occurrences of w in u.

We have not yet talked about how to add terms with intensional types to our lexicon. There are several strategies for doing this: the simplest method is to change the type map, the morphism ".*" from syntactic types to semantic types, and let it introduce type $\mathtt{s}$. For example, we can replace the basic type $\mathbf{t}$ by $\mathtt{s} \rightarrow \mathbf{t}$, in other words, we replace truth values by propositions, as proposed by Montague (1970a) van Benthem (see also 1991, chaper 12), we will call this the Montague/EFL type map — compare this to Muskens' strategy for interpreting Discourse Representation Structures in type theory as discussed in Section 3.6, which replaces truth values $\mathbf{t}$ by state changes $i \rightarrow i \rightarrow \mathbf{t}$. The Montague/EFL map produces $S^* = \mathtt{s} \rightarrow \mathbf{t}$ and $n^* = \mathbf{e} \rightarrow \mathtt{s} \rightarrow \mathbf{t}$. As another possibility, we can change the type map for the complex types and leave the translations of the atomic types unchanged, translating $(a \setminus b)^*$ and $(b \,/\, a)^*$ as $(\mathtt{s} \rightarrow a^*) \rightarrow b^*$, as proposed by Montague (1973) (see also Gamut, 1991; Hendriks, 1993), we will call this the Montague/PTQ type map. This second map replaces individuals (of type $\mathbf{e}$) by *individual concepts*: functions from possible worlds to entities. This allows expressions such as "the president of France" and "the temperature" to have different denotations in different worlds (Montague requires proper names to always refer to the same individual by means of a meaning postulate).

So according to van Montague's EFL type map, "believes that" of syntactic type $((np \setminus S) \,/\, S)$ would be of semantic type $(\mathtt{s} \rightarrow \mathbf{t}) \rightarrow \mathbf{e} \rightarrow \mathtt{s} \rightarrow \mathbf{t}$, whereas according to Montague's PTQ type map, it would be $(\mathtt{s} \rightarrow \mathbf{t}) \rightarrow (\mathtt{s} \rightarrow \mathbf{e}) \rightarrow \mathbf{t}$.

A final strategy is to add new connectives which are interpreted intensionally by the type map. Morrill (1990) (see also Morrill, 2011, Chapter 8) proposes an extension of the Lambek calculus which adds a modal connective '$\Box$', using the logical rules of S4 for this modal operator[11], where $(\Box a)^*$ is mapped to $\mathtt{s} \rightarrow a^*$.

[11] Not to be confused with the '$\Box$' connective of Section 5.2.2, but close to the '!' connective of Section 5.2.1, which is an S4 modality as well. Morrill (1994, Chapter 5) proposes to use an S5 modality instead.

In Morrill's calculus, the introduction rule for this modal operator corresponds to $^\wedge$ and the elimination rule to $^\vee$, adapting an idea from van Benthem (1986). In Morrill's intensional system "believes that" is assigned syntactic type $\Box((np\backslash S)/\Box S)$, which corresponds to the semantic type $\mathrm{s} \rightarrow (\mathrm{s} \rightarrow \mathbf{t}) \rightarrow \mathbf{e} \rightarrow \mathbf{t}$. Note how we mark explicitly that the sentential argument of believes is intensional by assigning it the syntactic type $\Box S$.

Let us conclude this very brief indication of the relation between intensional logic and categorial grammars by giving the lexical lambda terms corresponding to "believes that" for each of the three semantic types we have seen.

Montague/EFL	$(\mathrm{s} \rightarrow \mathbf{t}) \rightarrow \mathbf{e} \rightarrow \mathrm{s} \rightarrow \mathbf{t}$	$\lambda p^{\mathrm{s}\rightarrow\mathbf{t}} \lambda x^{\mathbf{e}}\ {}^\wedge\mathrm{believe}(x,p)$
Montague/PTQ	$(\mathrm{s} \rightarrow \mathbf{t}) \rightarrow (\mathrm{s} \rightarrow \mathbf{e}) \rightarrow \mathbf{t}$	$\lambda p^{\mathrm{s}\rightarrow\mathbf{t}} \lambda y^{\mathrm{s}\rightarrow\mathbf{e}}\ \mathrm{believe}({}^\vee y,p)$
Morrill	$\mathrm{s} \rightarrow (\mathrm{s} \rightarrow \mathbf{t}) \rightarrow \mathbf{e} \rightarrow \mathbf{t}$	${}^\wedge\lambda p^{\mathrm{s}\rightarrow\mathbf{t}} \lambda x^{\mathbf{e}}\ \mathrm{believe}(x,p)$

3.8 Concluding Remarks

Though we have touched only very briefly on many of the topics which we have discussed in this chapter, they include many of the "classic" topics in semantics: generalized quantifiers, intensionality, anaphora and Discourse Representation Theory. Carpenter (1996) discusses many more. So while the idea of combining Montague semantics and the Lambek calculus is very simple, its applications cover a wide range of semantically interesting phenomena.

Montague grammar and Discourse Representation Semantics are active and independent fields and we recommend (Portner and Partee, 2002; Partee and Hendriks, 2011; van Eijck and Kamp, 2011) as a starting point for the reader interesed in exploring these topics further.

Exercises for Chapter 3

Exercise 3.1. Reduce/expand the following lambda-terms into β normal η long form.

1. $f^{e\to(e\to t)}$
2. $(\lambda x^{e\to t} f^{(e\to t)\to(e\to t)} x) y^{e\to t}$
3. $((\lambda x^e \lambda y^e f^{e\to(e\to t)}(x,y))\ (h^{e\to e}(y)))\ c^e$
4. $(\lambda y^e f^{e\to(e\to t)}(y,y))((\lambda x^e g^{e\to(e\to e)}(x,x))\ c^e)$
5. $(\lambda x^e f^{e\to(e\to t)} y^e)(g^{e\to e} z^e)$

Exercise 3.2. Compute the types corresponding to the following formulae.

1. $n\,/\,n$
2. $(n\,\backslash\,n)\,/\,n$
3. $(n\,\backslash\,n)\,\backslash\,(S\,/\,np)$
4. $((np\,\backslash\,S)\,/\,np)\,\backslash\,(np\,\backslash\,S)$

Exercise 3.3. Using the terms and types described in the paragraph on *Systematic Type Raising* in Section 3.5 on page 84, derive "the cat sleeps" and calculate its lambda term.

Exercise 3.4. Assign a lambda-term to "is" of syntactic type $(np\,\backslash\,S)\,/\,(n\,/\,n)$ in such a way that "Leon is vegetarian" will produce a semantics equivalent to *vegetarian*(L).

Similarly, give a lambda-term to "is" of syntactic type $(np\,\backslash\,S)\,/\,np$ in such a way that "Leon is a vegetarian" will produce a semantics equivalent to *vegetarian*(L). Use the standard semantics for "a" and assign "vegetarian" the syntactic type n. Justify your answer.

Hint: assume a constant "=" of type $\mathbf{e}\to\mathbf{e}\to\mathbf{t}$ which evaluates "$=(x,y)$" (or $x=y$ in more convenient infix notation) to true iff x and y have the same denotation.

Exercise 3.5. For the semantics of adjectives, they are often classed into different groups according to the inference patterns they allow. Look at the following examples, all of which are of syntactic type n.

(3.17) vegetarian assassin
(3.18) efficient assassin
(3.19) alleged assassin

Example sentence 3.17 is an example of what is often called an *intersective* adjective. It is intersective because a vegetarian assassin is someone who is both a vegetarian and an assassin.

Example sentence 3.18 is different: suppose Leon is not just an efficient assassin but also an amateur gardener. If we know Leon is an efficient assassin, we want to be able to infer that Leon is an assassin. However, we do *not* want to conclude that he is an efficient gardener (an assassin's life may not leave him with enough time to water his plants....).

For the final example sentence, if someone is an “alleged assassin”, then he may or may not be an assassin.

Using the same syntactic type $n \,/\, n$ for all adjectives, assign semantic terms of the right type to each word. Verify that the three different inference patterns hold for the different lambda-terms.

Note: strictly speaking, inferences are between sentences. However, the notion extends rather naturally to other expressions (Keenan and Faltz, 1985). If you have trouble convincing yourself that the given inference patterns hold, use your solution for Exercise 3.4 and prefix the nouns and adjectives with “Leon is a” (resp. “Leon is”).

Exercise 3.6. Extend the lexicon of Example 3.5 with the appropriate syntactic types and lambda-terms for the generalized quantifiers “at least two” (for the current example, treat it as a single word *at_least_two*) and “no”. Add simple lexical entries for “pizzas” and “children”.

Compute the lambda-terms which correspond to the following two sentences.

(3.20) Every child ate at least two pizzas.
(3.21) No child ate at least two pizzas.

Comment on the difference between the subject wide scope and object wide scope readings. Is one reading more plausible than another?

Exercise 3.7. Give a lexicon which assigns “John” the syntactic type np, the word “and” an instantiation of the scheme $(X \setminus X) \,/\, X$ for some X.

(3.22) Every man and every child .
(3.23) Every man and child
(3.24) John and every child

Exercise 3.8. Following Example 3.4, assign formulas and lambda terms of the correct semantic types to the lexicon in order to derive the following sentence.

(3.25) *Chaque chat que je connais dort beaucoup.*
Every cat that I know sleeps a lot.
‘Every cat I know sleeps a lot.’

Exercise 3.9. Use the lexicon of Example 3.9 to compute the DRSs corresponding to the two readings of “every child ate a pizza”

Exercise 3.10. Extend the lexicon of Example 3.9 in order to treat the following two example sentences from (Muskens, 1996).

(3.26) A cat catches a fish and eats it.
(3.27) # A cat catches no fish and eats it.

Compute the DRSs for both sentences and resolve the anaphor for “it”. Make sure that your semantics for “no” — though it must introduce a discourse referent, of course — does not introduce a discourse referent which can corefer with “it”

References

Abramsky, S.: Computational interpretations of linear logic. Theoretical Computer Science 111, 3–57 (1993)

Amblard, M., Lecomte, A., Retoré, C.: Categorial minimalist grammars: From generative syntax to logical form. In: van Benthem, J., Moortgat, M. (eds.) Linguistic Analysis — Festschrift for Joachim Lambek. Linguistic Analysis, vol. 36, pp. 273–306 (2010)

van Benthem, J.: Categorial grammar. In: Essays in Logical Semantics, ch. 7, pp. 123–150. Reidel, Dordrecht (1986)

van Benthem, J.: Language in Action: Categories, Lambdas and Dynamic Logic. Sudies in logic and the foundation of mathematics, vol. 130. North-Holland, Amsterdam (1991)

van Benthem, J., ter Meulen, A. (eds.): Handbook of Logic and Language, 2nd edn. North-Holland Elsevier, Amsterdam (2011)

Carpenter, B.: Lectures on Type-Logical Semantics. MIT Press, Cambridge (1996)

van Dalen, D.: Logic and Structure, 4th edn., Universitext. Springer (1983)

Dowty, D., Wall, R.E., Peters, S.: Introduction to Montague Semantics. In: Classic Titles in Linguistics. Springer (1981)

van Eijck, J., Kamp, H.: Representing discourse in context. In: Van Benthem and ter Meulen, ch. 3, pp. 181–252 (2011)

Gallin, D.: Intensional and Higher-Order Logic: With Applications to Montague Semantics. Elsevier (1975)

Gamut, L.T.F.: Logic, Language and Meaning, vol. 2. The University of Chicago Press (1991)

Girard, J.Y., Lafont, Y., Taylor, P.: Proofs and Types. Cambridge Tracts in Theoretical Computer Science, vol. 7. Cambridge University Press (1988)

Heim, I., Kratzer, A.: Semantics in generative grammar. Blackwell textbooks in linguistics. Blackwell, Oxford (1997)

Hendriks, H.: Studied flexibility: Categories and types in syntax and semantics. PhD thesis, University of Amsterdam, ILLC Dissertation Series (1993)

Huet, G.: Résolution d'équations dans des langages d'ordre 1,2,...,ω. PhD thesis, Université Paris VII (1976)

Jackendoff, R.: The Architecture of the Language Faculty. Linguistic Inquiry Monographs, vol. 28. MIT Press, Cambridge (1995)

Kamp, H., Reyle, U.: From Discourse to Logic. D. Reidel, Dordrecht (1993)

Kamp, H., van Genabith, J., Reyle, U.: Discourse representation theory. In: Gabbay, D., Guenthner, F. (eds.) Handbook of Philosophical Logic, vol. 15, pp. 125–394. Springer (2011)

Keenan, E., Faltz, L.: Boolean Semantics for Natural Language, Synthese Language Library, vol. 23. D. Reidel (1985)

Krivine, J.L.: Lambda Calcul — Types et Modèles. Etudes et Recherches en Informatique, Masson, Paris (1990)

Montague, R.: English as a formal language. In: Visentini, B. (ed.) Linguaggi nella società e nella tecnica (1970a); reprinted as Chapter 6 of Thomason (1974)

Montague, R.: Universal grammar. Theoria 36, 373–398 (1970b); reprinted as Chapter 7 of Thomason (1974)

Montague, R.: The proper treatment of quantification in ordinary English. In: Hintikka, J., Moravcsik, J., Suppes, P. (eds.) Approaches to Natural Language: Proceedings of the 1970 Stanford Workshop on Grammar and Semantics. Reidel, Dordrecht (1973); reprinted as Chapter 8 of Thomason (1974)

Moortgat, M.: Categorial type logics. In: van Benthem, J., ter Meulen, A. (eds.) Handbook of Logic and Language, ch. 2, pp. 93–177. North-Holland Elsevier, Amsterdam (1997)

Morrill, G.: Intensionality and boundedness. Linguistics and Philosophy 13(6), 699–726 (1990)
Morrill, G.: Type Logical Grammar. Kluwer Academic Publishers, Dordrecht (1994)
Morrill, G.: Categorial Grammar: Logical Syntax, Semantics, and Processing. Oxford University Press (2011)
Muskens, R.: Categorial Grammar and Discourse Representation Theory. In: Proceedings of COLING 1994, Kyoto, pp. 508–514 (1994)
Muskens, R.: Combining Montague Semantics and Discourse Representation. Linguistics and Philosophy 19, 143–186 (1996)
Muskens, R., van Benthem, J., Visser, A.: Dynamics. In: Van Benthem and ter Meulen, ch. 10, pp. 587–688 (2011)
Partee, B., Hendriks, H.: Montague Grammar. In: Van Benthem and ter Meulen, ch. 1, pp. 3–94 (2011)
Portner, P., Partee, B.H. (eds.): Formal Semantics: The Essential Readings. Blackwell Publishers (2002)
Pustejovsky, J.: The generative lexicon. MIT Press (1995)
Seldin, J.P., Hindley, J.R.: To H. B. Curry: Essays on Combinatory Logic, Lambda Calculus and Formalism. Academic Press (1980)
Thomason, R. (ed.): Formal Philosophy: Selected papers of Richard Montague. Yale University Press (1974)

4

The Non-associative Lambek Calculus

Summary. In this chapter we will look at NL, the non-associative Lambek calculus, which was introduced by Lambek a few years after the Syntactic Calculus, L. The Gentzen-style presentation of the Lambek calculus uses a list of formulae as antecedents, whereas NL uses binary branching trees instead.

We will start this chapter with a brief introduction of the sequent calculus for NL, then illustrate why non-associativity is sometime desirable by presenting some ungrammatical sentences which — though derivable in L — are underivable in NL.

We will then revisit some of the results of L from the perspective of the non-associative calculus NL: we will reprove cut elimination for NL and present a natural deduction version of the calculus.

In addition, we will give a new type of model for Lambek calculi in the form of Kripke models, for which we prove soundness and completeness. We will also show how we can add an explicit rule of associativity to NL to obtain an alternative formulation of L and see how this corresponds to a constraint on the Kripke models.

Perhaps surprisingly, dropping the structural rule of associativity makes a big computational difference: whereas the Lambek calculus has been shown to be NP complete by Pentus, the non-associative Lambek calculus has a polynomial time decision algorithm. We will finish our investigation of NL by presenting the polynomial time algorithms of Aarts & Trautwein and of de Groote.

4.1 Introduction

The non-associative Lambek calculus NL is obtained from the Lambek calculus by dropping the (implicit) rule of associativity. Instead of using *lists* of formulae as antecedents, like we did for L, NL uses binary branching *trees* of formulae.

Since in many linguistic frameworks, the basic units of linguistic description are considered to be trees — notably in the Chomskyan tradition (Chomsky, 1982, 1995; Stabler, 1997) but also in several alternative frameworks such as tree adjoining grammars (Joshi and Schabes, 1997) or HPSG (Pollard and Sag, 1994), where the *daughters* feature encodes the tree information — it makes sense to investigate the non-associative Lambek calculus and see if it offers any advantages over the Lambek calculus.

4.2 Proof Theory

In order not to overburden the notation, we will use the same symbols for the connectives of NL as for L. Unless otherwise indicated in the text, the formulae we will talk about in this chapter will be the formulae of NL.

$$\mathsf{Lp} \quad ::= \quad \mathsf{P} \ \mid \ (\mathsf{Lp}\backslash\mathsf{Lp}) \ \mid \ (\mathsf{Lp}/\mathsf{Lp}) \ \mid \ (\mathsf{Lp}\bullet\mathsf{Lp})$$

In L, the antecedent of a sequent was a (non-empty) list of formulae, which had the convenience of making the rule of associativity implicit. For NL, we want to drop this implicit rule of associativity and this means using a binary-branching *tree*, with formulae as its leaves, as antecedents. We will call these binary-branching trees *antecedent terms*.

Definition 4.1. *The antecedent terms $\mathscr{A}$ are defined as follows.*

$$\mathscr{A} \quad ::= \quad \mathsf{Lp} \ \mid \ (\mathscr{A},\mathscr{A})$$

So for example $(np,np\backslash S)$ and $(((a,b/c),d),e)$ are antecedent terms (if the atomic formulae include np and S in the first case and a, b, c, d and e in the second).

In the following Γ,Δ will denote antecedent terms.

We define some basic functions which transform the tree-structured antecedent terms to lists (simply by removing the brackets) and to multisets (keeping only the formulae and the number of occurrences of each formula, but forgetting the order).

Definition 4.2. *Let Γ be an antecedent term, the* yield *of Γ is a list which is obtained as follows.*

$$\begin{array}{ll} yield(F) = F & \text{if } F \text{ is a formula} \\ yield(\Gamma,\Delta) = yield(\Gamma), yield(\Delta) & \end{array}$$

The comma is deliberately overloaded here, so that *yield* transforms an NL antecedent term into a valid (and non-empty) L list.

Definition 4.3. *Let Γ be an antecedent term, the* multiset of formulae *of Γ is defined by the function $formulae(\Gamma)$ as follows.*

$$\begin{array}{ll} formulae(F) = \{F\} & \text{if } F \text{ is a formula} \\ formulae(\Gamma,\Delta) = formulae(\Gamma) \cup formulae(\Delta) & \end{array}$$

where $\cup$ is the multiset union operation.

So $formulae((np,np\backslash S)) = \{np,np\backslash S\}$ and $formulae(((a,b),(c,(b,a)))) = \{a,a,b,b,c\}$.

For the definition of the sequent rules, it is necessary to refer to *contexts*, which are defined as follows.

Definition 4.4. *A* context *is defined as follows.*

$$\mathscr{C} \quad ::= \quad [] \quad | \quad (\mathscr{C}, \mathscr{A}) \quad | \quad (\mathscr{A}, \mathscr{C})$$

where $\mathscr{A}$ is an antecedent term according to Definition 4.1.

A context is an antecedent term with a single occurrence of a 'hole' denoted by '[]'; seen this way, the inductive definition defines a path to the hole, with the three cases corresponding to 'here', 'on the left branch' and 'on the right branch' respectively.

We will write $\Gamma[]$, $\Delta[]$ to denote contexts.

Definition 4.5. *The* substitution *of an antecedent term Δ in a context $\Gamma[]$, $subst(\Gamma[], \Delta)$ (which we will normally write simply as $\Gamma[\Delta]$) is defined as follows.*

$$\begin{aligned} subst([], \Delta) &= \Delta \\ subst((\Gamma, \Gamma'[]), \Delta) &= (\Gamma, subst(\Gamma'[], \Delta)) \\ subst((\Gamma[], \Gamma'), \Delta) &= (subst(\Gamma[], \Delta), \Gamma') \end{aligned}$$

Note that the substitution of an antecedent term into a context produces a valid antecedent term. We can define the substitution of a context $\Delta[]$ in a context $\Gamma[]$ analogously to Definition 4.5 above, which gives a context $\Gamma[\Delta[]]$ after substitution.

4.2.1 Sequent Calculus

We now have everything in place for giving the sequent calculus formulation of NL, which is shown in Figure 4.1.

$$\frac{\Gamma[B] \vdash C \quad \Delta \vdash A}{\Gamma[(\Delta, A \backslash B)] \vdash C} \backslash_h \qquad \frac{(A, \Gamma) \vdash C}{\Gamma \vdash A \backslash C} \backslash_i$$

$$\frac{\Gamma[B] \vdash C \quad \Delta \vdash A}{\Gamma[(B / A, \Delta)] \vdash C} /_h \qquad \frac{(\Gamma, A) \vdash C}{\Gamma \vdash C / A} /_i$$

$$\frac{\Gamma[(A, B)] \vdash C}{\Gamma[A \bullet B] \vdash C} \bullet_h \qquad \frac{\Delta \vdash A \quad \Gamma \vdash B}{(\Delta, \Gamma) \vdash A \bullet B} \bullet_i$$

$$\frac{\Gamma \vdash A \quad \Delta[A] \vdash B}{\Delta[\Gamma] \vdash B} \mathit{cut} \qquad \frac{}{A \vdash A} \mathit{axiom}$$

Fig. 4.1. Sequent calculus rule for NL, the non-associative Lambek calculus

NL does not allow empty antecedent derivations, which we have identified as undesirable in Section 2.5: the antecedent term Γ in the $\backslash_i$ and the $/_i$ rules is non-empty according to Definition 4.1.

When talking about the derivability of a sequent in the non-associative Lambek calculus, we can ask two different questions:

1. Given a list of formulae L and a goal formula C, for which antecedent terms Γ such that $yield(\Gamma) = L$, is $\Gamma \vdash C$ derivable?
2. Given an antecedent term Γ and a goal formula C, is $\Gamma \vdash C$ derivable?

That is, when we look at sequent proof search, do we consider the structure of the antecedent term to be part of the *output* of the proof search algorithm (as in option 1 above) or as part of its *input* (as in option 2). In other words, do we compute the brackets of the antecedent terms or just verify them? In what follows, unless otherwise noted — for example in Section 4.6 — when talking about parsing or proof search, we will consider the input of the parsing or proof search algorithm to be a list of formulae (or a list of *words*, when using a lexicon which assigns sets of formulae to these words). When there is a need to emphasize that the structure of the antecedent term of a sequent is unknown, we will write the sequent as $A_1,\ldots,A_n \vdash C$, just like we did for Lambek calculus sequents.

If we define the yield of an antecedent term of the form $\Gamma[\Delta]$ (for some Δ) to be $\Gamma, yield(\Delta), \Gamma'$ (with Γ' an unused antecedent term variable), then it is easy to verify that the following proposition holds.

Proposition 4.6. *If R is an NL sequent rule and $\Gamma_1,\ldots,\Gamma_n$ are the antecedent terms mentioned in R then replacing each Γ_i by $yield(\Gamma_i)$ will give an L sequent rule.*

Corollary 4.7. *If $\Gamma \vdash A$ is derivable in NL then $yield(\Gamma) \vdash A$ is derivable in L.*

The inverse does not hold however: some characteristic theorems of L are underivable in NL.

Example 4.8. An example is the transitivity of $/$, as shown by the following failed derivation.

$$\dfrac{\dfrac{(A,C) \vdash A \quad B\,/\,C \vdash B}{((A\,/\,B, B\,/\,C), C) \vdash A}\,/_h}{(A\,/\,B, B\,/\,C) \vdash A\,/\,C}\,/_i$$

Note how the parentheses prevent application of the $/_h$ rule to $B\,/\,C$ and C, as they are not sisters in the tree. Showing a failed proof attempt is not that same as showing underivability of a sequent, however! We need to show that *all* proof attempts fail. Exercise 4.1 asks you to verify by means of an exhaustive proof search that the sequent of this example has no proof in NL.

Showing non-derivability by exhaustive enumeration of proof attempts can be tedious and error-prone. We will see methods requiring less bookkeeping to show underivability in Section 4.6 and in Chapter 7. In Section 4.5.2 we will see another method to show a sequent is not derivable: the construction of a countermodel.

The count check of Proposition 2.6 is useful but incomplete. For example $(A / B, B / C) \vdash A / C$ satisfies the count check, but — as we have seen above — it is underivable nonetheless.

4.2.2 Arguments against Associativity

As an illustration of why associativity is sometimes undesirable, look at the following lexicon (after Lambek (1961)).

Word	Type(s)
the	np / n
Hulk	n
is	$(np \backslash s) / (n / n)$
incredible	n / n
green	n / n

Using this lexicon, both NL and L allow us the derive the following phrases.

(4.1) The Hulk is green.
(4.2) The Hulk is incredible.

However, L allows us the derive the following ungrammatical phrase as well.

(4.3) $*$ The Hulk is green incredible.

Exercise 4.3 at the end of this chapter asks you to show the non-derivability of

$$n / n, n / n \vdash n / n$$

in NL and to show that only the valid example sentences above are derivable in NL, whereas the invalid sentence 4.3 is derivable in L.

There is another type of example to show that associativity can lead to some very strange sentences. In order to present this argument, we need some introduction to the use of so-called *polymorphic* types, which can be used to conjoin expression which are assigned different formulae, as demonstrated by the examples below (Exercise 1.4.4 gives several other examples).

(4.4) Bill left the party and returned home.
(4.5) Bill gave flowers to Mary and a toy to the children.

Example 4.4 above shows that "left the party" and "returned home", both expressions of type $np \backslash S$, can be conjoined. This means that "and" can be assigned the formula $((np \backslash S) \backslash (np \backslash S)) / (np \backslash S)$, as well as many other instances of the general scheme $(X \backslash X) / X$.[1]

Example 4.5 shows that the items conjoined can be complex expressions such as "flowers to Mary" and "a toy to the children", which in the context of a formula assignment of $((np \backslash S) / pp) / np$ to "gave" makes "flowers to Mary" an expression of type $np \bullet pp$.

Now, with these examples in mind, look at the following sentence.

(4.6) *The mother of and Bill thought John arrived.

This sentence, as Paul Dekker was the first to notice, is not only clearly ungrammatical, but also — though this may seem surprising (and even shocking!) at first glance — derivable in L: there is an instantiation of the polymorphic type scheme $(X \backslash X) / X$ for "and" which makes the sentence derivable in the Lambek calculus: both "the mother of" and "Bill thought" can be shown to be of type $(S / (np \backslash S)) / np$ using the following lexicon, which is without surprises,

Word	Type(s)
Bill	np
John	np
the	np / n
mother	n / pp
of	pp / np
thought	$((np \backslash S) / S)$
arrived	$np \backslash S$

as shown by the following two natural deduction proofs in the associative Lambek calculus (using the Prawitz-style rules of Section 2.2.1 and the rule Lex to indicate the conclusion of the rule is a lexical entry for the word which serves as its premise)

$$\dfrac{\dfrac{\dfrac{\dfrac{\text{the}}{np / n}\,\text{Lex} \qquad \dfrac{\dfrac{\text{mother}}{n / pp}\,\text{Lex} \qquad \dfrac{\dfrac{\text{of}}{pp / np}\,\text{Lex} \qquad [np]^1}{pp}\,/e}{n}\,/e}{np}\,/e \qquad [np \backslash S]^2}{S}\,\backslash e}{\dfrac{S / (np \backslash S)}{(S / (np \backslash S)) / np}\,/i_1}\,/i_2$$

[1] This scheme corresponds to a quantification over formulas and it would be more correct to write $\forall X.(X \backslash X) / X$. Though a natural extension of L, the resulting calculus is undecidable, so this extension is not as innocent as it may appear (see Emms, 1993, 1995).

$$\dfrac{\dfrac{\dfrac{\dfrac{\text{Bill}}{np}\,\text{Lex} \qquad \dfrac{\dfrac{\text{thought}}{(np\backslash S)/S}\,\text{Lex} \qquad \dfrac{[np]^1 \quad [np\backslash S]^2}{S}\,\backslash e}{np\backslash S}\,/e}{S}\,\backslash e}{S/(np\backslash S)}\,/i_2}{(S/(np\backslash S))/np}\,/i_1$$

Neither proof can be transformed into a valid NL proof.

In some cases, however, the absence of associativity excludes some *grammatical* sentences as well. For example, the elegant treatment of peripheral extraction, as exemplified by the Italian sentence from Example 2.2 and the sentences of Exercise 2.7, is invalid in NL. Exercises 4.5 and 4.6 at the end of this chapter ask you to prove this. Similar remarks can be made about the quantifier scope ambiguities of the previous chapter (see Example 3.5). Exercise 4.9 asks you to verify that the type for object quantifiers in L cannot be used for NL.

So, summing up, we have seen that in some cases — in combination with coordination of in combination with the most natural type assignments for the verb "to be" and adjectives — it is desirable to restrict associativity, whereas in other cases — in the case of quantifier scope and of peripheral extraction — the easiest solution would be to permit associativity. In the next chapter, we will see how to combine an associative and a non-associative logic into one system. For the rest of this chapter, we will study the non-associative calculus (though, as a preamble to this, we will show in Section 4.3 how to add structural rules to NL and recover L).

4.2.3 Cut Elimination for the NL Sequent Calculus

In order to verify that cut elimination is valid for NL, following (Lambek, 1961; Kandulski, 1988), we revisit the different cases of the proof for L and verify that the bracketing is respected. This is just a simple exercise, but something we need to do to verify our logic is formulated correctly.

Remember that we are in the following general case for a cut rule of depth r and degree d and that we are looking at a cut rule of smallest depth in the proof.

$$\dfrac{\dfrac{\vdots\,\gamma}{\Gamma\vdash D}\,R^a \qquad \dfrac{\vdots\,\delta}{\Delta[D]\vdash C}\,R^f}{\Delta[\Gamma]\vdash C}\,\mathit{cut}\ d$$

We look at rule R^f and R^a. Since this is the rule with the smallest depth in the proof, there are no other cut rules in either γ or δ. We look at the other cases, which are the same as before: 1) at least one of the rules is an axiom, 2) R^a is not the rule which creates the cut formula, 3) R^f is not the rule which creates the cut formula or 4) both R^a and R^f create the cut formula.

1. If at least one of the rules is an axiom, we can remove the cut as follows.

$$\dfrac{D \vdash D \qquad \overset{\vdots\,\delta}{\Gamma[D] \vdash C}}{\Gamma[D] \vdash C}\,cut \qquad \text{reduces to} \qquad \overset{\vdots\,\delta}{\Gamma[D] \vdash C}$$

2. If R^a does not create the cut formula D, then we move the rule up past R^a. We know R^a must have been one of $\backslash_h$, $/_h$, $\bullet_h$, since an introduction rule would have necessarily introduced the cut formula. The table below lists the different cases (the implications are symmetric, so only $\backslash$ is shown).

2 R^a **does not create** D**, the cut formula**		
R^a	Before reduction	After reduction
$\bullet_h$	$\dfrac{\dfrac{\overset{\vdots\,\gamma}{\Delta[(A,B)] \vdash D}}{\Delta[A \bullet B] \vdash D}\bullet_h \qquad \overset{\vdots\,\delta}{\Gamma[D] \vdash C}}{\Gamma[\Delta[A \bullet B]] \vdash C}\,cut\ d$	$\dfrac{\dfrac{\overset{\vdots\,\gamma}{\Delta[(A,B)] \vdash D} \quad \overset{\vdots\,\delta}{\Gamma[D] \vdash C}}{\Gamma[\Delta[(A,B)]] \vdash C}\,cut\ d}{\Gamma[\Delta[A \bullet B]] \vdash C}\bullet_h$
$\backslash_h$	$\dfrac{\dfrac{\overset{\vdots\,\delta}{\Delta'[B] \vdash D} \quad \overset{\vdots\,\delta'}{\Delta \vdash A}}{\Delta'[(\Delta,A \backslash B)] \vdash D}\backslash_h \qquad \overset{\vdots\,\gamma}{\Gamma[D] \vdash C}}{\Gamma[\Delta'[(\Delta,A \backslash B)]] \vdash C}\,cut\ d$	$\dfrac{\dfrac{\overset{\vdots\,\delta}{\Delta'[B] \vdash D} \quad \overset{\vdots\,\gamma}{\Gamma[D] \vdash C}}{\Gamma[\Delta'[B]] \vdash C}\,cut\ d \qquad \overset{\vdots\,\delta'}{\Delta \vdash A}}{\Gamma[\Delta'[(\Delta,A \backslash B)]] \vdash C}\backslash_h$

3. If R^f does not create the cut formula, then we move the rule up past R^f. There are rather many cases to consider.

   For the left rules $\backslash_h$, $/_h$, $\bullet_h$, since the cut formula is not the main formula of the rule, the cut formula is a formula in $\Gamma[]$ (in the case of $\backslash_h$ and $/_h$ it can be a formula of Θ as well, hence the alternative case in the cut elimination below), which is already a context. Instead of introducing a new type of context with two distinguished formulae — D and the main formula of the rule — we simply indicate that D is a formula in $\Gamma[]$ and write $\Gamma^{\{D:=\Delta\}}[]$ for the context $\Gamma[]$ where D has been replaced by Δ in the reductions for $\backslash_h$ and $\bullet_h$ (the reduction for $/_h$ is symmetric to the reduction for $\backslash_h$ and has been omitted).

3 R^f does not create D, the cut formula		
R^f	Before reduction	After reduction
$\bullet_h$	$\dfrac{\overset{\vdots\,\delta}{\Delta \vdash D} \quad \dfrac{\overset{\vdots\,\gamma}{\Gamma[(A,B)] \vdash C}}{\Gamma[A\bullet B]\vdash C}\,\bullet_h}{\Gamma^{\{D:=\Delta\}}[A\bullet B]\vdash C}\,cut\ d$	$\dfrac{\dfrac{\overset{\vdots\,\delta}{\Delta\vdash D}\quad \overset{\vdots\,\gamma}{\Gamma[(A,B)]\vdash C}}{\Gamma^{\{D:=\Delta\}}[(A,B)]\vdash C}\,cut\ d}{\Gamma^{\{D:=\Delta\}}[A\bullet B]\vdash C}\,\bullet_h$
$\backslash_h$	$\dfrac{\overset{\vdots\,\delta}{\Delta \vdash D} \quad \dfrac{\overset{\vdots\,\gamma}{\Gamma[B] \vdash C}\quad \overset{\vdots\,\theta}{\Theta\vdash A}}{\Gamma[(\Theta,A\backslash B)]\vdash C}\,\backslash_h}{\Gamma^{\{D:=\Delta\}}[(\Theta,A\backslash B)]\vdash C}\,cut\ d$	$\dfrac{\dfrac{\overset{\vdots\,\delta}{\Delta\vdash D}\quad \overset{\vdots\,\gamma}{\Gamma[B]\vdash C}}{\Gamma^{\{D:=\Delta\}}[B]\vdash C}\,cut\ d \quad \overset{\vdots\,\theta}{\Theta\vdash A}}{\Gamma^{\{D:=\Delta\}}[(\Theta,A\backslash B)]\vdash C}\,\backslash_h$
$\backslash_h$	$\dfrac{\overset{\vdots\,\delta}{\Delta \vdash D} \quad \dfrac{\overset{\vdots\,\gamma}{\Gamma[B] \vdash C}\quad \overset{\vdots\,\theta}{\Theta[D]\vdash A}}{\Gamma[(\Theta[D],A\backslash B)]\vdash C}\,\backslash_h}{\Gamma[(\Theta[\Delta],A\backslash B)]\vdash C}\,cut\ d$	$\dfrac{\overset{\vdots\,\gamma}{\Gamma[B]\vdash C}\quad \dfrac{\overset{\vdots\,\delta}{\Delta\vdash D}\quad \overset{\vdots\,\theta}{\Theta[D]\vdash A}}{\Theta[\Delta]\vdash A}\,cut\ d}{\Gamma[(\Theta[\Delta],A\backslash B)]\vdash C}\,\backslash_h$
$\bullet_i$	$\dfrac{\overset{\vdots\,\delta}{\Delta \vdash D} \quad \dfrac{\overset{\vdots\,\gamma}{\Gamma[D] \vdash A}\quad \overset{\vdots\,\theta}{\Theta\vdash B}}{(\Gamma[D],\Theta)\vdash A\bullet B}\,\bullet_i}{(\Gamma[\Delta],\Theta)\vdash A\bullet B}\,cut\ d$	$\dfrac{\dfrac{\overset{\vdots\,\delta}{\Delta\vdash D}\quad \overset{\vdots\,\gamma}{\Gamma[D]\vdash A}}{\Gamma[\Delta]\vdash A}\,cut\ d \quad \overset{\vdots\,\theta}{\Theta\vdash B}}{(\Gamma[\Delta],\Theta)\vdash A\bullet B}\,\bullet_i$
$\bullet_i$	$\dfrac{\overset{\vdots\,\delta}{\Delta \vdash D} \quad \dfrac{\overset{\vdots\,\gamma}{\Gamma \vdash A}\quad \overset{\vdots\,\theta}{\Theta[D]\vdash B}}{(\Gamma,\Theta[D])\vdash A\bullet B}\,\bullet_i}{(\Gamma,\Theta[\Delta])\vdash A\bullet B}\,cut\ d$	$\dfrac{\overset{\vdots\,\gamma}{\Gamma\vdash A}\quad \dfrac{\overset{\vdots\,\delta}{\Delta\vdash D}\quad \overset{\vdots\,\theta}{\Theta[D]\vdash B}}{\Theta[\Delta]\vdash B}\,cut\ d}{(\Gamma,\Theta[\Delta])\vdash A\bullet B}\,\bullet_i$
$\backslash_i$	$\dfrac{\overset{\vdots\,\delta}{\Delta \vdash D} \quad \dfrac{\overset{\vdots\,\gamma}{(A,\Gamma[D]) \vdash B}}{\Gamma[D]\vdash A\backslash B}\,\backslash_i}{\Gamma[\Delta]\vdash A\backslash B}\,cut\ d$	$\dfrac{\dfrac{\overset{\vdots\,\delta}{\Delta\vdash D}\quad \overset{\vdots\,\gamma}{(A,\Gamma[D])\vdash B}}{(A,\Gamma[\Delta])\vdash B}\,cut\ d}{\Gamma[\Delta]\vdash A\backslash B}\,\backslash_i$

4. Finally, in the crucial case, both rules introduce the cut formula d and we replace it by two cuts of lesser degree.

4	Both R^a and R^f create the cut-formula	
	Before reduction	After reduction
$\bullet$	$\dfrac{\dfrac{\overset{\vdots\,\delta}{\Delta\vdash A} \quad \overset{\vdots\,\theta}{\Theta\vdash B}}{(\Delta,\Theta)\vdash A\bullet B}\bullet_i \quad \dfrac{\overset{\vdots\,\gamma}{\Gamma[(A,B)]\vdash C}}{\Gamma[(A\bullet B)]\vdash C}\bullet_h}{\Gamma[(\Delta,\Theta)]\vdash C}\,cut\ d$	$\dfrac{\overset{\vdots\,\delta}{\Delta\vdash A} \quad \dfrac{\overset{\vdots\,\theta}{\Theta\vdash B} \quad \overset{\vdots\,\gamma}{\Gamma[(A,B)]\vdash C}}{\Gamma[(A,\Theta)]\vdash C}\,cut<d}{\Gamma[(\Delta,\Theta)]\vdash C}\,cut<d$
$\backslash$	$\dfrac{\dfrac{\overset{\vdots\,\delta}{(A,\Delta)\vdash B}}{\Delta\vdash A\backslash B}\backslash_i \quad \dfrac{\overset{\vdots\,\gamma}{\Gamma[B]\vdash C} \quad \overset{\vdots\,\theta}{\Theta\vdash A}}{\Gamma[(\Theta,A\backslash B)]\vdash C}\backslash_h}{\Gamma[(\Theta,\Delta)]\vdash C}\,cut\ d$	$\dfrac{\dfrac{\overset{\vdots\,\theta}{\Theta\vdash A} \quad \overset{\vdots\,\delta}{(A,\Delta)\vdash B}}{(\Theta,\Delta)\vdash B}\,cut<d \quad \overset{\vdots\,\gamma}{\Gamma[B]\vdash C}}{\Gamma[(\Theta,\Delta)]\vdash C}\,cut<d$

4.2.4 Natural Deduction

Like L, NL permits a natural deduction formulation. However, given that for NL it no longer suffices to demand that the hypothesis which is withdrawn by the introduction rules is the leftmost or rightmost hypothesis, the more explicit Gentzen

$$\frac{\Gamma\vdash A \quad \Delta\vdash A\backslash B}{(\Gamma,\Delta)\vdash B}\backslash_e \qquad \frac{(A,\Gamma)\vdash C}{\Gamma\vdash A\backslash C}\backslash_i$$

$$\frac{\Delta\vdash B/A \quad \Gamma\vdash A}{(\Delta,\Gamma)\vdash B}/_e \qquad \frac{(\Gamma,A)\vdash C}{\Gamma\vdash C/A}/_i$$

$$\frac{\Delta\vdash A\bullet B \quad \Gamma[(A,B)]\vdash C}{\Gamma[\Delta]\vdash C}\bullet_e \qquad \frac{\Delta\vdash A \quad \Gamma\vdash B}{(\Delta,\Gamma)\vdash A\bullet B}\bullet_i$$

$$\frac{}{A\vdash A}\,axiom$$

Fig. 4.2. Natural deduction rules for NL

style formulation introduced in Section 2.2.2 is preferred. Figure 4.2 shows the natural deduction rules for NL.

Again, with the exception of the parentheses in the current rules, the natural deduction rules for NL are the same as those of L.

4.3 Structural Rules

From NL we can recover L simply by adding the two structural rules of associativity, shown in Figure 4.3 on the left and middle, to the logic. By adding commutativity as well, as shown below to the right, two new logics become available: adding just commutativity to NL gives us the non-associative Lambek calculus with permutation NLP, whereas adding both associativity and commutativity gives the Lambek-van Benthem calculus LP.

$$\frac{\Gamma[(\Delta_1,(\Delta_2,\Delta_3))]\vdash C}{\Gamma[((\Delta_1,\Delta_2),\Delta_3)]\vdash C}\,ass1 \qquad \frac{\Gamma[((\Delta_1,\Delta_2),\Delta_3)]\vdash C}{\Gamma[(\Delta_1,(\Delta_2,\Delta_3))]\vdash C}\,ass2 \qquad \frac{\Gamma[(\Delta_2,\Delta_1)]\vdash C}{\Gamma[(\Delta_1,\Delta_2)]\vdash C}\,com$$

Fig. 4.3. The structural rules of associativity and commutativity

Figure 4.4 lists the four possible combinations of the structural rules and the corresponding logics.

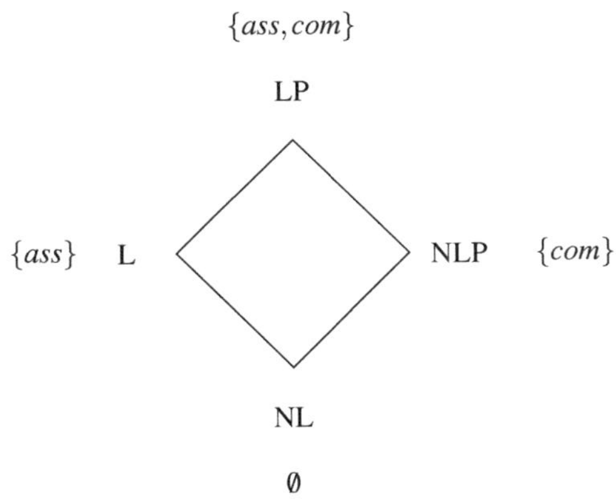

Fig. 4.4. The four logics NL, L, NLP and LP

We've already talked about the differences between NL and L and the benefits of both systems. A consequence of adding commutativity to our logic is that we can no longer distinguish between the two implications. The natural deduction proof below shows one direction, the other direction is symmetric.

$$\dfrac{\dfrac{\dfrac{\dfrac{}{A\,/\,B \vdash A\,/\,B}\;axiom \qquad \dfrac{}{B \vdash B}\;axiom}{(A\,/\,B, B) \vdash A}\;/_e}{(B, A\,/\,B) \vdash A}\;com}{A\,/\,B \vdash B \backslash A}\;\backslash_i$$

An example of where this property would be useful is for the treatment of English adverbs: some adverbs like 'completely' and 'carefully' can appear both before and after the verb phrase. The following two pairs of sentences, for example, should all be derivable.

(4.7) Loren carefully read Neuromancer.
(4.8) Loren read Neuromancer carefully.
(4.9) Stewart completely destroyed his credibility.
(4.10) Stewart destroyed his credibility completely.

The following lexicon allows us to derive example sentences 4.8 and 4.10 in NL. It assigns the formula $(np \backslash S) \backslash (np \backslash S)$ to the adverbs, which allows them to appear to the right of a verb phrase $np \backslash S$. We can add additional lexical entries to allow us to derive sentences 4.7 and 4.9 as well: the formula $(np \backslash S) / (np \backslash S)$ will do exactly that. So there is a trade-off to be made: do we add additional lexical entries, or do we try to generalize by adding structural rules? There is no easy answer to this question: both a small lexicon and a small set of structural rules are desirable.

Word	Type(s)
Loren	np
Stewart	np
Neuromancer	np
credibility	n
his	np / n
read	$(np \backslash S) / np$
destroyed	$(np \backslash S) / np$
carefully	$(np \backslash S) \backslash (np \backslash S)$
completely	$(np \backslash S) \backslash (np \backslash S)$

In this case, using NLP to model the behavior of adverbs and reduce the size of the lexicon has a serious drawback: it also permits the derivation of a number of ungrammatical sentences, like the following.

(4.11) $*$ Loren Neuromancer read.
(4.12) $*$ Destroyed credibility his Stewart.

These derivations are made possible, simply by the fact that NLP allows us to change the order of *any* two sister formulae in the tree.

LP, which has associativity as well as commutativity, allows us to reorder and rebracket our antecedent formulae in any way we want. While this would allow us to treat languages which have (nearly) free word order, like Latin, even in these languages word order is not completely free. For example, Latin sentences tend to have the preposition occurring closely *before* its argument noun phrase.

What we would like is to have some sort of *controlled* access to the structural rules of associativity and commutativity. We will see a number of solutions to this problem in the next chapter.

4.4 Combinator Calculi for NL

In this section, we will look at three combinator calculi for NL and show that all three are equivalent to the sequent calculus for NL. This has two goals: first, in Section 4.5 we will use one of these calculi for our soundness and completeness result with respect to the Kripke models for NL, and second, we will use one of the other calculi to give a polynomial algorithm in Section 4.6.

The first combinator calculus (Došen, 1988, 1989, 1992) consists of a set of axioms (identity, application and its dual "co-application") and a set of rules (monotonicity for all connectives and transitivity) as shown in Figure 4.5.

Axioms

$$\frac{}{A \vdash A}\ (Id)$$

$$\frac{}{A \bullet (A \backslash B) \vdash B}\ (Appl\backslash) \qquad \frac{}{(B / A) \bullet A \vdash B}\ (Appl/)$$

$$\frac{}{A \vdash B \backslash (B \bullet A)}\ (Co\text{-}appl\backslash) \qquad \frac{}{A \vdash (A \bullet B) / B}\ (Co\text{-}appl/)$$

Rules

$$\frac{A \vdash B \quad C \vdash D}{A \bullet C \vdash B \bullet D}\ (Mon\bullet) \qquad \frac{A \vdash B \quad C \vdash D}{B \backslash C \vdash A \backslash D}\ (Mon\backslash) \qquad \frac{A \vdash B \quad C \vdash D}{C / B \vdash D / A}\ (Mon/)$$

$$\frac{A \vdash B \quad B \vdash C}{A \vdash C}\ (Trans)$$

Fig. 4.5. Došen's axiomatic or combinator presentation of NL

The axiomatic calculus is a bit more tricky to use than either the sequent calculus or natural deduction (try Exercise 4.11 at the end of the chapter to get an idea). This is because the transitivity rule (Trans) plays a role similar to the cut rule in the sequent calculus, but, unlike for the sequent calculus, where the cut rule can be eliminated, the transitivity rule is a necessary component of the axiomatic calculus and finding the intermediate *B* formula for this rule is not always easy.

4.4.1 Alternative Axiomatic Presentations

Though not directly relevant to the soundness and completeness proofs which follow, it is worthwhile to spend some time discussing two alternative axiomatic presentations of NL, one proposed by Lambek (1988), which is useful to reinforce the links with what we have seen in Chapter 2 — notably the principle of *residuation* — and one proposed by Moortgat and Oehrle (1999), which is a system without the transitivity rules and which we will use in Section 4.6 to prove that we can find NL derivations in polynomial time. Note that when we speak about the axiomatic calculus without further qualification we will mean Došen's formulation of Figure 4.5.

Figure 4.6 shows Lambek's formulation. The only rules in this calculus, besides the (*Axiom*) and (*Trans*) rules are the residuation rules. We will sometimes call this calculus the residuation-based calculus.

Axiom

$$\frac{}{A \vdash A}\ (Id)$$

Rules

$$\frac{B \vdash A \backslash C}{A \bullet B \vdash C}\ (Res_{\backslash\bullet}) \qquad \frac{A \bullet B \vdash C}{B \vdash A \backslash C}\ (Res_{\bullet\backslash}) \qquad \frac{A \vdash C / B}{A \bullet B \vdash C}\ (Res_{/\bullet}) \qquad \frac{A \bullet B \vdash C}{A \vdash C / B}\ (Res_{\bullet/})$$

$$\frac{A \vdash B \quad B \vdash C}{A \vdash C}\ (Trans)$$

Fig. 4.6. Lambek's residuation-based combinatorial presentation

Figure 4.7 shows Moortgat and Oehrle's calculus: it has the application and co-application axioms of the Došen presentation and the residuation rules of the Lambek presentation. This formulation has the important advantage that the (*Trans*) rule is admissible, making the calculus more appropriate for proof search, as we will show in Section 4.6.

Axiom

$$\frac{}{A \vdash A}\ (Id)$$

Rules

$$\frac{A \vdash B \quad C \vdash D}{A \bullet C \vdash B \bullet D}\ (Mon\bullet) \qquad \frac{A \vdash B \quad C \vdash D}{B \backslash C \vdash A \backslash D}\ (Mon\backslash) \qquad \frac{A \vdash B \quad C \vdash D}{C / B \vdash D / A}\ (Mon/)$$

$$\frac{B \vdash A \backslash C}{A \bullet B \vdash C}\ (Res_{\backslash\bullet}) \qquad \frac{A \bullet B \vdash C}{B \vdash A \backslash C}\ (Res_{\bullet\backslash}) \qquad \frac{A \vdash C / B}{A \bullet B \vdash C}\ (Res_{/\bullet}) \qquad \frac{A \bullet B \vdash C}{A \vdash C / B}\ (Res_{\bullet/})$$

Fig. 4.7. Moortgat & Oehrle's presentation using residuation and monotonicity

Lemma 4.9. *Došen's combinator calculus (Figure 4.5) and Lambek's combination calculus (Figure 4.6) are equivalent.*

Proof. This is fairly easy to see. We show that the different *Res* rules are derived rules of the first calculus and that the *Appl* and *Co-appl* axioms and the monotonicity rules are derivable in the second calculus.

$\Longrightarrow$

We show the *Res* rules are derivable in Došen's axiomatic calculus. We show only the cases for $\backslash$, those for $/$ are symmetric.

$$\dfrac{\dfrac{\dfrac{}{A \vdash A}\,(\mathit{Id}) \qquad \begin{matrix}\vdots\\ B \vdash A \backslash C\end{matrix}}{A \bullet B \vdash A \bullet (A \backslash C)}\,(\mathit{Mon}\bullet) \qquad \dfrac{}{A \bullet (A \backslash C) \vdash C}\,(\mathit{Appl}\backslash)}{A \bullet B \vdash C}\,(\mathit{Trans})$$

$$\dfrac{\dfrac{}{B \vdash A \backslash (A \bullet B)}\,(\mathit{Coappl}\backslash) \qquad \dfrac{\dfrac{}{A \vdash A}\,(\mathit{Id}) \qquad \begin{matrix}\vdots\\ A \bullet B \vdash C\end{matrix}}{A \backslash (A \bullet B) \vdash A \backslash C}\,(\mathit{Mon}\backslash)}{B \vdash A \backslash C}\,(\mathit{Trans})$$

$\Longleftarrow$

We first show that *Appl*$\backslash$ and *Co-appl*$\backslash$ are derivable. The cases for $/$ are again symmetric.

$$\dfrac{\dfrac{}{A \backslash B \vdash A \backslash B}\,(\mathit{Id})}{A \bullet (A \backslash B) \vdash B}\,(\mathit{Res}_{\backslash\bullet}) \qquad \dfrac{\dfrac{}{B \bullet A \vdash B \bullet A}\,(\mathit{Id})}{A \vdash B \backslash (B \bullet A)}\,(\mathit{Res}_{\bullet\backslash})$$

To conclude, we show that the monotonicity rules are derived rules of the residuation calculus. We only show $(\mathit{Mon}\bullet)$ and $(\mathit{Mon}\backslash)$, the case for $(\mathit{Mon}/)$ is symmetric to the case for $(\mathit{Mon}\backslash)$.

$$\dfrac{\dfrac{\begin{matrix}\vdots\\ A \vdash B\end{matrix} \qquad \dfrac{\dfrac{\dfrac{\begin{matrix}\vdots\\ C \vdash D\end{matrix} \qquad \dfrac{\dfrac{}{B \bullet D \vdash B \bullet D}\,(\mathit{Id})}{D \vdash B \backslash (B \bullet D)}\,(\mathit{Res}_{\bullet\backslash})}{C \vdash B \backslash (B \bullet D)}\,(\mathit{Trans})}{B \bullet C \vdash B \bullet D}\,(\mathit{Res}_{\backslash\bullet})}{B \vdash (B \bullet D) / C}\,(\mathit{Res}_{\bullet/})}{A \vdash (B \bullet D) / C}\,(\mathit{Trans})}{A \bullet C \vdash B \bullet D}\,(\mathit{Res}_{/\bullet})$$

$$\dfrac{\dfrac{\dfrac{\dfrac{\begin{matrix}\vdots\\ A\vdash B\end{matrix}\quad \dfrac{\dfrac{\dfrac{}{B\backslash C\vdash B\backslash C}\,(Id)}{B\bullet(B\backslash C)\vdash C}\,(Res_{\backslash\bullet})}{B\vdash C/(B\backslash C)}\,(Res_{\bullet/})}{A\vdash C/(B\backslash C)}\,(Trans)}{A\bullet(B\backslash C)\vdash C}\,(Res_{/\bullet})\quad \begin{matrix}\vdots\\ C\vdash D\end{matrix}}{A\bullet(B\backslash C)\vdash D}\,(Trans)}{B\backslash C\vdash A\backslash D}\,(Res_{\bullet\backslash})$$

This completes the equivalence proof of the two combinator calculi. □

For the equivalence of the third calculus, it is a simple corollary of the proof of Lemma 4.9 and of Lemma 4.14 in the next section (though see Moortgat and Oehrle, 1999, for a direct proof).

Corollary 4.10. *All derivable sequents of the monotonicity-residuation calculus of Moortgat and Oehrle shown in Figure 4.7 are derivable sequents of the two other calculi as well.*

Proof. Since by Lemma 4.9 the two other calculi are equivalent, it suffices to show that the residuation calculus generates all theorems of the residuation-monotonicity calculus. By the proof of Lemma 4.9, the monotonicity rules are admissible in the residuation calculus, therefore all theorems of Moortgat and Oehrle's calculus are theorems of Lambek's calculus. □

4.4.2 Equivalence between the Axiomatic Representation and Sequent Calculus

Since the axiomatic formulation of NL doesn't have the commas and parentheses we used to construct antecedent terms in NL, we introduce a simple translation from antecedent terms of NL to formulae, which replaces all commas by products.

Definition 4.11. *Let $\mathscr{A}$ be an antecedent term, we define the function $\|\mathscr{A}\|^{\bullet}$, which translates an antecedent term into a formula, as follows.*

$$\|\mathsf{Lp}\|^{\bullet} = \mathsf{Lp}$$
$$\|(\mathscr{A},\mathscr{A})\|^{\bullet} = \|\mathscr{A}\|^{\bullet}\bullet\|\mathscr{A}\|^{\bullet}$$

Before starting the equivalence proof, we first prove a useful substitution lemma, which shows how we can replace a term B by a less general term A in a context Γ. This lemma is an easy combination of rules (*Mon*•) and (*Trans*). The intuition behind the Lemma is that it allows us to strengthen the transitivity rule into something which corresponds to (the translation of) the cut rule.

Lemma 4.12. *If* $\|\Gamma[B]\|^{\bullet} \vdash C$ *and* $A \vdash B$ *then* $\|\Gamma[A]\|^{\bullet} \vdash C$

Proof. Assume $A \vdash B$ and $\|\Gamma[B]\|^{\bullet} \vdash C$. In order to prove $\|\Gamma[A]\|^{\bullet} \vdash C$, it suffices to show that $A \vdash B$ implies $\|\Gamma[A]\|^{\bullet} \vdash \|\Gamma[B]\|^{\bullet}$, since this allows us to combine the hypotheses as follows.

$$\frac{\|\Gamma[A]\|^{\bullet} \vdash \|\Gamma[B]\|^{\bullet} \qquad \|\Gamma[B]\|^{\bullet} \vdash C}{\|\Gamma[A]\|^{\bullet} \vdash C}\ (Trans)$$

Induction of the length l of the unique path in $\Gamma[A]$ to A (which is of course the same as the path in $\Gamma[B]$ to B).

If $l = 0$ then the context is empty and we have $\Gamma[A] = A$, $\Gamma[B] = B$ and therefore we can conclude that $A \vdash B$ implies $\|\Gamma[A]\|^{\bullet} \vdash \|\Gamma[B]\|^{\bullet}$ because they are identical.

If $l > 0$ then, since Γ is not the empty context, Γ is either of the form $(\Delta[B], \Delta')$ or of the form $(\Delta', \Delta[B])$.

If the first case $\|(\Delta[B], \Delta')\|^{\bullet} = \|\Delta[B]\|^{\bullet} \bullet \|\Delta'\|^{\bullet}$. Given that we know by induction hypothesis that $\|\Delta[A]\|^{\bullet} \vdash \|\Delta[B]\|^{\bullet}$, we can simply apply the monotonicity rule for the product formula as follows.

$$\frac{\begin{array}{c}\vdots\ IH \\ \|\Delta[A]\|^{\bullet} \vdash \|\Delta[B]\|^{\bullet}\end{array} \qquad \dfrac{}{\|\Delta'\|^{\bullet} \vdash \|\Delta'\|^{\bullet}}\ (Id)}{\|(\Delta[A], \Delta')\|^{\bullet} \vdash \|(\Delta[B], \Delta')\|^{\bullet}}\ (Mon\bullet)$$

The case where Γ is of the form $(\Delta', \Delta[B])$ is symmetric. □

Lemma 4.13. $\Gamma \vdash A$ *is derivable in the sequent calculus iff* $\|\Gamma\|^{\bullet} \vdash A$ *is derivable in the axiomatic representation.*

Proof

$\Longrightarrow$

We proceed by induction on the length l of the sequent calculus proof.

If $l = 1$ the sequent calculus proof contains a single axiom rule and the axiomatic proof is the same, justified by axiom (Id).

If $l > 1$ then induction hypothesis gives us an axiomatic proof of length $l - 1$ and depending on the last rule, we extend it as follows.

- $(\bullet_h)$ Note that $\|\Gamma[(A,B)]\|^{\bullet}$ is equal to $\|\Gamma[A \bullet B]\|^{\bullet}$, so the axiomatic proof we have by induction hypothesis for $\|\Gamma[(A,B)]\|^{\bullet}$ is a proof of $\|\Gamma[A \bullet B]\|^{\bullet}$ as well.
- $(\bullet_i)$ Induction hypothesis gives us an axiomatic proof of $\|\Delta\|^{\bullet} \vdash A$ and an axiomatic proof of $\|\Gamma\|^{\bullet} \vdash B$, which we can combine into a proof of $\|(\Delta, \Gamma)\|^{\bullet} \vdash A \bullet B$ as shown below ("*Def* $\|.\|^{\bullet}$" is not a rule of the calculus, but simply denotes the two antecedent terms translate to the same formula according to Definition 4.11).

$$\dfrac{\dfrac{\begin{matrix}\vdots\, IH \\ \|\Delta\|^\bullet \vdash A\end{matrix} \qquad \begin{matrix}\vdots\, IH \\ \|\Gamma\|^\bullet \vdash B\end{matrix}}{\|\Delta\|^\bullet \bullet \|\Gamma\|^\bullet \vdash A \bullet B}\,(Mon\bullet)}{\|(\Delta,\Gamma)\|^\bullet \vdash A \bullet B}\,(Def\,\|.\|^\bullet)$$

($\backslash_h$) Induction hypothesis gives us an axiomatic proof of $\|\Gamma[B]\|^\bullet \vdash C$ and of $\|\Delta\|^\bullet \vdash A$. We need to show that $\|\Gamma[(\Delta, A \backslash B)]\|^\bullet \vdash C$.

With $\|\Delta\|^\bullet \vdash A$ (induction hypothesis) and $A \backslash B \vdash A \backslash B$ (*Id*) we apply rule (*Mon*$\bullet$) to obtain $\|\Delta\|^\bullet \bullet A \backslash B \vdash A \bullet (A \backslash B)$. Using axiom (*Appl*$\backslash$) and the transitivity rule (*Trans*) we obtain $\|\Delta\|^\bullet \bullet A \backslash B \vdash B$. Combining this with the other induction hypothesis $\|\Gamma[B]\|^\bullet \vdash C$ using Lemma 4.12 gives us $\|\Gamma[(\Delta \bullet (A \backslash B))]\|^\bullet \vdash C$ which is equivalent to $\|\Gamma[(\Delta, A \backslash B)]\|^\bullet \vdash C$.

The proof schema below displays the different steps used.

$$\dfrac{\dfrac{\dfrac{\dfrac{\begin{matrix}\vdots\, IH \\ \|\Delta\|^\bullet \vdash A\end{matrix} \qquad \dfrac{}{A \backslash B \vdash A \backslash B}\,(Id)}{\|\Delta\|^\bullet \bullet A \backslash B \vdash A \bullet (A \backslash B)}\,(Mon\bullet) \qquad \dfrac{}{A \bullet (A \backslash B) \vdash B}\,(Appl\backslash)}{\|\Delta\|^\bullet \bullet A \backslash B \vdash B}\,(Trans)}{\|\Delta \bullet A \backslash B\|^\bullet \vdash B}\,(Def\|.\|^\bullet) \qquad \begin{matrix}\vdots\, IH \\ \|\Gamma[B]\|^\bullet \vdash C\end{matrix}}{\dfrac{\|\Gamma[\Delta \bullet A \backslash B]\|^\bullet \vdash C}{\|\Gamma[(\Delta, A \backslash B)]\|^\bullet \vdash C}\,(Def\|.\|^\bullet)}\,Lem\ 4.12$$

($\backslash_i$) Induction hypothesis gives us an axiomatic proof of $\|(A,\Gamma)\|^\bullet \vdash C$, which is equal to $A \bullet \|\Gamma\|^\bullet \vdash C$. Using the axiom (*Id*) to obtain $A \vdash A$ we apply rule (*Mon*$\backslash$) to obtain $A \backslash (A \bullet \|\Gamma\|^\bullet) \vdash A \backslash C$. Now we use axiom (*Co-appl*$\backslash$) to obtain $\|\Gamma\|^\bullet \vdash A \backslash (A \bullet \|\Gamma\|^\bullet)$. We combine this with the previous statement using rule (*Trans*) to obtain $\|\Gamma\|^\bullet \vdash A \backslash C$.

The proof schema below shows the different steps.

$$\dfrac{\dfrac{}{\|\Gamma\|^\bullet \vdash A \backslash (A \bullet \|\Gamma\|^\bullet)}\,(Co\text{-}appl\backslash) \qquad \dfrac{\dfrac{}{A \vdash A}\,(Id) \qquad \dfrac{\begin{matrix}\vdots\, IH \\ \|(A,\Gamma)\|^\bullet \vdash C\end{matrix}}{A \bullet \|\Gamma\|^\bullet \vdash C}\,(Def\|.\|^\bullet)}{A \backslash (A \bullet \|\Gamma\|^\bullet) \vdash A \backslash C}\,(Mon\backslash)}{\|\Gamma\|^\bullet \vdash A \backslash C}\,(Trans)$$

($/_h$), ($/_i$) Symmetric to the cases for ($\backslash_h$) and ($\backslash_i$).

(*cut*) Induction hypothesis give us a proof of $\|\Gamma\|^\bullet \vdash A$ and of $\|\Delta[A]\|^\bullet \vdash B$. We apply Lemma 4.12 directly to obtain $\|\Delta[\Gamma]\|^\bullet \vdash B$.

$\Longleftarrow$

This part of the proof is easy, since all rules and axioms have simple proofs in the sequent calculus.

$(Mon\bullet)$

$$\dfrac{\dfrac{\begin{array}{c}\vdots\, IH\\ A \vdash B\end{array} \qquad \begin{array}{c}\vdots\, IH\\ C \vdash D\end{array}}{(A,C) \vdash B \bullet D}\,\bullet_h}{A \bullet C \vdash B \bullet D}\,\bullet_i$$

$(Mon\backslash)$

$$\dfrac{\dfrac{\begin{array}{c}\vdots\, IH\\ A \vdash B\end{array} \qquad \begin{array}{c}\vdots\, IH\\ C \vdash D\end{array}}{(A, B \backslash C) \vdash D}\,\backslash_h}{B \backslash C \vdash A \backslash D}\,\backslash_i$$

$(Trans)$

$$\dfrac{\begin{array}{c}\vdots\, IH\\ A \vdash B\end{array} \qquad \begin{array}{c}\vdots\, IH\\ B \vdash C\end{array}}{A \vdash C}\,cut$$

This completes the equivalence proof of the sequent and the combinator calculus. □

Lemma 4.14. *The residuation-monotonicity calculus of Figure 4.7 is equivalent to the cut-free sequent calculus of NL.*

Proof. We only need to prove that if δ is a cut free NL sequent derivation of $\Gamma \vdash C$, then there is a residuation-monotonicity (ResMon) derivation δ' of $\|\Gamma\|^{\bullet} \vdash C$ which does not use the rule $(Trans)$. The other direction is a direct consequence Corollary 4.10, which shows derivability in the residuation-monotonicity calculus implies derivability in the Došen-style axiomatic calculus, and Lemma 4.13, which shows derivability in the axiomatic calculus implies derivability in the sequent calculus for NL.

We prove the lemma by induction on the length l of δ. If $l = 1$, the we have an axiom rule in the sequent calculus and the (Id) rule in the axiomatic calculus.

If $l > 1$ we do a case analysis on the last rule of the proof.

The only cases which require a little work are when the last rule is $\backslash_h$ or $/_h$. We only sketch the basic idea for the case of $\backslash_h$ here, Exercise 4.12 and 4.14 ask you to give a more formal proof. By induction hypothesis, we have an axiomatic proof of $\|\Gamma[B]\|^{\bullet} \vdash C$. Now, using a series of $Res_{\bullet/}$ and $Res_{\bullet\backslash}$ rules, we can obtain a proof of $B \vdash D$, where D has C as a subformula inside a number of $\backslash$ and $/$ connectives, then we can apply the monotonicity rule for $\backslash$ followed by the residuation rule to change $\backslash$ into $\bullet$, after which we "put back" the context $\Gamma[]$ around the antecedent, by using $Res_{/\bullet}$ and $Res_{\backslash\bullet}$ to obtain $\|\Gamma[(\Delta, A \backslash B)]\|^{\bullet} \vdash C$. This gives us the following schematic proof.

$$
\dfrac{\dfrac{\begin{array}{c}\vdots\; IH\\ \|\Gamma[B]\|^{\bullet} \vdash C\\ \vdots\; Res_{\bullet/}, Res_{\bullet\backslash}\\ B \vdash D\end{array} \qquad \begin{array}{c}\vdots\; IH\\ \|\Delta\|^{\bullet} \vdash A\end{array}}{A \backslash B \vdash \|\Delta\|^{\bullet} \backslash D}\,Mon\backslash}{\begin{array}{c}\|\Delta\|^{\bullet} \bullet A \backslash B \vdash D\\ \vdots\; Res_{/\bullet}, Res_{\backslash\bullet}\\ \|\Gamma[(\Delta, A \backslash B)]\|^{\bullet} \vdash C\end{array}}\,Res_{\backslash\bullet}
$$

The case for $/_h$ is symmetric.

The other cases are easily checked. In case the last rule is $\backslash_i$, we can simulate it using $Res_{\bullet\backslash}$ as follows.

$$
\dfrac{\dfrac{\begin{array}{c}\vdots\; IH\\ \|(A,\Gamma)\|^{\bullet} \vdash B\end{array}}{A \bullet \|\Gamma\|^{\bullet} \vdash B}\,Def\|.\|^{\bullet}}{\|\Gamma\|^{\bullet} \vdash A \backslash B}\,Res_{\bullet\backslash}
$$

Given that $\|\Gamma(A,B)\|^{\bullet}$ is equal to $\|\Gamma[A \bullet B]\|^{\bullet}$, the rule $\bullet_h$ is trivial.

Finally, the rule $\bullet_i$ corresponds to an application of the monotonicity rule for the product as follows.

$$
\dfrac{\dfrac{\begin{array}{c}\vdots\; IH\\ \|\Delta\|^{\bullet} \vdash A\end{array} \qquad \begin{array}{c}\vdots\; IH\\ \|\Gamma\|^{\bullet} \vdash B\end{array}}{\|\Delta\|^{\bullet} \bullet \|\Gamma\|^{\bullet} \vdash A \bullet B}\,Mon_{\bullet}}{\|(\Delta,\Gamma)\|^{\bullet} \vdash A \bullet B}\,Def_{\|.\|^{\bullet}}
$$

□

4.5 Model Theory

We have seen group models for the Lambek calculus in the previous chapter. Even though it is possible to adapt these models to the non-associative Lambek calculus, we will introduce a different kind of models for the non-associative Lambek calculus in this chapter: Kripke models. This is another important group of models for the Lambek calculus and they have the advantage of easily accommodating the multimodal extensions to NL we will see in the next chapter.

Kripke models were originally introduced for modal logics and they used possible worlds and a binary accessibility relation R^2 between worlds to give models for logical necessity and possibility (for many other applications and a very good general overview of modal logics, see Blackburn et al, 2001).

For our current purposes, however, our 'worlds' are simply the linguistic structures we use: formulae and their (structured) combinations. In the following text we will use the words 'world' and 'linguistic resource' interchangeably.

We use a *ternary* accessibility relation R^3 to "merge" these formulae and structures: R^3abc holds between a b and c if and only if resource a is the result of merging resource b and resource c. We can represent R^3abc in the form of a picture as shown below.

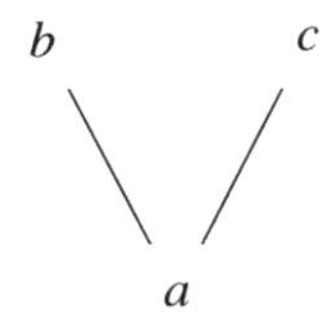

We intend to interpret this ternary relation in such a way that if, for example, b is in the interpretation of np and c is in the interpretation of $np\backslash S$ then we want to conclude that there exists a world a such that a is in the interpretation of S.

Though it is often useful to draw the accessibility relation as a tree-like structure, as shown above, it is important to remember that the structures we define using this ternary relation do not necessarily represent trees. In particular, if we want to interpret R^3abc as saying that a is the mother of its daughters b and c, we have to keep in mind that unicity of this mother for a given b and c is *not* necessary, or, more precisely, the formula $\forall a \forall b \forall c \forall d.(R^3abc \wedge R^3dbc) \rightarrow d = a$ does not hold. However, to make the intuitions behind the use of the accessibility relation R^3 clear, we will sometime refer to a trio of nodes a, b, c such that R^3abc using the vocabulary of binary trees, saying the a is the parent of b and c or that b is the left sister of c.

Before giving the definitions of the connectives, we first give some standard definitions in modal logic: frames, models, etc. In what follows, we will assume the set of atomic formulae P to be fixed.

Definition 4.15. *A* Kripke frame *F is a pair $\langle W, R^3 \rangle$ where W is a non-empty set of worlds and R^3 is a ternary accessibility relation over triples of elements from W.*

Definition 4.16. *A* Kripke model *M is a triple $\langle W, R^3, V \rangle$ where $\langle W, R^3 \rangle$ is a Kripke frame and where V is an evaluation function from elements of P to subsets of W. We say that $\langle W, R^3 \rangle$ is the* underlying frame *of M.*

The following definitions are also standard.

Definition 4.17. *A sequent $A \vdash B$ is* true in a model M at world a *— we will write $M, a \models (A \vdash B)$ — in case if $M, a \models A$ then $M, a \models B$ as well.*

A sequent $A \vdash B$ *is* true in a model M *— we will write* $M \models (A \vdash B)$ *— iff it is true at all worlds in that model.*

A sequent $A \vdash B$ *is* valid on a frame F *— we will write* $F \models (A \vdash B)$ *— iff it is true at all worlds and under all valuations of that frame.*

The notion of Kripke frame is often useful when we want to study properties of the accessibility relation without considering the assignments of the atomic formulae. We will return to this point in Section 4.5.2, where we see that adding structural rules to NL corresponds to restricting the frame we use for interpreting the logic.

As is clear from Definition 4.16, a Kripke model is simply a Kripke frame with an evaluation which maps atomic formulae to sets of worlds in our model, so, for example, it would tell us at which worlds the atomic formula np is true and at which worlds the atomic formula S is true. The interpretation of the atomic formulae can vary from one model to another.

So, though the interpretation of the atomic formulae is fixed by the model, the interpretation of the complex formulae defined as follows.

$$
\begin{array}{rl}
M,a \models p & \textit{iff } a \in V(p) \\
M,a \models A \bullet B & \textit{iff } \exists b \exists c (R^3abc \;\&\; M,b \models A \;\&\; M,c \models B) \\
M,a \models A \setminus B & \textit{iff } \forall b \forall c ((R^3cba \;\&\; M,b \models A) \Rightarrow M,c \models B) \\
M,a \models A \,/\, B & \textit{iff } \forall b \forall c ((R^3cab \;\&\; M,b \models A) \Rightarrow M,c \models B)
\end{array}
$$

Again, the picture below helps us to better see the different worlds, the formulae which hold at them and the way they are related by the accessibility relation.

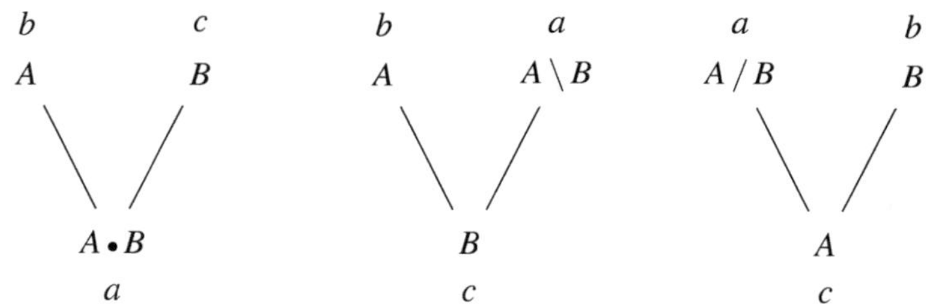

Seen this way, the semantics of the connectives is quite close to their intuitive meanings: if $A \bullet B$ holds at world a then this world is the merger by R^3abc of two worlds b and c such that A holds at world b and B holds at world c. Similarly, if $A \,/\, B$ holds at world a then for any world b which is "to the right" of a via R^3cab and which is in the interpretation of B, then the world c, which represents the "parent" or the merger of a and b, is in the interpretation of A.

4.5.1 Soundness and Completeness

Lemma 4.18 (Soundness). *If* $A \vdash B$ *is derivable then for all models* M *and for all words* a*,* $M,a \models (A \vdash B)$.

Proof. Soundness is relatively simple. We prove by induction on the depth of the axiomatic proof of $A \vdash B$ that for all models M and worlds a, whenever $M,a \models A$ then $M,a \models B$. This is just a matter of writing out the definitions for the different connectives.

First, we look at the axioms.

(Id) Axiom (Id) is a simple tautology.

($Appl\backslash$) Suppose we obtained $A \bullet (A \backslash B) \vdash B$ by application of axiom ($Appl\backslash$). In that case, we need to prove that for every world a such that $M,a \models A \bullet (A \backslash B)$, B is true at world a in M as well. Given the definition of the product, this means there are worlds b and c such that R^3abc and $M,b \models A$ and $M,c \models A \backslash B$. The picture below on the left shows these three worlds and their relation.
We can then unfold the definition of $A \backslash B$ at world c, obtaining that for every world d and e such that R^3edc and $M,d \models A$, $M,e \models B$ as well. But if this holds for any worlds, then it also holds for $d = b$, for which we already knew that $M,b \models A$ and for world $e = a$, which means R^3edc holds by virtue of being equal to R^3abc. $M,b \models A$ and R^3abc together then imply $M,a \models B$ which we needed to prove.

The picture below shows the unfolding of the product formula on the left, then the unfolding of the implication in the middle and finally the composed figure on the right.

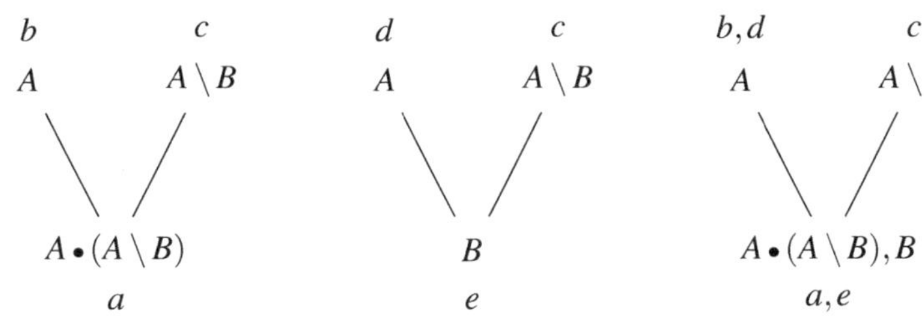

($Appl/$) Symmetric.

($Co\text{-}appl\backslash$) If we obtained $A \vdash B \backslash (B \bullet A)$ as an instantiation of axiom ($Co\text{-}appl\backslash$), we verify that for every model M and world a, if $M,a \models A$ then $M,a \models B \backslash (B \bullet A)$, that is, we verify that for all worlds b and c such that R^3cba and $M,b \models B$ we have $M,c \models B \bullet A$. Take any worlds b and c such that R^3cba and $M,b \models B$. We need to show that $M,c \models B \bullet A$. In other words, we need to show that there exists a d and an e such that R^3cde and $M,d \models B$ and $M,e \models A$. Take $b = d$ and $a = e$, then $R^3cde = R^3cba$, $M,b \models B$ and $M,a \models A$ all hold by assumption. Therefore, we have shown that if $M,a \models A$ then $M,a \models B \backslash (B \bullet A)$.

($Co\text{-}appl/$) Symmetric.

Next for the rules. For rules ($Mon\bullet$), ($Mon\backslash$) and ($Mon/$) we know by induction hypothesis that for all models M and worlds a, if $M,a \models A$ then $M,a \models B$ and if $M,a \models C$ then $M,a \models D$.

($Mon\bullet$) For rule ($Mon\bullet$) we have to show that for every model M and world a if $M,a \models A \bullet C$ then $M,a \models B \bullet D$. Take an arbitrary world and model satisfying $M,a \models A \bullet C$. The definition of $A \bullet C$ tells us there are worlds b and c such that R^3abc, $M,b \models A$ and $M,c \models C$. Induction hypothesis gives us that $M,b \models B$ and $M,c \models D$ and then, given that we already knew R^3abc we can apply the definition of the product in the opposite direction to obtain $M,a \models B \bullet D$.

($Mon\backslash$) We have to show that for every model M and world a, if $M,a \models B \backslash C$ then $M,a \models A \backslash D$. Given that $M,a \models B \backslash C$ we know that for all worlds b and c if R^3cba and $M,b \models B$ then $M,c \models C$. But induction hypothesis also gives us $M,b \models A$ and $M,c \models D$ for all b and c, allowing us to apply the definition of $\backslash$ again to obtain $M,a \models A \backslash D$.

($Mon/$) Symmetric.

(*Trans*) Induction hypothesis gives us that for all models M and worlds a, if $M,a \models A$ then $M,a \models B$ and if $M,a \models B$ then $M,a \models C$. We just have to show that for every model M and every world a if $M,a \models A$ then $M,a \models C$. Suppose $M,a \models A$. By induction hypothesis we know that $M,a \models A$ implies $M,a \models B$. Induction hypothesis also gives us that $M,a \models B$ implies $M,a \models C$, which is all we needed to show. □

For the completeness result, we first define a canonical model for NL and prove a strong result: the Truth Lemma which states that truth in this model corresponds exactly with provability in the axiomatic calculus. This means that for *any* underivable statement $A \vdash B$, our canonical model will be a countermodel but also that for any statement $A \vdash B$ which is true in the canonical model, $A \vdash B$ is derivable.

Definition 4.19. *The canonical model* $\mathscr{M} = \langle W, R^3, V \rangle$ *for NL is defined as follows:*

- W *is the set of all formulae,*
- R^3abc *iff* $a \vdash b \bullet c$, *and*
- $a \in V(p)$ *iff* $a \vdash p$.

Let's look at the different clauses for the canonical model in a bit more detail. First, the worlds of the canonical model are just the formulae of NL. Second, the accessibility relation is linked directly to the derivability relation of the product formula. Finally, for all atomic formulae p and all formulae (worlds) a such that $a \vdash p$ we have $a \in V(p)$. In other words, $V(p)$ is true at world p and at all words a such that we can derive p from a.

Lemma 4.20. *The* Truth Lemma, $\mathscr{M}, a \models A$ *iff* $a \vdash A$

Proof.

$\Longleftarrow$

Soundness is a simple corollary of Lemma 4.18 which proves the stronger claim that for *any* model M and word a we have $M,a \models A$ implies $a \vdash A$.

$\Longrightarrow$

For the completeness part we show that if $\mathcal{M}, a \models A$ then $a \vdash A$. In other words, when $a \models A$ is true in the canonical model for formulae a and A, then there is a derivation of $a \vdash A$ in the axiomatic calculus of Figure 4.5 as well. We prove the completeness part of the Truth Lemma by induction on A.

[$A = p$] For atomic formulae p the definition of $\models$ for $\mathcal{M}$ gives us $a \vdash p$ directly, by construction.

[$A = B \bullet C$] A is of the form $B \bullet C$. Assume that a is a world such that $\mathcal{M}, a \models B \bullet C$. By the definition of $M, a \models B \bullet C$ there are worlds b and c such that R^3abc where $\mathcal{M}, b \models B$ and $\mathcal{M}, c \models C$. Given that B and C are subformulae of A, we can apply the induction hypothesis and obtain proofs $b \vdash B$ and $c \vdash C$. Using rule ($Mon\bullet$) we obtain

$$b \bullet c \vdash B \bullet C$$

But, given that R^3abc, the definition of R^3 in the canonical frame gives us.

$$a \vdash b \bullet c$$

Finally, using the transitivity rule (*Trans*) we obtain

$$a \vdash B \bullet C$$

as required.

[$A = B \backslash C$] Assume that a is a world such that $\mathcal{M}, a \models B \backslash C$, we have to show $a \vdash B \backslash C$. Given that $\mathcal{M}, a \models B \backslash C$ this means that for all e and f if R^3efa and $\mathcal{M}, f \models B$ then $\mathcal{M}, e \models C$. Choose $e = B \bullet a$ and $f = B$. The picture below summarizes the worlds we have named so far and their relations.

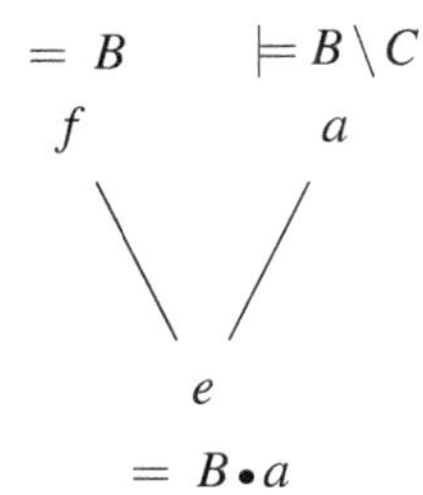

Now we can use the fact that the derivability of $B \bullet a \vdash B \bullet a$ (which is equal to $e \vdash f \bullet a$) and the definition of R^3 in the canonical frame give us R^3efa.
Given that $\mathcal{M}, f \models B$ and R^3efa are satisfied, the definition of $B \backslash C$ gives us $\mathcal{M}, e \models C$, with $e = B \bullet a$. By induction hypothesis, we have a proof of $B \bullet a \vdash C$. We can transform this proof into a proof of $a \vdash B \backslash C$ as follows.

$$\dfrac{\dfrac{}{a \vdash B\backslash(B\bullet a)}\,(Co\text{-}appl\backslash) \qquad \dfrac{\dfrac{}{B\vdash B}\,(Id) \qquad \overset{\vdots\ IH}{B\bullet a\vdash C}}{B\backslash(B\bullet a)\vdash B\backslash C}\,(Mon\backslash)}{a\vdash B\backslash C}\,(Trans)$$

$[A = C\,/\,B]$ Symmetric. □

Corollary 4.21. *An important corollary of the Truth Lemma is that the canonical model* $\mathscr{M}$ *is the* most general model *for NL, that is, for every model M if* $\mathscr{M} \models A \vdash B$ *then* $M \models A \vdash B$.

Proof. The corollary has a very easy proof by contraposition. We show that if there is a model M such that $M \not\models A \vdash B$ then $\mathscr{M} \not\models A \vdash B$. But if the statement $A \vdash B$ has a countermodel then $A \nvdash B$ (the sequent is underivable), which by the Truth Lemma means $\mathscr{M} \not\models A \vdash B$. □

Theorem 4.22. *NL is sound and complete with respect to all models.*

Proof. Soundness was proved as Lemma 4.18 and with the Truth Lemma in place, the completeness proof is trivial.

For completeness, we need to show that if a statement $A \vdash B$ is true in all models, then $A \vdash B$ is derivable. But if $A \vdash B$ is true in all models, then it is true in the canonical model as well, so we have $\mathscr{M} \models (A \vdash B)$. In other words, for any world a we have that if $\mathscr{M}, a \models A$ then $\mathscr{M}, a \models B$ (1). If this holds for any world a then in particular for world A, and since $A \vdash A$ is derivable, by the Truth Lemma we have $\mathscr{M}, A \models A$ and therefore $\mathscr{M}, A \models B$, by (1). Now by the Truth Lemma $\mathscr{M}, A \models B$ implies that $A \vdash B$ is derivable. □

4.5.2 Adding Structural Rules

An interesting benefit of the current formulation is that, like in modal logic, we can add *restrictions* to our frame to change the properties of the connectives in our logic. This branch of modal logic is called *correspondence theory*: a well-know example is that the modal logic S4 corresponds to the reflexive, transitive frames. Chapter 3 of Blackburn et al (2001) gives a detailed overview of correspondence theory for modal logic.

Many of the standard techniques which apply to frames for modal logics can be adapted to categorial grammars. In this section we will be interested in finding first-order formulae which express constraints on modal frames (see Došen, 1992; Kurtonina, 1995, 1998). Kurtonina (1995, 1998) shows that first-order constraints allow us to define a large class of categorial logics; we will return to her results briefly in Section 5.5.1.

For now, we will only treat the structural rules of associativity and commutativity, as we have seen them in Section 4.3.

The structural rules for associativity and commutativity can be added to our models by adding the following restrictions on the accessibility relation R^3.

$$\forall a \forall b \forall c.(R^3abc \Rightarrow R^3acb) \quad \text{(com)}$$
$$\forall a \forall b \forall c \forall d \forall f.((R^3afd \;\&\; R^3fbc) \Rightarrow \exists e.(R^3abe \;\&\; R^3ecd)) \quad \text{(ass1)}$$
$$\forall a \forall b \forall c \forall d \forall e.((R^3abe \;\&\; R^3ecd) \Rightarrow \exists f.(R^3afd \;\&\; R^3fbc)) \quad \text{(ass2)}$$

If we want to add several of these constraints to the accessibility relation, this corresponds simply to adding the conjunction of the corresponding formula as a frame constraint.

As usual, these principles are best shown in picture form, as done in Figure 4.8.

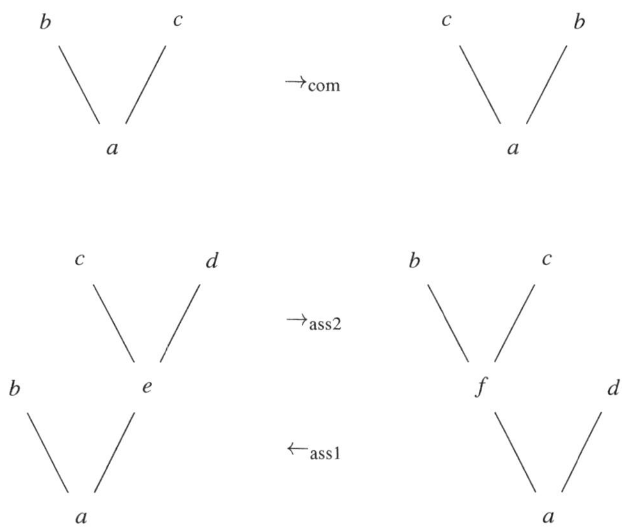

Fig. 4.8. Visual representation of the constraints on the accessibility relation R^3 corresponding to associativity and commutativity

These frame constraints correspond to adding the following axioms to the axiomatic calculus for NL.

$$A \bullet B \vdash B \bullet A \quad \text{(com)}$$
$$A \bullet (B \bullet C) \vdash (A \bullet B) \bullet C \quad \text{(ass1)}$$
$$(A \bullet B) \bullet C \vdash A \bullet (B \bullet C) \quad \text{(ass2)}$$

Now, the following is easy to see.

Proposition 4.23. *The axiomatic calculus with a subset of the additional axioms* { *(comm), (ass1), (ass2)*} *corresponds to the sequent calculus with the additional structural rules of the same name (these are shown in Figure 4.3)*

Proof. This is an easy extension of Lemma 4.13. We show only the case for (ass1), the other cases are similar.

Showing that $A \bullet (B \bullet C) \vdash (A \bullet B) \bullet C$ is derivable in the sequent calculus using the structural rule (ass1) of Figure 4.3 (repeated below) is trivial.

$$\frac{\Gamma[((\Delta_1,\Delta_2),\Delta_3)] \vdash D}{\Gamma[(\Delta_1,(\Delta_2,\Delta_3))] \vdash D}\ (\text{ass1})$$

Now, for the inverse direction, we only need to treat the new case (the rest of the proof follows from Lemma 4.13). Suppose we have an axiomatic proof of $\|\Gamma[((\Delta_1,\Delta_2),\Delta_3)]\|^{\bullet} \vdash D$, then we need to show we can transform it into an axiomatic proof of $\|\Gamma[(\Delta_1,(\Delta_2,\Delta_3))]\|^{\bullet} \vdash D$. By the definition of $\|.\|^{\bullet}$, this is equivalent to showing we can transform a proof of $\|\Gamma[(\Delta_1 \bullet \Delta_2) \bullet \Delta_3]\|^{\bullet} \vdash D$ into a proof of $\|\Gamma[\Delta_1 \bullet (\Delta_2 \bullet \Delta_3)]\|^{\bullet} \vdash D$.

According to Lemma 4.12, whenever $\|\Gamma[F]\|^{\bullet} \vdash D$ and $E \vdash F$ are derivable in the axiomatic calculus, then $\|\Gamma[E]\|^{\bullet} \vdash D$ is derivable as well. Taking $E = \|\Delta_1\|^{\bullet} \bullet (\|\Delta_2\|^{\bullet} \bullet \|\Delta_3\|^{\bullet})$ and $F = (\|\Delta_1\|^{\bullet} \bullet \|\Delta_2\|^{\bullet}) \bullet \|\Delta_3\|^{\bullet}$, makes $E \vdash F$ an instantiation of the axiom (ass1) and gives us $\|\Gamma[(\Delta_1 \bullet \Delta_2) \bullet \Delta_3]\|^{\bullet} \vdash D$ implies $\|\Gamma[\Delta_1 \bullet (\Delta_2 \bullet \Delta_3)]\|^{\bullet} \vdash D$ as required. □

Lemma 4.24. *Let F be a frame and FO be the first-order formula expressing one of the frame constraints { (com), (ass1), (ass2) } and $A \vdash B$ the corresponding sequent. F satisfies FO if and only if $F \models (A \vdash B)$.*

Proof
We verify only the case for (com), the cases for associativity are similar.

For (com) this means we need to show that F satisfies $\forall a \forall b \forall c.(R^3abc \Rightarrow R^3acb)$ iff $F \models (A \bullet B \vdash B \bullet A)$.

$\Longrightarrow$

We need to show that if F satisfies the frame constraint, then F satisfies $A \bullet B \vdash B \bullet A$ as well. Let M be any model $\langle W, R^3, V\rangle$ such that its underlying frame $F = \langle W, R^3\rangle$ satisfies the constraint $\forall a \forall b \forall c.(R^3abc \Rightarrow R^3acb)$ and let d be any world in this model. Suppose $M,d \models A \bullet B$. We need to show that $M,d \models B \bullet A$. If $M,d \models A \bullet B$ then writing out the definition of the valuation for $A \bullet B$ we conclude that there are worlds e and f in the model such that R^3def, $M,e \models A$ and $M,f \models B$. Applying the frame constraint with $a = d$, $b = e$ and $c = f$ gives us R^3dfe. Therefore, at world d the following are true: R^3dfe, $M,e \models A$ and $M,f \models B$. This means that $M,d \models B \bullet A$ as required.

$\Longleftarrow$

We prove the other direction by contraposition. Suppose our frame does *not* satisfy the frame constraint, that is F satisfies its negation, $\exists c \exists a \exists b.(R^3cab \,\&\, \neg R^3cba)$. In other words, there are worlds a, b and c such that R^3cab but not R^3cba. Given this frame F, we can construct a countermodel by giving a valuation v making $A \bullet B \vdash B \bullet A$ invalid: set $v(A) = \{a\}$, $v(B) = \{b\}$. We claim that $M,c \models A \bullet B$ but that $M,c \not\models B \bullet A$. In order to show that $A \bullet B$ is true at c, we need to show that

$\exists x \exists y. R^3cxy \;\&\; M,x \models A \;\&\; M,y \models B$. Choosing $x = a$ and $y = b$ makes this statement true by our chosen validation. What remains to be done is show that $M,c \not\models B \bullet A$, that is $\neg\exists x \exists y.(R^3cxy \;\&\; M,x \models B \;\&\; M,y \models A)$. We prove this by contradiction: suppose there are x and y which satisfy the three conjuncts. Since $M,x \models B$ and $M,y \models A$ our valuation forces $x = b$ and $y = a$. This means R^3cba must hold as well, contradicting $\neg R^3cba$ from our assumption that F does not satisfy the frame constraint. □

The important point of this section is that Kripke models give fairly easy soundness and completeness results not just for the non-associative Lambek calculus but also — with the appropriate frame constraints, for the Lambek calculus L — as well as for the commutative versions both calculi, NLP and LP.

4.6 Polynomial Complexity

In this section, we will talk about the complexity of parsing sentences and proving theorems for the non-associative Lambek calculus. It is important in this context to distinguish between the complexity of theorem proving and the complexity of parsing: theorem proving asks the question whether a given sequent is derivable in a calculus, whereas parsing asks the question of whether, given a sentence $w_1, \ldots, w_n$ and a lexicon Lex mapping words to formulas, there is a sequence $f_1, \ldots f_n$ such that each $f_i \in \mathrm{Lex}(w_i)$ and an antecedent term Γ with yield $f_1, \ldots, f_n$ such that $\Gamma \vdash S$ is derivable. From this, it is easy to see that parsing is at least as difficult as theorem proving.

4.6.1 Complexity

Before talking about NL, we will very briefly discuss some of the known results for other Lambek calculi: the associative Lambek calculus L and the associative, commutative Lambek-van Benthem calculus LP. We have a rather complete picture of the complexity of the different Lambek calculi NL, L and LP, both with the product • and without it.

Figure 4.9 summarizes the results. For LP, its relationship with linear logic, which we will discuss in more detail in Chapter 6, makes it possible to apply the complexity results for the multiplicative fragment of linear logic (Kanovich, 1994) directly to LP to obtain NP completeness results, both for the logic which contains only implication and for the logic with implication and conjunction. NP-completeness for the Lambek calculus with product was proved by Pentus (2006); Savateev (2009) showed that NP-completeness holds even in the case without the product formula. All these results are for theorem proving, but it is easy to see that the parsing problem has the same complexity: given that parsing is at least as difficult as theorem proving and that theorem proving is NP complete, we only need to show that parsing is in NP to show NP-completeness and it is as easy to verify whether or not a parse is successful as it it to verify whether or not a proof is valid.

Though L and LP are NP complete both for parsing and for theorem proving, there are polynomial algorithms for NL. However, there is an important difference between NL with product, for which we show *theorem proving* is polynomial and product-free NL, for which *parsing* is polynomial. We will discuss these polynomial algorithms in what follows.

Polynomial	*NP complete*	
$\mathrm{NL}^{\backslash/\bullet}$ de Groote (1999)	$\mathrm{L}^{\backslash/\bullet}$ Pentus (2006)	$\mathrm{LP}^{\backslash/\bullet}$ Kanovich (1994)
$\mathrm{NL}^{\backslash/}$ Aarts and Trautwein (1995)	$\mathrm{L}^{\backslash/}$ Savateev (2009)	$\mathrm{LP}^{\backslash/}$ Kanovich (1994)

Fig. 4.9. The complexity of different variants of the Lambek calculus

4.6.2 De Groote's Context Calculus SC

Philippe de Groote has shown that theorem proving for NL takes polynomial time (de Groote, 1999), generalizing an earlier result from Aarts and Trautwein (1995) for product-free NL which we will discuss in Section 4.6.4. In this section, we will first present de Groote's result and then talk about the special case without product.

The context calculus SC is defined using two-formula sequents and *contexts* which are defined as follows.

Definition 4.25. *A formula with a hole or a* context *is defined as follows:*

$$\mathscr{F}[] \quad ::= \quad [] \mid \mathscr{F}[] \backslash \mathsf{Lp} \mid \mathsf{Lp} \backslash \mathscr{F}[] \mid \mathscr{F}[] / \mathsf{Lp} \mid \mathsf{Lp} / \mathscr{F}[] \mid \mathscr{F}[] \bullet \mathsf{Lp} \mid \mathsf{Lp} \bullet \mathscr{F}[]$$

Though similar in spirit to the definition of context in the beginning of this chapter (Definition 4.4), where a context is an antecedent term with a hole as opposed to a formula with a hole here, the two notions are of course distinct. We will use $\Gamma[], \Delta[], \Theta[], \ldots$ to range over contexts. $\Gamma[A]$ is the formula obtained by filling the hole in $\Gamma[]$ by A.

The context calculus SC is shown in Figure 4.10.

As can be seen in the rules, we need to distinguish between positive and negative contexts. The context rules $\bullet / N$ and $\bullet \backslash N$ should be compared — modulo some manipulation of contexts — to the (*Appl*) axioms of the axiomatic presentation of Figure 4.5 on page 113, the $/ \bullet P$ and $\backslash \bullet P$ rules correspond to the (*Co-appl*) axioms, whereas the $/ \backslash P$ and $\backslash / P$ rules correspond to the type lifting rules.

Sequent Rules

$$\frac{}{A \vdash A}\ axiom$$

$$\frac{A \vdash B \quad C \vdash D}{A \bullet C \vdash B \bullet D}\ \bullet mon \qquad \frac{A \vdash B \quad C \vdash D}{B \backslash C \vdash A \backslash D}\ \backslash mon \qquad \frac{A \vdash B \quad C \vdash D}{A / D \vdash B / C}\ / mon$$

$$\frac{A \vdash B \quad \vdash_N \Gamma[\,]}{\Gamma[A] \vdash B}\ cont_N \qquad \frac{A \vdash B \quad \vdash_P \Gamma[\,]}{A \vdash \Gamma[B]}\ cont_P$$

Negative Context Rules

$$\frac{}{\vdash_N [\,]}\ [\,]N$$

$$\frac{A \vdash B \quad \vdash_N \Gamma[\,] \quad \vdash_N \Delta[\,]}{\vdash_N (A \bullet \Gamma[(B \backslash \Delta[\,])])}\ \bullet\backslash N \qquad \frac{A \vdash B \quad \vdash_N \Gamma[\,] \quad \vdash_N \Delta[\,]}{\vdash_N (\Gamma[(\Delta[\,] / B)] \bullet A)}\ \bullet / N$$

Positive Context Rules

$$\frac{}{\vdash_P [\,]}\ [\,]P$$

$$\frac{A \vdash B \quad \vdash_P \Gamma[\,] \quad \vdash_P \Delta[\,]}{\vdash_P (A \backslash \Gamma[(B \bullet \Delta[\,])])}\ \backslash\bullet P \qquad \frac{A \vdash B \quad \vdash_P \Gamma[\,] \quad \vdash_P \Delta[\,]}{\vdash_P (\Gamma[(\Delta[\,] \bullet B)] / A)}\ / \bullet P$$

$$\frac{B \vdash A \quad \vdash_N \Gamma[\,] \quad \vdash_P \Delta[\,]}{\vdash_P (A / \Gamma[(\Delta[\,] \backslash B)])}\ / \backslash P \qquad \frac{B \vdash A \quad \vdash_N \Gamma[\,] \quad \vdash_P \Delta[\,]}{\vdash_P (\Gamma[(B / \Delta[\,])] \backslash A)}\ \backslash / P$$

Fig. 4.10. The context calculus SC from de Groote (1999)

Before analyzing the complexity of proof search in the calculus SC, we first need to show that it is equivalent to NL. That is. we need to show that SC derives all and only the theorems of NL. To do this, we follow the structure of the proof of de Groote (1999) and first introduce some auxiliary notions.

Definition 4.26. *A negative context* $\Gamma[\,]$ *is* correct *iff* $A \vdash B$ *implies* $\Gamma[A] \vdash B$.
A positive context $\Delta[\,]$ *is* correct *iff* $A \vdash B$ *implies* $A \vdash \Delta[B]$.

The following lemma shows that we can "nest" contexts.

Lemma 4.27. *If* $\vdash_N \Gamma[]$ *and* $\vdash_N \Delta[]$ *are derivable using the negative context rules, then* $\vdash_N \Gamma[\Delta[]]$ *is derivable as well.*

If $\vdash_P \Gamma[]$ *and* $\vdash_P \Delta[]$ *are derivable using the positive context rules, then* $\vdash_P \Gamma[\Delta[]]$ *is derivable as well.*

Proof. Both parts of the lemma are proved by a simple induction on $\vdash_N \Gamma[]$ resp. $\vdash_P \Gamma[]$. □

For the completeness proof, it is useful to restrict the SC derivation into those that have a certain form.

Definition 4.28. *An SC derivation is* normal *if the following conditions are satisfied.*

1. *The rule axiom is restricted to atomic formulae.*
2. *No rule* $cont_N$ *or* $cont_P$ *has an axiom* $[\,]N$ *resp.* $[\,]P$ *as its right premise.*
3. *No rule* $cont_N$ *or* $cont_P$ *has another rule* $cont_N$ *or* $cont_P$ *as its left premise.*

Lemma 4.29. *Every SC derivation can be transformed into a normal SC derivation.*

Proof. Condition 1 corresponds to the possibility to restrict NL to atomic axioms. In the calculus SC, the proof is even easier than in the sequent calculus, and follows directly from the monotonicity rules for the three connectives of the calculus.

Condition 2 is trivial as well. If the right premise on a *cont* rule is a context axiom, then the context is empty and the conclusion of the rule is equal to the left premise and the *cont* rule can be eliminated from the proof.

Condition 3, finally, is a direct corollary of Lemma 4.27. □

Lemma 4.30. *SC is equivalent to the monotonicity-residuation presentation of NL shown in Figure 4.7 on page 114.*

Proof. We show that a sequent is derivable in the context calculus SC if and only if the corresponding sequent is derivable in the combinatorial presentation ResMon. Our work is made easier by the fact that the two calculi share the monotonicity and axiom rules.

$\Longrightarrow$

We show by induction on the length of the SC proof of $A \vdash B$ that there is a derivation in the residuation-monotonicity calculus of $A \vdash B$ as well.

The *axiom* rule and the three *mon* rules are identical to the *Id* and monotonicity rules of the residuation-monotonicity calculus.

The negative and positive context rules are valid by the definition of correct contexts.

So, what remains to be shown is that the context rules allow us to derive only correct contexts.

Negative contexts

For a negative context $\Gamma[]$, being correct means that if $A \vdash B$, then $\Gamma[A] \vdash B$.

- If $\vdash_N \Gamma[]$ is obtained by the axiom for negative contexts $[\,]N$, the result holds trivially.
- If $\vdash_N \Theta[]$ is obtained by the $\bullet\backslash N$ rule, then Θ is of the form $A \bullet \Gamma[(B \backslash \Delta[])]$ and, by induction hypothesis, we have a ResMon proof of $A \vdash B$ and we know that $\Gamma[]$ and $\Delta[]$ are correct negative contexts. We need to show that $\Theta[]$ is a correct negative context as well, that is, if $C \vdash D$ then $A \bullet \Gamma[(B \backslash \Delta[C])] \vdash D$.

  Now, if $C \vdash D$, then, given that $\Delta[]$ is a correct negative context by induction hypothesis, we can conclude $C \vdash \Delta[D]$. $C \vdash \Delta[D]$ and $A \vdash B$ (induction hypothesis) together allow us to conclude $B \backslash \Delta[C] \vdash A \backslash D$ by $(Mon\backslash)$. This, together with the fact that $\Gamma[]$ is a correct positive context, allows us to conclude $\Gamma[(B \backslash \Delta[C])] \vdash A \backslash D$, followed by $(Res_{\backslash\bullet})$ to conclude $A \bullet \Gamma[(B \backslash \Delta[C])] \vdash D$ as required.

$$\dfrac{\dfrac{\dfrac{\overset{\vdots\, IH}{A \vdash B} \qquad \dfrac{C \vdash D}{\Delta[C] \vdash D}\,\Delta[]\ correct}{B \backslash \Delta[C] \vdash A \backslash D}\,(Mon_{\backslash})}{\Gamma[(B \backslash \Delta[C])] \vdash A \backslash D}\,\Gamma[]\ correct}{A \bullet \Gamma[(B \backslash \Delta[C])] \vdash D}\,(Res_{\backslash\bullet})$$

- The case for $\bullet / N$ is symmetric.

Positive contexts

If $\Delta[]$ is a positive context such that $\vdash_P \Delta[]$ and $A \vdash B$ is derivable in ResMon, then $A \vdash \Delta[B]$ is derivable in ResMon as well. We use induction on the proof of $\vdash_P \Delta[]$.

- If $\vdash_P \Delta[]$ is obtained by the axiom for negative contexts $[\,]P$, the result holds trivially.
- If $\vdash_P \Theta[]$ is obtained by the $\backslash\bullet P$ rule, then $\Theta[]$ is of the form $A \backslash \Gamma[(B \bullet \Delta[])]$, we know by induction hypothesis that $\Gamma[]$ and $\Delta[]$ are correct positive contexts and that $A \vdash B$ is derivable. In order to show that $\Theta[]$ is a correct context, we need to show that if $C \vdash D$ then $C \vdash A \backslash \Gamma[(B \bullet \Delta[D])]$. Given that we know by induction hypothesis that $\Delta[]$ is a correct positive context, $C \vdash D$ allows us to conclude $C \vdash \Delta[D]$, which together with $A \vdash B$ (induction hypothesis) gives us $A \bullet C \vdash B \bullet \Delta[D]$ using $Mon_\bullet$. Given that $\Gamma[]$ is a correct positive context as well, this allows us to conclude $A \bullet C \vdash \Gamma[(B \bullet \Delta[D])]$. Finally, the residuation rule $Res_{\bullet\backslash}$ allows us to conclude $C \vdash A \backslash \Gamma[(B \bullet \Delta[D])]$ as required. The figure below summarizes the proof.

$$\dfrac{\dfrac{\vdots\, IH}{A \vdash B} \quad \dfrac{C \vdash D}{C \vdash \Delta[D]}\,\Delta[]\ correct}{\dfrac{A \bullet C \vdash B \bullet \Delta[D]}{\dfrac{A \bullet C \vdash \Gamma[(B \bullet \Delta[D])]}{C \vdash A \setminus \Gamma[(B \bullet \Delta[D])]}\,Res_{\bullet\setminus}}\,\Gamma[]\ correct}\,Mon_{\bullet}$$

- The case for $/\bullet P$ is symmetric.
- In case $\vdash_P \Theta[]$ is obtained by the $/\setminus P$ rule, $\Theta[]$ is of the form $A / \Gamma[(\Delta[] \setminus B)]$ and we need to show that $\Theta[]$ is a correct positive context, that is, if $C \vdash D$, then $C \vdash A / \Gamma[(\Delta[D] \setminus B)]$ given that $\Gamma[]$ is a correct negative context, $\Delta[]$ is a correct positive context and $B \vdash A$, all by induction hypothesis.

  If $C \vdash D$ and if $\Delta[]$ is a valid positive context, then $C \vdash \Delta[D]$, which together with $B \vdash A$ and $mon_{\setminus}$ gives $\Delta[D] \setminus B \vdash C \setminus A$. This, together with the fact that $\Gamma[]$ is a correct negative context, allows us to conclude $\Gamma[(\Delta[D] \setminus B)] \vdash C \setminus A$. From this, using $Res_{\setminus\bullet}$ we can conclude $C \bullet \Gamma[(\Delta[D] \setminus B)] \vdash A$ and finally, using $Res_{\bullet/}$, $C \vdash A / \Gamma[(\Delta[D] \setminus B)]$ as required.

$$\dfrac{\dfrac{\dfrac{\dfrac{\dfrac{C \vdash D}{C \vdash \Delta[D]}\,\Delta[]\ correct \quad \dfrac{\vdots\, IH}{B \vdash A}}{\Delta[D] \setminus B \vdash C \setminus A}\,mon_{\setminus}}{\Gamma[(\Delta[D] \setminus B)] \vdash C \setminus A}\,\Gamma[]\ correct}{C \bullet \Gamma[(\Delta[D] \setminus B)] \vdash A}\,Res_{\setminus\bullet}}{C \vdash A / \Gamma[(\Delta[D] \setminus B)]}\,Res_{\bullet/}$$

- The case for $\setminus / P$ is symmetric.

$\Longleftarrow$

Assume we have a ResMon derivation of $A \vdash B$, we show by induction on the length of the proof that there is a corresponding normal SC derivation.

We proceed by a case analysis on the last rule of the proof. The only cases which need some work are the four *Res* rules.

($Res_{\setminus\bullet}$) Suppose the last rule in the ResMon derivation is the $Res_{\setminus\bullet}$ rule.

$$\frac{B \vdash A \setminus C}{A \bullet B \vdash C}\,(Res_{\setminus\bullet})$$

By induction hypothesis, we know there is a normal SC derivation δ of $B \vdash A \setminus C$. Looking at the form of the rules, only the three following rules can have produced the sequent $B \vdash A \setminus C$: $mon\setminus$, $cont_P$ or $cont_N$. For the two *cont* rules, a further case analysis is necessary.

($mon\backslash$) If the last rule of the normal SC derivation was the monotonicity rule for $\backslash$, then we are in the following situation

$$\dfrac{\dfrac{\vdots\ \delta_1}{A \vdash B} \quad \dfrac{\vdots\ \delta_2}{D \vdash C}}{B \backslash D \vdash A \backslash C}\ mon\backslash$$

We can combine proofs δ_1 and δ_2 to produce a proof of $A \bullet (B \backslash D) \vdash C$ as follows.

$$\dfrac{\dfrac{\vdots\ \delta_2}{D \vdash C} \quad \dfrac{\dfrac{\vdots\ \delta_1}{A \vdash B} \quad \dfrac{}{\vdash_N [\,]}\,[\,]N \quad \dfrac{}{\vdash_N [\,]}\,[\,]N}{\vdash_N A \bullet (B \backslash [\,])}}{A \bullet (B \backslash D) \vdash C}\ cont_N$$

Though this new proof is not necessarily normal (the last rule of δ_2 may be $cont_N$), we can transform it into a normal proof by Lemma 4.29, which completes this case.

($cont_N$) If the last rule of the SC derivation is $cont_N$, then we are schematically in the following situation.

$$\dfrac{\dfrac{\vdots\ \delta_1}{B \vdash A \backslash C} \quad \dfrac{\vdots\ \delta_2}{\vdash_N \Gamma[]}}{\Gamma[B] \vdash A \backslash C}\ cont_N$$

We do a further case analysis on the last rule of δ_1: it can be either $mon\backslash$ or $cont_P$ ($cont_N$ and *axiom* are excluded because the proof is normal).

- If the last rule is $cont_P$, the proof is of the following form

$$\dfrac{\dfrac{\dfrac{\vdots\ \delta_1'}{A \vdash B} \quad \dfrac{\vdots\ \delta_1''}{\vdash_P \Delta[]}}{A \vdash \Delta[B]}\ cont_P \quad \dfrac{\vdots\ \delta_2}{\vdash_N \Gamma[]}}{\Gamma[A] \vdash \Delta[B]}\ cont_N$$

  and we simply swap the $cont_N$ and $cont_P$ rules as follows,

$$\dfrac{\dfrac{\dfrac{\vdots\ \delta_1'}{A \vdash B} \quad \dfrac{\vdots\ \delta_2}{\vdash_N \Gamma[]}}{\Gamma[A] \vdash B}\ cont_N \quad \dfrac{\vdots\ \delta_1''}{\vdash_P \Delta[]}}{\Gamma[A] \vdash \Delta[B]}\ cont_P$$

  normalize the resulting derivation if necessary and apply the appropriate case for $cont_P$ below.

- If the last rule of δ_1 is $mon\backslash$, then we are in the following situation

$$\dfrac{\dfrac{\begin{array}{c}\vdots\,\delta_1'\\ A\vdash B\end{array}\quad\begin{array}{c}\vdots\,\delta_1''\\ D\vdash C\end{array}}{B\backslash D\vdash A\backslash C}\,mon\backslash\qquad\begin{array}{c}\vdots\,\delta_2\\ \vdash_N\Gamma[]\end{array}}{\Gamma[B\backslash D]\vdash A\backslash C}\,cont_N$$

and need to prove $A\bullet\Gamma[B\backslash D]\vdash C$, which we can do as follows.

$$\dfrac{\begin{array}{c}\vdots\,\delta_1''\\ D\vdash C\end{array}\qquad\dfrac{\begin{array}{c}\vdots\,\delta_1'\\ A\vdash B\end{array}\quad\begin{array}{c}\vdots\,\delta_2\\ \vdash_N\Gamma[]\end{array}\quad\dfrac{}{\vdash_N[\,]}\,[\,]N}{\vdash_N A\bullet\Gamma[(B\backslash[\,])]}\,\bullet\backslash N}{A\bullet\Gamma[B\backslash D]\vdash C}\,cont_N$$

($cont_P$) If the last rule of the SC derivation is $cont_P$, then since the positive context $\Gamma[]$ will have $\backslash$ as its main connective, derivation of the positive context $\Gamma[]$ will have as its last rule either $\backslash\bullet P$ or $\backslash/P$. We will consider each case in turn.

- In the first case, we are in are in the following situation

$$\dfrac{\begin{array}{c}\vdots\,\delta_1\\ B\vdash C\end{array}\qquad\dfrac{\begin{array}{c}\vdots\,\delta_2\\ A\vdash D\end{array}\quad\begin{array}{c}\vdots\,\delta_3\\ \vdash_P\Gamma[]\end{array}\quad\begin{array}{c}\vdots\,\delta_4\\ \vdash_P\Delta[]\end{array}}{\vdash_P A\backslash\Gamma[(D\bullet\Delta[])]}\,\backslash\bullet P}{B\vdash A\backslash\Gamma[(D\bullet\Delta[C])]}\,cont_P$$

and we need to prove $A\bullet B\vdash\Gamma[(D\bullet\Delta[C])]$, which we can do as follows.

$$\dfrac{\dfrac{\begin{array}{c}\vdots\,\delta_2\\ A\vdash D\end{array}\qquad\dfrac{\begin{array}{c}\vdots\,\delta_1\\ B\vdash C\end{array}\quad\begin{array}{c}\vdots\,\delta_4\\ \vdash_P\Delta[]\end{array}}{B\vdash\Delta[C]}\,cont_P}{A\bullet B\vdash D\bullet\Delta[C]}\,mon\bullet\qquad\begin{array}{c}\vdots\,\delta_3\\ \vdash_P\Gamma[]\end{array}}{A\bullet B\vdash\Gamma[(D\bullet\Delta[C])]}\,cont_P$$

- In the second case, we have $cont_P$ followed by $\backslash/P$ and are in the following situation.

$$\dfrac{\begin{array}{c}\vdots\,\delta_1\\ B\vdash A\end{array}\qquad\dfrac{\begin{array}{c}\vdots\,\delta_2\\ D\vdash C\end{array}\quad\begin{array}{c}\vdots\,\delta_3\\ \vdash_N\Gamma[]\end{array}\quad\begin{array}{c}\vdots\,\delta_4\\ \vdash_P\Delta[]\end{array}}{\vdash_P\Gamma[(D/\Delta[])]\backslash C}\,\backslash/P}{B\vdash\Gamma[(D/\Delta[A])]\backslash C}\,cont_P$$

We need to give a proof of $\Gamma[(D\,/\,\Delta[A])]\bullet B\vdash C$, which we can do as follows.

$$\dfrac{\begin{array}{c}\vdots\,\delta_2\\ D\vdash C\end{array} \quad \dfrac{\dfrac{\begin{array}{c}\vdots\,\delta_1\\ B\vdash C\end{array}\quad \begin{array}{c}\vdots\,\delta_4\\ \vdash_P \Delta[]\end{array}}{B\vdash \Delta[A]}\,cont_P \quad \begin{array}{c}\vdots\,\delta_3\\ \vdash_N \Gamma[]\end{array}\quad \dfrac{}{\vdash_N []}\,[]N}{\vdash_N \Gamma[([]/\Delta[A])\bullet B}\,\bullet/N}{\Gamma[(D\,/\,\Delta[A])]\bullet B\vdash C}\,cont_N$$

($Res_{/\bullet}$) Symmetric to the case for $Res_{\backslash\bullet}$

($Res_{\bullet\backslash}$) We need to show that whenever we have a normal proof δ of $A\bullet B\vdash C$, we can transform it into a proof of $B\vdash A\backslash C$. We do a case analysis on the last rule of δ as before. The only rules which can have applied to form a sequent $A\bullet B\vdash C$ are the monotonicity rule for the product, the positive context rule and the negative context rule, with the context rules requiring a further case analysis. We treat the cases in the indicated order.

($\bullet mon$) In case the last rule was $\bullet mon$, we have a proof of the following form.

$$\dfrac{\begin{array}{c}\vdots\,\delta_1\\ A\vdash C\end{array}\quad \begin{array}{c}\vdots\,\delta_2\\ B\vdash D\end{array}}{A\bullet B\vdash C\bullet D}\,\bullet mon$$

We can combine the subproofs into a proof of $B\vdash A\backslash(C\bullet D)$ as follows.

$$\dfrac{\begin{array}{c}\vdots\,\delta_2\\ B\vdash D\end{array}\quad \dfrac{\begin{array}{c}\vdots\,\delta_1\\ A\vdash C\end{array}\quad \dfrac{}{\vdash_P []}\,[]P \quad \dfrac{}{\vdash_P []}\,[]P}{\vdash_P A\backslash(C\bullet[])}\,\backslash\bullet P}{B\vdash A\backslash(C\bullet D)}\,cont_P$$

($cont_P$) If the last rule of the proof is $cont_P$, then we are in the following situation.

$$\dfrac{\begin{array}{c}\vdots\,\delta_1\\ A\bullet B\vdash C\end{array}\quad \begin{array}{c}\vdots\,\delta_2\\ \vdash_P \Delta[]\end{array}}{A\bullet B\vdash \Delta[C]}\,cont_P$$

We do a further case analysis on the last rule of the subproof δ_1, which can be either $\bullet mon$ or $cont_N$ ($cont_P$ is excluded because the proof is normal). We treat both subcases.

- If the last rule of δ_1 is $\bullet mon$, then the proof looks as follows.

$$\dfrac{\dfrac{\dfrac{\vdots\,\delta_1'}{A \vdash C} \quad \dfrac{\vdots\,\delta_1''}{B \vdash D}}{A \bullet B \vdash C \bullet D}\,\bullet mon \qquad \dfrac{\vdots\,\delta_2}{\vdash_P \Delta[]}}{A \bullet B \vdash \Delta[C \bullet D]}\,cont_P$$

We need to combine the subproofs to create a proof of $B \vdash A \backslash \Gamma[C \bullet D]$, which we can do as follows.

$$\dfrac{\dfrac{\vdots\,\delta_1''}{B \vdash D} \qquad \dfrac{\dfrac{\vdots\,\delta_1'}{A \vdash C} \quad \dfrac{\vdots\,\delta_2}{\vdash_P \Delta[]} \quad \dfrac{}{\vdash_P []}\,[]P}{\vdash_P A \backslash \Delta[(C \bullet [])]}\,\backslash \bullet P}{B \vdash A \backslash \Delta[C \bullet D]}\,cont_P$$

- If the last rule of δ_1 is $cont_N$, then the proof looks as follows.

$$\dfrac{\dfrac{\dfrac{\vdots\,\delta_1'}{A \vdash B} \quad \dfrac{\vdots\,\delta_1''}{\vdash_N \Gamma[]}}{\Gamma[A] \vdash B}\,cont_N \qquad \dfrac{\vdots\,\delta_2}{\vdash_P \Delta[]}}{\Gamma[A] \vdash \Delta[B]}\,cont_P$$

and we swap the $cont_P$ and $cont_N$ rules, similar to the way in which we treated the situation in the $Res \backslash \bullet$ section of the proof, as follows

$$\dfrac{\dfrac{\dfrac{\vdots\,\delta_1'}{A \vdash B} \quad \dfrac{\vdots\,\delta_2}{\vdash_P \Delta[]}}{A \vdash \Delta[B]}\,cont_P \qquad \dfrac{\vdots\,\delta_1''}{\vdash_N \Gamma[]}}{\Gamma[A] \vdash \Delta[B]}\,cont_N$$

normalize, then continue with case $cont_N$ described below.

($cont_N$) For the case $cont_N$, we know the negative context must have $\bullet$ as its outermost symbol. This means only rules $\bullet \backslash N$ and $\backslash \bullet N$ can apply.

- Suppose our proof ends with the combination $\bullet \backslash N$, $cont_N$ as shown below

$$\dfrac{\dfrac{\vdots\,\delta_1}{D \vdash C} \qquad \dfrac{\dfrac{\vdots\,\delta_2}{A \vdash B} \quad \dfrac{\vdots\,\delta_3}{\vdash_N \Gamma[]} \quad \dfrac{\vdots\,\delta_4}{\vdash_N \Delta[]}}{\vdash_N A \bullet \Gamma[(B \backslash \Delta[])]}\,\bullet \backslash N}{A \bullet \Gamma[(B \backslash \Delta[D])] \vdash C}\,cont_N$$

then we can transform it into a proof of $\Gamma[(B \backslash \Delta[D])] \vdash A \backslash C$ as follows.

$$\dfrac{\dfrac{\begin{matrix}\vdots\,\delta_2\\ A\vdash B\end{matrix} \qquad \dfrac{\begin{matrix}\vdots\,\delta_1\\ D\vdash C\end{matrix} \quad \begin{matrix}\vdots\,\delta_4\\ \vdash_N \Delta[]\end{matrix}}{\Delta[D]\vdash C}\,cont_N}{B\backslash\Delta[D]\vdash A\backslash C}\,Mon\backslash \qquad \begin{matrix}\vdots\,\delta_3\\ \vdash_N \Gamma[]\end{matrix}}{\Gamma[(B\backslash\Delta[D])]\vdash A\backslash C}\,cont_N$$

- Finally, suppose our proof ends with the combination $\bullet/N$, $cont_N$. This means the proof ends as follows.

$$\dfrac{\begin{matrix}\vdots\,\delta_1\\ A\vdash C\end{matrix} \qquad \dfrac{\begin{matrix}\vdots\,\delta_2\\ B\vdash D\end{matrix} \quad \begin{matrix}\vdots\,\delta_3\\ \vdash_N \Gamma[]\end{matrix} \quad \begin{matrix}\vdots\,\delta_4\\ \vdash_N \Delta[]\end{matrix}}{\vdash_N \Gamma[(\Delta[]\,/\,D)]\bullet B}\,\bullet/N}{\Gamma[(\Delta[A]\,/\,D)]\bullet B\vdash C}\,cont_N$$

We transform it, as required, into a proof of $B\vdash \Gamma[(\Delta[A]\,/\,D)]\backslash C$ as follows.

$$\dfrac{\begin{matrix}\vdots\,\delta_2\\ B\vdash D\end{matrix} \qquad \dfrac{\dfrac{\begin{matrix}\vdots\,\delta_1\\ A\vdash C\end{matrix} \quad \begin{matrix}\vdots\,\delta_4\\ \vdash_N \Delta\end{matrix}}{\Delta[A]\vdash C}\,cont_N \quad \begin{matrix}\vdots\,\delta_3\\ \vdash_N \Gamma[]\end{matrix} \quad \dfrac{}{\vdash_P []}\,[]P}{\vdash_P \Gamma[(\Delta[A]\,/\,[])]\backslash C}\,\backslash/P}{B\vdash \Gamma[(\Delta[A]\,/\,D)]\backslash C}\,cont_P$$

$(Res_{\bullet/})$ Symmetric to the case for $Res_{\bullet\backslash}$ □

4.6.3 A Theorem Proving Algorithm

Now that we have established the equivalence of the context calculus SC with the combinator presentation of NL, we will take some time to give a very rough upper bound on the complexity of proof search for SC.

We first define the size of formulae, contexts and sequents. This corresponds simply to the number of symbols other than '(' and ')' we use to write it down. The context marker '[]' is counted as a single symbol.

Definition 4.31. *Let F be a formula or a context, the* size *of F is defined as follows.*

$$\begin{aligned} size([]) &= 1\\ size(p) &= 1\\ size(A\bullet B) &= size(A)+size(B)+1\\ size(A\backslash B) &= size(A)+size(B)+1\\ size(A\,/\,B) &= size(A)+size(B)+1 \end{aligned}$$

The size *of a sequent* $A\vdash B$ *is defined as* $size(A)+size(B)$.

By inspection of the rules, it is clear that in any SC proof of a sequent $A \vdash B$, the formulae and the contexts which can appear in the proof are all subformulae or subcontexts of the sequent $A \vdash B$.

Now for a given sequent $A \vdash B$ of size n, the number of subformulae is bounded by n and the number of subcontexts is bounded by n^2: if we write the sequent as a tree, this tree will have n nodes. Apart from the root node, which corresponds to the sequent symbol, each node in the tree corresponds to a subformula and each pair of nodes such that the second node is a descendant of this first corresponds to a context.

So a proof search algorithm will require $O(n^2)$ space to store all contexts and pairs of formulae which can appear in an SC proof. The following is a very naive tabular search algorithm which decides whether or not a sequent $A \vdash B$ is derivable. It is not difficult to make the algorithm a bit smarter, for example by taking into account the polarities of formulae and contexts and by storing only sequents and contexts which have an equal number of positive and negative occurrences of all atomic formulae.

1. Given a sequent $A \vdash B$, store all pairs of subformulae and all subcontexts of $A \vdash B$.
2. Mark all instantiations of the axioms and all empty contexts as derivable.
3. For each rule which has all its premises marked as derivable but which does not have its conclusion sequent/context marked as derivable, mark its conclusion sequent or its conclusion context as derivable.
4. Repeat step 3 until no further derivable sequents/contexts are added.
5. Answer "yes" if the sequent $A \vdash B$ is marked as derivable and "no" otherwise.

For each rule, except the $cont_N$ and $cont_P$ rules, the size of the conclusion is equal to the size of its premises plus two. This gives us a maximum of $\frac{1}{2}n$ rule applications other than *cont* to produce a sequent of size n. In addition, in a normal proof there are never two consecutive $cont_N$ rules and never two consecutive $cont_P$ rules, bounding the number of *cont* rules by two thirds of the total number of rules, giving us a maximum of $\frac{3}{2}n$ iterations of step 3 of the algorithm. Now there are at most $O(n^2)$ sequents and contexts which are marked as derivable and 11 rules (all rules except the three axioms) for which we have to verify if:

1. this context is one of its premises,
2. the other premises are marked as derivable,
3. the conclusion is in the search space but not yet marked as derivable.

All of the above steps take $O(1)$. So we can conclude that it takes $O(\frac{3}{2}n * 11 * n^2) = O(n^3)$ steps to decide whether or not a given sequent is derivable using the context calculus.

4.6.4 NL without Product

Aarts and Trautwein (1995) have given an earlier proof of polynomial *parsing* for product-free NL. Figure 4.11 shows the calculus proposed by Aarts and Trautwein.

$$\frac{\Gamma, B, \Gamma' \vdash D \quad C \vdash A}{\Gamma, C, A \backslash B, \Gamma' \vdash D} \backslash'_h \qquad \frac{\Gamma, B, \Gamma' \vdash D \quad C \vdash A}{\Gamma, B / A, C, \Gamma' \vdash D} /'_h$$

$$\frac{A \vdash B \quad C \vdash D}{A \vdash (D / B) \backslash C} \backslash lift \qquad \frac{A \vdash B \quad C \vdash D}{A \vdash D / (B \backslash C)} / lift$$

$$\frac{A \vdash B \quad C \vdash D}{B \backslash C \vdash A \backslash D} \backslash mon \qquad \frac{A \vdash B \quad C \vdash D}{A / D \vdash B / C} / mon$$

$$\frac{}{A \vdash A} \; axiom$$

Fig. 4.11. Calculus for product-free NL (from Aarts and Trautwein (1995))

A remarkable feature of this calculus is that, as is clear from the formulation of the $\backslash'_h$ and $/'_h$ rules, it works using *lists* of formulae: it does not require the bracketing of the antecedent term as its input and, given a parse in the calculus of Aarts and Trautwein, we can easily extract the antecedent term simply by adding parentheses around the two formulae for each application of an $\backslash'_h$ and $/'_h$ rule.

From the perspective of the context calculus we have seen in the previous section, the product-free calculus corresponds to the following restriction on formulae

$$C \quad ::= \quad (C \bullet C) \;\Big|\; F$$

$$F \quad ::= \quad \mathsf{P} \;\Big|\; F \backslash F \;\Big|\; F / F$$

with sequents being of the form $C \vdash F$. C is simply an antecedent term written using product formulae instead of parentheses and comma's in order to make the comparison with SC more evident. So the resulting calculus still has product formulae, but only on the outside (not inside the scope of one of the implications).

Keeping this formula restriction in mind when looking at the rules for the calculus SC (Figure 4.10 on page 131), we notice the following:

- the $\bullet mon$ rule can never apply, since we cannot have a product formula as the succedent,
- neither the $\backslash \bullet P$ nor the $/ \bullet P$ rule can apply, since they necessarily have a product formula as a subformula of an implication
- for the $\bullet \backslash N$ and the $\bullet / N$ rules, the context $\Delta[]$ must be empty, since a negative context always has a product formula as its main connective and $\Delta[]$ occurs as a subformula of an implication

Sequent Rules

$$\frac{}{A \vdash A}\,axiom$$

$$\frac{A \vdash B \quad C \vdash D}{B \backslash C \vdash A \backslash D}\,\backslash mon \qquad \frac{A \vdash B \quad C \vdash D}{A / D \vdash B / C}\,/mon$$

$$\frac{A \vdash B \quad \vdash_N \Gamma[\,]}{\Gamma[A] \vdash B}\,cont_N \qquad \frac{A \vdash B \quad \vdash_P \Gamma[\,]}{A \vdash \Gamma[B]}\,cont_P$$

Negative Context Rules

$$\frac{}{\vdash_N [\,]}\,[\,]N$$

$$\frac{A \vdash B \quad \vdash_N \Gamma[\,]}{\vdash_N (A \bullet \Gamma[(B \backslash [\,])])}\,\bullet\backslash N \qquad \frac{A \vdash B \quad \vdash_N \Gamma[\,]}{\vdash_N (\Gamma[([\,] / B)] \bullet A)}\,\bullet/N$$

Positive Context Rules

$$\frac{}{\vdash_P [\,]}\,[\,]P$$

$$\frac{B \vdash A \quad \vdash_P \Delta[\,]}{\vdash_P (A / (\Delta[\,] \backslash B))}\,/\backslash P \qquad \frac{B \vdash A \quad \vdash_N \vdash_P \Delta[\,]}{\vdash_P ((B / \Delta[\,]) \backslash A)}\,\backslash/P$$

Fig. 4.12. The context calculus SC from Figure 4.10 on page 131 with product formulae occurring only on the outside

- for the $\backslash/P$ and $/\backslash P$ rules, the negative context $\Gamma[\,]$ must be empty by the same reasoning.

If we take all of these restrictions into account, the reduced calculus SC for NL without product is shown in Figure 4.12.

In order to show the equivalence of the two systems, we only need to show the following:

1. the positive context rules correspond to $\backslash$*lift* and /*lift*,
2. the negative context rules correspond to $\backslash_h'$ and $/_h'$.

Item 1 is trivial; item 2 is easy to see once we realize that the two calculi work in the opposite direction: in de Groote's SC, the argument of the implication which is reduced first is always one of the outermost formulae A of the sequent, whereas in

Aarts and Trautwein's calculus the argument which is reduced first is always of of the innermost arguments C.

From the Aarts and Trautwein calculus, it is rather easy to see that (product-free) NL generates only context-free languages. Using a strategy similar to the one used for parsing AB grammars in Section 1.4, we can generate a context-free grammar which has all words in the lexicon as terminal symbols, all formulas in our grammar as non-terminal symbols and which has as its rules:

- a rule $F \to w$ if $F \in \mathrm{Lex}(w)$,
- all instances of the proof rules in Aarts and Trautwein's calculus which the formulas in the lexicon allow, eg. if $(S\,/\,(np \setminus S)) \setminus S$ is a member of $\mathrm{Lex}(\mathrm{is\_missing})$, then the following rules will be generated.

$$
\begin{array}{l}
((S\,/\,(np \setminus S)) \setminus S) \to \mathrm{is\_missing} \\
(np \setminus S) \to ((S\,/\,(np \setminus S)) \setminus S) \\
S \to (S\,/\,(np \setminus S))\ ((S\,/\,(np \setminus S)) \setminus S) \\
S \to np\ (np \setminus S)
\end{array}
$$

Though the *lift* and *mon* rules mean that the resulting context-free grammar is not in Chomsky Normal Form (the second rule shown above is an example), we can apply the standard conversion to obtain a context-free grammar in Chomsky Normal Form and apply the Cocke Kasami Younger algorithm to parse the resulting grammar.

Kandulski (1988) shows that NL with product generates only context-free languages as well.

4.7 Concluding Remarks

This completes our overview of NL. Compared to the Lambek calculus, it offers both advantages and disadvantages: there are some cases where associativity is undesirable, but other cases where it seems necessary. From a computational point of view, parsing NL grammars is simpler than parsing L grammars (though Pentus' result (Pentus, 1997), which we treated in detail in Section 2.11, shows that for a *fixed* L grammar, we can convert it to a context-free grammar and benefit from polynomial parsing).

The next chapter shows how we can combine NL and L into a single logic, to obtain a *multimodal* grammar. In addition to allowing us to specify lexically whether or not we want associativity to apply, multimodal grammars allow us to give an account of phenomena which do not have a satisfactory treatment in context-free grammars.

Exercises for Chapter 4

Exercise 4.1. Verify yourself that any proof attempt of the sequent $(A\,/\,B, B\,/\,C) \vdash A\,/\,C$ which starts with a rule application other than the $/_i$ rule produces a sequent which is invalid according to Proposition 2.6. In other words, show that for each of these rule applications there is a subproof where the number of positive occurrences of one of the formulae A, B and C is not equal to the number of negative occurrences. In the derivation shown in Example 4.8 on page 104, the two failing subsequents $(A,C) \vdash A$ has negative occurrence of C and no positive occurrences, whereas $B\,/\,C \vdash B$ has one positive occurrence of C and no negative occurrences.

Exercise 4.2. Which of the following sequents — all derivable in L — are derivable in NL as well?

1. $A\,/\,B \vdash (A\,/\,C)\,/\,(B\,/\,C)$
2. $A \vdash B\,/\,(A\,\backslash\,B)$
3. $(A\,/\,B)\bullet B \vdash A$
4. $A\,\backslash\,(B\,\backslash\,C) \vdash (B\bullet A)\,\backslash\,C$

Give a proof of all derivable sequents and show in case of underivability how all proof attempts fail.

Exercise 4.3. Using the lexicon in Section 4.2.2 on page 105, derive sentences 4.1 and 4.2 in NL and sentence 4.3 in L.

Exercise 4.4. In Section 2.5, we have seen that L requires us to state explicitly that none of the antecedents are empty. Take the following lexicon.

Word	Type(s)
very	$(n\,/\,n)\,/\,(n\,/\,n)$
interesting	$n\,/\,n$
book	n

Show that in NL "very interesting book" is derivable as an expression of type n, but "very book" isn't.

Exercise 4.5. The Italian lexicon we've seen in Example 2.2 — the relevant part is repeated below —

Word	Type(s)
cosa	$(S\,/\,(S\,/\,np))$
guarda	$(S\,/\,inf)$
passare	$(inf\,/\,np)$

allows us to derive "cosa guarda passare" in L. Show that this sentence is underivable in NL.

Exercise 4.6. Exercise 4.5 has shown that we cannot treat peripheral extraction in the same simple and elegant way we used for the Lambek calculus. Give a type assignment to *cosa* in the previous exercise which makes to sentence "cose guarda passare" derivable. Comment on this type assignment. Would it allow you to treat the sentences of Exercise 2.7?

Exercise 4.7. Example 4.5 "Bill gave flowers to Mary and a toy to the children" (on page 105) is not derivable in NL under the type assignment of $((np \backslash S) / pp) / np$ to "gave". Give a type assignment to "gave" and a type assignment to "and" which is an instantiation of $(X \backslash X) / X$ which allow us to derive Example 4.5 in NL.

Exercise 4.8. In Section 4.3 we have seen the following patterns of adverbs.

(4.13) Loren carefully read Neuromancer.
(4.14) Loren read Neuromancer carefully.
(4.15) Stewart completely destroyed his credibility.
(4.16) Stewart destroyed his credibility completely.

together with the lexicon repeated below.

Word	Type(s)
Loren	np
Stewart	np
Neuromancer	np
credibility	n
his	np / n
read	$(np \backslash S) / np$
destroyed	$(np \backslash S) / np$
carefully	$(np \backslash S) \backslash (np \backslash S)$
completely	$(np \backslash S) \backslash (np \backslash S)$

1. Give natural deduction proofs for sentences 4.14 and 4.16 in NL.
2. Give natural deduction proofs for sentences 4.13 and 4.15 in NLP.

Exercise 4.9. We have seen in Exercise 2.6 that the sentences "Someone loves everyone" and "Someone is_missing" both have two normal form natural deduction derivations in L. The lexicon is repeated below. How many normal form natural deduction derivations does each of these sentences have in NL?

Word	Type(s)
someone	$(S / (np \backslash S))$
loves	$((np \backslash S) / np)$
is_missing	$((S / (np \backslash S)) \backslash S)$
everyone	$((S / np) \backslash S)$

Comment on the difference.

Exercise 4.10. Associativity corresponds to a set of two separate rules. There are some cases where it suffices to have only one of the two rules. A canonical example are cases of what is often called *right node raising*. The sentence below is a typical example.

(4.17) Loren loved but Stewart hated Neuromancer.

"loved" and "hated" are both transitive verbs and "Neuromancer" is the objet of both of them. This is another case of polymorphic coordination, of which we have already seen examples in Exercise 1.4.4 and Section 4.2.1: "but", like' "and" can be assigned the formula $(X \setminus X) / X$ for several instantiations of X. We have seen several examples in Exercise 1.4.4.

1. What is the instantiation of X which is appropriate for the sentence above?
2. Which of the two associativity rules do we need to derive the sentence?

Exercise 4.11. To familiarize yourself with the axiomatic calculus, prove $A \vdash B / (A \setminus B)$. Though this statement has a trivial proof in the sequent calculus as well as in natural deduction, you'll find that it requires a bit more effort here.

Exercise 4.12. Prove the following lemma for the residuation calculus of Figure 4.6 on page 114 without using the (*Trans*) rule.

Lemma 4.13. *If from* $A \vdash C$ *we can derive* $B \vdash C$*, then from* $\Gamma[A] \vdash C$ *we can derive* $\Gamma[B] \vdash C$*.*

Exercise 4.14. Reprove the case for $\setminus_h$ of Lemma 4.14 (on page 119) using Lemma 4.13 you proved for Exercise 4.12.

Exercise 4.15. Find all different proofs of the sequent

$$S / (np \setminus S), (S / (np \setminus S)) \setminus S \vdash S$$

using both de Groote's context calculus SC, as shown on Figure 4.10 on page 131, and Aarts and Trautwein's calculus, shown on Figure 4.11 on page 141.

Exercise 4.16. Prove the example sequent (from de Groote, 1999)

$$a \vdash (c / ((a \bullet b) \setminus c)) / b$$

in the context calculus SC from Figure 4.10.

References

Aarts, E., Trautwein, K.: Non-associative Lambek categorial grammar in polynomial time. Mathematical Logic Quarterly 41, 476–484 (1995)

Blackburn, P., de Rijke, M., Venema, Y.: Modal Logic. Cambridge Tracts in Theoretical Computer Science, vol. 53. Cambridge University Press (2001)

Chomsky, N.: Some concepts and consequences of the theory of Government and Binding. MIT Press, Cambridge (1982)

Chomsky, N.: The minimalist program. MIT Press, Cambridge (1995)

Došen, K.: Sequent-systems and groupoid models I. Studia Logica 47(4), 353–385 (1988)

Došen, K.: Sequent-systems and groupoid models II. Studia Logica 48(1), 41–65 (1989)

Došen, K.: A brief survey of frames for the Lambek calculus. Zeitschrift für Mathematische Logic und Grundlagen der Mathematik 38, 179–187 (1992)

Emms, M.: Parsing with polymorphism. In: Proceedings of the Sixth Conference of the European Association of Computational Linguistics, pp. 120–129 (1993)

Emms, M.: An undecidability result for polymorphic Lambek calculus. In: Dekker, P., Stokhof, M. (eds.) Proceedings 10th Amsterdam Colloquium, pp. 539–549 (1995)

de Groote, P.: The Non-associative Lambek Calculus with Product in Polynomial Time. In: Murray, N.V. (ed.) TABLEAUX 1999. LNCS (LNAI), vol. 1617, pp. 128–139. Springer, Heidelberg (1999)

Joshi, A., Schabes, Y.: Tree adjoining grammars. In: Rozenberg, G., Salomaa, A. (eds.) Handbook of Formal Languages, vol. 3, ch. 2. Springer, Berlin (1997)

Kandulski, M.: The equivalence of nonassociative Lambek categorial grammars and context free grammars. Zeitschrift für Mathematische Logic und Grundlagen der Mathematik 34, 41–52 (1988)

Kanovich, M.: The complexity of horn fragments of linear logic. Ann. Pure Appl. Logic 69(2-3), 195–241 (1994)

Kurtonina, N.: Frames and labels. A modal analysis of categorial inference. PhD thesis, OTS Utrecht, ILLC Amsterdam (1995)

Kurtonina, N.: Categorial inference and modal logic. Journal of Logic, Language and Information 7(4), 399–411 (1998)

Lambek, J.: On the calculus of syntactic types. In: Jakobson, R. (ed.) Structure of Language and its Mathematical Aspects, pp. 166–178. American Mathematical Society (1961)

Lambek, J.: Categorial and categorical grammars. In: Oehrle, R.T., Bach, E., Wheeler, D. (eds.) Categorial Grammars and Natural Language Structures. Reidel, Dordrecht (1988)

Moortgat, M., Oehrle, R.T.: Proof nets for the grammatical base logic. In: Abrusci, V.M., Casadio, C., Sandri, G. (eds.) Dynamic Perspectives in Logic and Linguistics. Cooperativa Libraria Universitaria Editrice Bologna (1999)

Pentus, M.: Product-free Lambek calculus and context-free grammars. Journal of Symbolic Logic 62(2), 648–660 (1997)

Pentus, M.: Lambek calculus is NP-complete. Theoretical Computer Science 357(1), 186–201 (2006)

Pollard, C., Sag, I.A.: Head-Driven Phrase Structure Grammar. Center for the Study of Language and Information, Stanford (1994) (distributed by Cambridge University Press)

Savateev, Y.: Product-Free Lambek Calculus Is NP-Complete. In: Artemov, S., Nerode, A. (eds.) LFCS 2009. LNCS, vol. 5407, pp. 380–394. Springer, Heidelberg (2008)

Stabler, E.P.: Derivational Minimalism. In: Retoré, C. (ed.) LACL 1996. LNCS (LNAI), vol. 1328, pp. 68–95. Springer, Heidelberg (1997)

5

The Multimodal Lambek Calculus

Summary. The multimodal Lambek calculus extends the (non-associative) Lambek calculus in two ways. First, it provides a way of mixing different resource management possibilities — for example associative and non-associative or commutative and non-commutative — without collapse, that is to say that associativity can be valid for certain formulae but not for others. Second, it introduces unary connectives, which introduce new derivability patterns and which provide us with another way to lexically anchor the structural rules of associativity and commutativity.

Since the multimodal extensions of the Lambek calculus have been motivated by the desire to give a better *linguistic* treatment of some grammatical phenomena, some of these linguistic analyses will be discussed as examples illustrating the use of the multimodal extensions.

5.1 Combining Different Calculi

While we discussed different calculi L, NL, NLP, LP in the previous chapters, none of them — taken by itself — is very suited for linguistic analysis. For example, as we have seen in Section 2.11, L generates only context-free languages and Shieber (1985) argues that linguistic phenomena like Swiss verb clusters are not context-free.

The fundamental idea of the multimodal Lambek calculus is that we use different families of connectives, which we will distinguish by means of indices or modes. Each family of connectives has it own structural punctuation. This logic was first introduced by Oehrle and Zhang (1989); Moortgat and Morrill (1991) and later extended by Moortgat and Oehrle (1993, 1994) and by Hepple (1993). The interest of this setup is that this allows us to have different structural rules which apply to different modes and thereby use each of the different calculi where it is most advantageous.

Formulae for the multimodal Lambek calculus are a simple extension of the formulae of L we have seen before. Given a set of primitive formulae P and for all elements i from the set of indices I (this set of indices I is defined by the grammar, though often implicitly), the set of multimodal formulae is defined as follows.

$$\mathsf{Lp} \quad ::= \quad \mathsf{P} \quad | \quad (\mathsf{Lp} \backslash_i \mathsf{Lp}) \quad | \quad (\mathsf{Lp} /_i \mathsf{Lp}) \quad | \quad (\mathsf{Lp} \bullet_i \mathsf{Lp})$$

Definition 5.1. *Multimodal antecedent terms $\mathscr{A}$ are like the antecedent terms for NL with the addition of mode information. We now have, for every mode i in the set of modes I a different pair of parentheses, which we distinguish by the mode information which is written as a superscript on the closing parenthesis.*

$$\mathscr{A} \quad ::= \quad \mathsf{Lp} \quad | \quad (\mathscr{A}, \mathscr{A})^i$$

As useful way of looking at multimodal antecedent terms is to see them as binary branching trees — with formulas as its leaves, just like for NL antecedent terms — but where, in addition, each internal node of the tree is labeled with an element of the set I. So where NL had unlabeled binary branching trees as antecedent terms, by multimodal antecedent terms are labeled binary branching trees. These labels do not correspond to the classic *np* and *s* of phrase structure trees (we can already derive that constituents are of the corresponding types) but to additional information orthogonal to these types.

Figure 5.1 shows the logical rules for the multimodal Lambek calculus; each mode $i \in I$ has its own set of logical rules, though they all follow the same rule scheme.

$$\frac{\Gamma[B] \vdash C \quad \Delta \vdash A}{\Gamma[(\Delta, A \backslash_i B)^i] \vdash C} \backslash h \qquad \frac{(A,\Gamma)^i \vdash C}{\Gamma \vdash A \backslash_i C} \backslash i$$

$$\frac{\Gamma[B] \vdash C \quad \Delta \vdash A}{\Gamma[(B /_i A, \Delta)^i] \vdash C} /h \qquad \frac{(\Gamma, A)^i \vdash C}{\Gamma \vdash C /_i A} /i$$

$$\frac{\Gamma[(A,B)^i] \vdash C}{\Gamma[A \bullet_i B] \vdash C} \bullet h \qquad \frac{\Delta \vdash A \quad \Gamma \vdash B}{(\Delta, \Gamma)^i \vdash A \bullet_i B} \bullet i$$

$$\frac{\Gamma \vdash A \quad \Delta[A] \vdash B}{\Delta[\Gamma] \vdash B} \, cut \qquad \frac{}{A \vdash A} \, axiom$$

Fig. 5.1. Binary logical rules for the sequent calculus for the multimodal Lambek calculus

Note how these rules are just the rules of NL with mode information added to the logical rules, where in each case the logical connective and the structural punctuation share the same index.

As for NL, when we talk about the derivability of a sequent $A_1, \ldots, A_n \vdash C$ we will in general mean: is there an antecedent term Γ with yield $A_1, \ldots, A_n$ such that $\Gamma \vdash C$ is derivable? Though this is often left implicit, we will often need a slightly more restricted variant of this condition in the multimodal Lambek calculus. Each grammar defines not only a set I of modes but also a set $E \subseteq I$ of *external* modes and when we look for a derivation of a sequent, we want to find an antecedent term Γ *containing only external modes* such that $\Gamma \vdash C$ is derivable. Modes in $I \setminus E$ are called *internal* modes, they are modes which are only used for grammar-internal calculations, or — to use a different metaphor — we can see them as a type of *features* which must be checked, as is done in minimalist grammars (Chomsky, 1995; Vermaat, 2004).

So the motivation for allowing only certain modes to appear in the antecedent term of the end-sequent is either to force certain structural rules to apply or, inversely, to prevent them from applying. We will see why this is useful in Example 5.2.

5.1.1 Multimodal Structural Rules

Modally Licensed Structural Rules and Inclusion Rules

Using a multimodal system allows us to reintroduce the structural rules of Section 4.5.2, but now indexed in such a way that they apply only to certain modes: in other words, the mode information on the formulae — which comes from the lexicon — *licenses* the application of structural rules. This gives us a *combined* logic, containing all the different systems we have seen before, with the lexicon indicating where the different structural rules can apply.

System	Mode	Structural Rules
NL	n	–
L	a	Associativity
NLP	nc	Commutativity
LP	c	Associativity, Commutativity

This gives us the following set of structural rules shown in Figure 5.2 for the four modes n, a, nc, c. The complete calculus also includes instances of the logical rules in Figure 5.1 for each of these four modes.

In addition to having these mode-indexed structural rules, it is often useful to have *inclusion* rules which relate different modes to each other. Here, Moortgat and Oehrle (1993) and Hepple (1993); de Groote (1996) differ in the direction of the inclusions. As shown by the inclusion rules in Figure 5.1, we follow Moortgat and Oehrle. Given these inclusions, we have the following relations between the implications.

$$\dfrac{\dfrac{\dfrac{\vdots}{(a /_c b, b)^c \vdash a}}{(a /_c b, b)^a \vdash a}\, incl_{c,a}}{a /_c b \vdash a /_a b}\, /_h \qquad \dfrac{\dfrac{???}{(a /_a b, b)^c \vdash a}}{a /_a b \vdash a /_c b}\, /_h$$

Associativity for a, c

$$\frac{\Gamma[(\Delta_1,(\Delta_2,\Delta_3)^a)^a] \vdash C}{\Gamma[((\Delta_1,\Delta_2)^a,\Delta_3)^a] \vdash C}\, ass1a \qquad \frac{\Gamma[((\Delta_1,\Delta_2)^a,\Delta_3)^a] \vdash C}{\Gamma[(\Delta_1,(\Delta_2,\Delta_3)^a)^a] \vdash C}\, ass2a$$

$$\frac{\Gamma[(\Delta_1,(\Delta_2,\Delta_3)^c)^c] \vdash C}{\Gamma[((\Delta_1,\Delta_2)^i,\Delta_3)^c] \vdash C}\, ass1c \qquad \frac{\Gamma[((\Delta_1,\Delta_2)^c,\Delta_3)^c] \vdash C}{\Gamma[(\Delta_1,(\Delta_2,\Delta_3)^c)^c] \vdash C}\, ass2c$$

Commutativity for nc, c

$$\frac{\Gamma[(\Delta_2,\Delta_1)^{nc}] \vdash C}{\Gamma[(\Delta_1,\Delta_2)^{nc}] \vdash C}\, comnc$$

$$\frac{\Gamma[(\Delta_2,\Delta_1)^{c}] \vdash C}{\Gamma[(\Delta_1,\Delta_2)^{c}] \vdash C}\, comc$$

Inclusion rules

$$\frac{\Gamma[(\Delta_1,\Delta_2)^a] \vdash C}{\Gamma[(\Delta_1,\Delta_2)^n] \vdash C}\, incl_{a,n} \qquad \frac{\Gamma[(\Delta_1,\Delta_2)^{nc}] \vdash C}{\Gamma[(\Delta_1,\Delta_2)^n] \vdash C}\, incl_{nc,n}$$

$$\frac{\Gamma[(\Delta_1,\Delta_2)^c] \vdash C}{\Gamma[(\Delta_1,\Delta_2)^a] \vdash C}\, incl_{c,a} \qquad \frac{\Gamma[(\Delta_1,\Delta_2)^{c}] \vdash C}{\Gamma[(\Delta_1,\Delta_2)^{nc}] \vdash C}\, incl_{c,nc}$$

Fig. 5.2. Structural rules for a mixed calculus containing NL, L, NLP and LP as subsystems

In the partial derivation shown above on the left, we can — reading from the bottom upwards — use the inclusion rule to change mode a into c and then we can continue the derivation using the $/_h$ rule now that we have obtained the correct mode. On the right, this strategy is not possible: we obtain an a implication in a c context and at this point no rules apply to the sequent, the $/_h$ rule being blocked by the fact that the mode of the implication does not correspond to the mode of the structural punctuation.

The interpretation of the inclusions in this way is that an $a /_c b$ formula can select its b argument both to the right as well as to the left and we can therefore derive $a /_a b$, which selects its argument just to its right, from it.

Interaction Rules

In addition to having modally licensed structural rules and inclusion rules, the last type of structural rules are structural rules of *interaction*, which open up new possibilities when two different modes appear together. We will illustrate this by two well-known examples, one due to Morrill (1994, 1995) and another due to Moortgat and Oehrle (1993, 1994).

Wrapping

As a first example, Morrill (1994, 1995) has proposed the set of "wrapping" structural rules shown in Figure 5.3, for a multimodal Lambek calculus containing an associative mode a, a non-associative mode n and a "wrapping" mode w. Morrill's formulation of the calculus includes rules for an identity element ε for the associative mode a, where ε corresponds to the empty antecedent term. For the reasons we discussed in Section 2.5, ε, $(\varepsilon,\varepsilon)^a$, ... (ie. any antecedent term equivalent to ε) are not valid antecedent terms, though Morrill admits $(\varepsilon,\varepsilon)^n$ as a valid antecedent term, which allows us to derive $(\varepsilon,\varepsilon)^n \vdash A /_w A$.

In this calculus, we have $I = \{a,n,w\}$ and $E = \{a,n\}$. In other words, the end-sequent can contain only the associative and the non-associative mode, but not the wrapping mode w.

ε is the identity element for a

$$\frac{\Gamma[(\Delta,\varepsilon)^a] \vdash C}{\Gamma[\Delta] \vdash C}\,\varepsilon_r \qquad \frac{\Gamma[\Delta] \vdash C}{\Gamma[(\Delta,\varepsilon)^a] \vdash C}\,\varepsilon_r \qquad \frac{\Gamma[(\varepsilon,\Delta)^a] \vdash C}{\Gamma[\Delta] \vdash C}\,\varepsilon_l \qquad \frac{\Gamma[\Delta] \vdash C}{\Gamma[(\varepsilon,\Delta)^a] \vdash C}\,\varepsilon_l$$

Associativity rules for a

$$\frac{\Gamma[(\Delta_1,(\Delta_2,\Delta_3)^a)^a] \vdash C}{\Gamma[((\Delta_1,\Delta_2)^a,\Delta_3)^a] \vdash C}\,ass \qquad \frac{\Gamma[((\Delta_1,\Delta_2)^a,\Delta_3)^a] \vdash C}{\Gamma[(\Delta_1,(\Delta_2,\Delta_3)^a)^a] \vdash C}\,ass$$

Wrapping rules

$$\frac{\Gamma[((\Delta_1,\Delta_2)^a,\Delta_3)^a] \vdash C}{\Gamma[((\Delta_1,\Delta_3)^n,\Delta_2)^w] \vdash C}\,wrap \qquad \frac{\Gamma[((\Delta_1,\Delta_3)^n,\Delta_2)^w] \vdash C}{\Gamma[((\Delta_1,\Delta_2)^a,\Delta_3)^a] \vdash C}\,wrap$$

Fig. 5.3. The set of wrapping structural rules from Morrill (1994)

The key to the calculus are the two wrapping rules. Figure 5.4 shows them in tree form. Read right-to-left, they allow us to "move out" a subcontext Δ_2 when it occurs between two other contexts Δ_1 and Δ_3 in the associative mode a, leaving behind this pair of contexts in the non-associative mode n. The other rule allows us to move Δ_2 back to its original position. In a sense, we can see the non-associative mode n as serving as a kind of marker saying that material can wrap or insert itself between the two direct daughter antecedent terms of the mode.

Example 5.2. Morrill gives several interesting applications of the wrapping operator. One of them is their use for generalized quantifiers, which we discussed in Chapter 3. We will see another in Exercise 5.4 at the end of this chapter.

In Chapter 3, we have used type assignments $S / (np \setminus S)$ and $(S / np) \setminus S$ for generalized quantifiers. Besides the fact that we need to use a different type for subject generalized quantifiers and for object generalized quantifiers, this analysis suffers from the problem that the quantifiers can only take their correct scope if they occur either in the leftmost or in the rightmost position of the sentence they occur in.

One of the classic examples of a generalized quantifier which can take scope from the middle of the sentence it occurs in is the following.

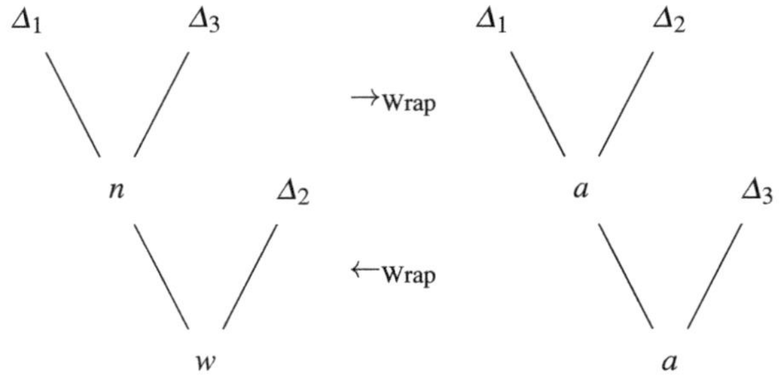

Fig. 5.4. Wrapping structural rules

(5.1) Bond believes someone left.

This sentences has two readings, a 'de dicto' reading where Bond believes there exists a person such that he left. This reading is true, for example, when Bond has counted the number of persons at two instances and inferred that there is one person missing. So this sentence can be true without Bond believing of any specific person that he left. In this sentence, the existential quantifier has scope only over the sentence "someone left". The 'de re' reading represents the other possibility. It would be true if it is the case that Bond is watching a specific person, say Blofeld, and then seeing him slip out the back door. In this reading, the existential quantifier has scope over the entire sentence. This last reading is not available with either of the type assignments for generalized quantifiers for the Lambek calculus.

The wrapping structural rules of Figure 5.3 allow us to handle cases where a generalized quantifier appears in the middle of the sentence. The lexicon below gives a wrapping analysis of generalized quantifiers.

Word	Type(s)
Bond	np
believes	$(np \backslash_a S) /_a S$
someone	$(S /_w np) \backslash_w S$
left	$np \backslash_a S$

Apart from the type assigned to "someone", this lexicon contains no surprises. The idea behind the generalized quantifier type $(S /_w np) \backslash_w S$ is that it wraps out of the structure to take its scope at the S level (using the *wrap* and $\backslash_h$ rule), then wraps its np subformula back to the original position of the generalized quantifier (using the $/_i$ rule followed by the *wrap* rule).

The two partial sequent proofs (the final part of the proofs has been left as an easy exercise) below show this type in action by providing a proof for each of the two different readings of the sentence. The first proof shows the reading where "someone" has wide scope. We use the wrap structural rule — with Δ_1 = "Bond believes", Δ_2 = "someone" and Δ_3 = "left" — to move the generalized quantifier to the rightmost position, where it is the rightmost daughter of a w bracket. This means it is in the correct configuration to apply the $\backslash_h$ rule for mode w. Then we use the $/_i$ rule for mode w to add an np with mode w to the right of the antecedent term. This is the correct configuration to wrap the np back to the position which was first occupied by the $(S /_w np) \backslash_w S$ formula. The non-associative mode n can be interpreted as a pair of terms Γ_1, Γ_3 which allows another term Γ_2 to wrap between them. The interplay between the $/_i$ and the $\backslash_h$ rule for mode w ensures that the np is inserted at the same place as the original generalized quantifier formula.

$$
\dfrac{\dfrac{\dfrac{\dfrac{\vdots}{(((np,(np\backslash_a S)/_a S)^a,np)^a,np\backslash_a S)^a \vdash S}}{(((np,(np\backslash_a S)/_a S)^a,np\backslash_a S)^n,np)^w \vdash S}\,wrap}{((np,(np\backslash_a S)/_a S)^a,np\backslash_a S)^n \vdash S/_w np}\,/_i \qquad \dfrac{}{S \vdash S}\,axiom}{\dfrac{(((np,(np\backslash_a S)/_a S)^a,np\backslash_a S)^n,(S/_w np)\backslash_w S)^w \vdash S}{(((np,(np\backslash_a S)/_a S)^a,(S/_w np)\backslash_w S)^a,np\backslash_a S)^a \vdash S}\,wrap}\,\backslash_h
$$

The second proof is the reading where the generalized quantifier "someone" has scope over the subordinate clause only. This is clear in the proof below since we need to derive $(S /_w np) \backslash_w S, np \backslash_a S$ as being of type S. Given that we have only two formulae in the antecedent, we cannot apply the wrap structural rule. However, we can apply the rules for the identity element to add the empty context ε to the left of the generalized quantifier type. After that, we can wrap the quantifier to the outside and apply the combination $/_i$, $\backslash_h$ to put the formula np in the place of the quantifier. Wrapping the np to the subject position, then eliminating the empty context leaves us with the sequent $(np, np \backslash_a S)^a \vdash S$ to prove.

$$
\dfrac{\dfrac{\vdots}{(np,np\backslash_a S)^a \vdash S} \qquad \dfrac{\dfrac{\dfrac{\dfrac{\dfrac{\dfrac{\dfrac{\vdots}{(np,np\backslash_a S)^a \vdash S}}{((\varepsilon,np)^a,np\backslash_a S)^a \vdash S}\,\varepsilon_l}{((\varepsilon,np\backslash_a S)^n,np)^w \vdash S}\,wrap}{(\varepsilon,np\backslash_a S)^n \vdash S/_w np}\,/_i \qquad \dfrac{}{S \vdash S}\,axiom}{((\varepsilon,np\backslash_a S)^n,(S/_w np)\backslash_w S)^w \vdash S}\,\backslash_h}{((\varepsilon,(S/_w np)\backslash_w S)^a,np\backslash_a S)^a \vdash S}\,wrap}{((S/_w np)\backslash_w S,np\backslash_a S)^a \vdash S}\,\varepsilon_l}{(np,((np\backslash_a S)/_a S)^a,((S/_w np)\backslash_w S,np\backslash_a S)^a)^a)^a \vdash S}\,/_h
$$

The reason for disallowing mode w to appear in the end-sequent by declaring it internal is the following: the end-sequent above is of the right form for the wrap structural rule, which allows us to continue the above proof to produce the sequent.

$$
\dfrac{(np,((np\backslash_a S)/_a S)^a,((S/_w np)\backslash_w S,np\backslash_a S)^a)^a)^a \vdash S}{(np,((S/_w np)\backslash_w S,np\backslash_a S)^a)^n(np\backslash_a S)/_a S)^w \vdash S}\,wrap
$$

However, this sequent would correspond to the ungrammatical sentence "∗Bond someone left believes". Disallowing mode w in the end-sequent prevents this grammar from generating this (and many other) ungrammatical sentences.

Mixed associativity and mixed commutativity

Figures 5.5 and 5.6 shows the structural rules from Moortgat and Oehrle (1993, 1994). In the rule descriptions below, we will interpret mode 0 as the basic, continuous composition and modes 1 to 4 as different types of discontinuous composition modes. These modes are organized along two parameters:

1. whether the moving constituent is on the left branch (modes 2 and 4) or on the right branch (modes 1 and 3) of the discontinuous mode,
2. whether the moving constituent moves up towards the root of the tree (modes 1 and 2) or down towards its leaves (modes 3 and 4).

Figure 5.5 shows the structural rules for when the moving element is on the right branch and Figure 5.6 shows the structural rules for the left branch. Each set of structural rules consists of a pair of rules (this pair of rules is in the same row in both figures) which sees modes i and j interacting.

The left branch rules in Figure 5.6 show — in the top row and reading the rules from top to bottom — that Δ_1 is on the left branch of mode 2. If this context of mode 2 is dominated by a context of mode 0, there are two cases to consider: one where mode 2 occurs as the left daughter of mode 0 (rule MA, for mixed associativity) and one where mode 2 occurs as its right daughter (rule MC, for mixed commutativity). In both cases, Δ_1 can move one step up towards to root, while staying on the left branch of mode 2. For the mixed associativity rule, Δ_2 and Δ_3 form a new constituent after application of the rule; for the mixed commutativity rule, this means that Δ_2 now occurs to the left of Δ_1, changing the order of the constituents.

The names used for the rules are somewhat complex, but this is just done to allow us to distinguish between the different possibilities by means of the notation: MX_YZ_{ij} signifies whether the rule is a mixed associativity $X = A$ or mixed commutativity rule $X = C$, whether the moving element is on the left branch $Y = l$ or on the right branch $Y = r$ and whether the rule is one of extraction $Z = e$ or infixation $Z = i$. Finally i and j denote the two modes for which the structural rule is defined, with the continuous mode mentioned first and the discontinuous mode second. This naming convention allows us to refer to the different instantiations of the mixed associativity and mixed commutativity rule scheme in a compact way.

The reader is advised to keep in mind that though the different versions of these rules are often simply called *MA* and *MC*, the names used for these rules can vary between different authors. The same is true for the different modes: the use of one continuous mode 0 and four discontinuous modes 1 to 4 is just a matter of listing the possibilities; in practice, authors use just one or two of these available possibilities and choose an index for the mode which does not necessarily correspond to those used here.

$$\frac{\Gamma[(\Delta_1,(\Delta_2,\Delta_3)^1)^0]\vdash C}{\Gamma[((\Delta_1,\Delta_2)^0,\Delta_3)^1]\vdash C}\,MA_re_{0,1} \qquad \frac{\Gamma[((\Delta_1,\Delta_3)^1,\Delta_2)^0]\vdash C}{\Gamma[((\Delta_1,\Delta_2)^0,\Delta_3)^1]\vdash C}\,MC_re_{0,1}$$

$$\frac{\Gamma[((\Delta_1,\Delta_2)^0,\Delta_3)^3]\vdash C}{\Gamma[(\Delta_1,(\Delta_2,\Delta_3)^3)^0]\vdash C}\,MA_ri_{0,3} \qquad \frac{\Gamma[((\Delta_1,\Delta_2)^0,\Delta_3)^3]\vdash C}{\Gamma[((\Delta_1,\Delta_3)^3,\Delta_2)^0]\vdash C}\,MC_ri_{0,3}$$

Fig. 5.5. The structural rules for mixed associativity and mixed commutativity (right branch)

$$\frac{\Gamma[((\Delta_1,\Delta_2)^2,\Delta_3)^0]\vdash C}{\Gamma[(\Delta_1,(\Delta_2,\Delta_3)^0)^2]\vdash C}\,MA_le_{0,2} \qquad \frac{\Gamma[(\Delta_2,(\Delta_1,\Delta_3)^2)^0]\vdash C}{\Gamma[(\Delta_1,(\Delta_2,\Delta_3)^0)^2]\vdash C}\,MC_le_{0,2}$$

$$\frac{\Gamma[(\Delta_1,(\Delta_2,\Delta_3)^0)^4]\vdash C}{\Gamma[((\Delta_1,\Delta_2)^4,\Delta_3)^0]\vdash C}\,MA_li_{0,4} \qquad \frac{\Gamma[(\Delta_1,(\Delta_2,\Delta_3)^0)^4]\vdash C}{\Gamma[(\Delta_2,(\Delta_1,\Delta_3)^4)^0]\vdash C}\,MC_li0,4$$

Fig. 5.6. The structural rules for mixed associativity and mixed commutativity (left branch)

Figure 5.8 shows the rules in a more easily readable tree format. In the context of the figure, the two rules of mixed associativity and mixed commutativity for mode 0 and mode 2 correspond to setting $j=2$ in the figure and looking at the $MC_le_{0,2}$ and $MA_le_{0,2}$ arrows: the arrows moving from the leftmost tree to the center tree and from the rightmost tree to the center tree respectively.

Another way of looking at the left branch extraction pattern is that they allow the following inference pattern (under the condition that there is a path in Γ to $(B,\Delta)^2$ that passes only through binary mode 0).

$$\begin{array}{c}\Gamma[(B,\Delta)^2]\vdash A\\ \vdots\,\delta\\ \dfrac{(B,\Gamma[\Delta])^2\vdash A}{\Gamma[\Delta]\vdash B\backslash_2 A}\,\backslash i\end{array}$$

We can move the B formula up through the context Γ — using the $MC_le_{0,2}$ and $MA_le_{0,2}$ structural rules — until it is on the left branch of the antecedent term and has $\Gamma[\Delta]$ as its sister. As noted by Vermaat (2004, 2005), the derivation pattern above corresponds rather naturally to the *move* operation which has been proposed for minimalist grammars (Chomsky, 1995).

The left branch infixation rules, shown in the bottom row of Figure 5.6 and the arrows $MC_li_{0,4}$ and $MA_li_{0,4}$ of Figure 5.8 (when setting $j=4$; these are the arrows moving from the tree in the center outwards to the left and right), allow us to move an antecedent term Δ_1 down when it is on the left of mode 4 and when there is a mode 0 to the right of it. This structural rule is non-deterministic: the structural rules allow us to choose whether to move Δ_1 to the left, where it becomes the left sister of Δ_2 or to the right, where it becomes the left sister of Δ_3.

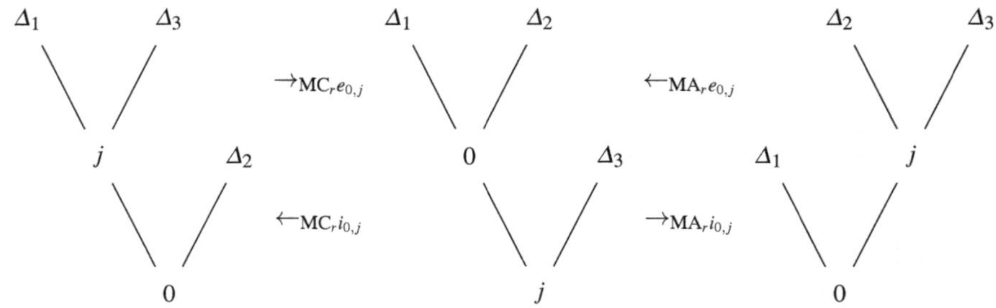

Fig. 5.7. Trees corresponding to the structural rules for right branch extraction/infixation

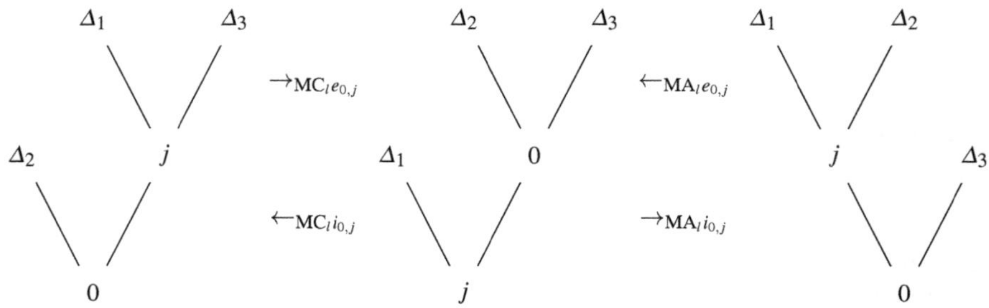

Fig. 5.8. Trees corresponding to the structural rules for left branch extraction/infixation

A deterministic version of these infixation rules exists: if we replace mode 0 by two different modes l (left) and r (right) then the mode information will decide if Δ_1 moves to the left with Δ_2 (MA, the arrow pointing to the right) or to the right with Δ_3 (MC, the arrow pointing to the left). We can see the modes l and r as coding a path to a node in the tree. In (Moortgat and Morrill, 1991), these "paths" represent the heads of phrases and are not linked to any infixation rules. A set of structural rules using extraction and infixation using modes to specify paths has been proposed by Moortgat (1996a) to give an account of quantifier scope. The deterministic infixation rules will feature in Exercise 5.8 at the end of this chapter.

The right branch extraction and infixation rules Figure 5.5 (Figure 5.7 shows them in tree form) are left-to-right symmetric to the left branch rules: they allow us to move an antecedent term Δ_3 which is on the right branch of a discontinuous mode either up towards the root of the tree (the extraction rules, which correspond to the arrows pointing towards the center in Figure 5.7) or down towards the leaves (the infixation rules, which correspond to the arrows pointing from the center outwards in the same figure).

We will see a variation of these rules in Section 5.2.3, where instead of two binary modes interacting, we will have a unary mode interacting with a binary mode. The general form of the structural rules we will consider is presented in Section 5.2.4.

Example 5.3 (Dutch verb clusters I)
Using these new structural rules, we will make a first step towards giving a treatment of verb clusters in Dutch subordinate clauses, following the analysis of Moortgat and Oehrle (1994).

The following examples illustrate a very simply case of this phenomenon (more complex cases can be found in Exercise 5.9).

(5.2)	*(dat)*	*Marie*	*Jan*	*plaagt.*	
	(that)	Mary	John	teases.	

'(that) Mary teases John.'

(5.3)	*(dat)*	*Marie*	*Jan*	*wil*	*plagen.*
	(that)	Mary	John	wants	to tease.

'(that) Mary wants to tease John.'

As shown in Sentence 5.2, the verb "plaagt" — when it occurs in a subordinate clause — selects its object "Jan" to its left. What is interesting is that in Sentence 5.3, the infinitive "plagen" selects its object to the left, but now the auxiliary verb "wil" is placed between the infinitive and its object.

Moortgat and Oehrle use two modes of composition, in our notation these are: 0 which corresponds to regular phrasal adjunction and 4 which corresponds to what they call *head adjunction*: it is the mode which allows a constituent on its left branch to descend downwards to the leaves of the antecedent term. According to the terminology we have used in this section, these are the left branch infixation rules $MC_l i_{0,4}$ and $MA_l i_{0,4}$ of Figure 5.8 (for $j = 4$). Look at the following lexicon.

Word	Type(s)	Translation
Marie	np	*Mary*
Jan	np	*John*
wil	$(np \backslash_0 S) /_4 inf$	*wants*
plagen	$np \backslash_0 inf$	*to tease*

The type assignment to the auxiliary verb "wil" selects in infinitival phrase to its right, but does so using the head adjunction mode 4. This means that if we derive "wil Jan plagen", where "plagen" selects its object to the left and "wil" selects the resulting infinitive to its right to form an S, we are in the situation shown in the middle of Figure 5.8 with $\Delta_1 = (np \backslash_0 S) /_4 inf$ "wil", $\Delta_2 = np$ "Jan" and $\Delta_3 = np \backslash_0 inf$ "plagen". The mixed commutativity rule $MC_l i_{0,4}$ allows us to reconfigure this tree into "Jan wil plagen" as shown in the following proof.

$$\dfrac{\dfrac{\dfrac{\dfrac{\overline{np \vdash np}\ axiom \quad \overline{S \vdash S}\ axiom}{(np, np \backslash_0 S)^0 \vdash S}\ \backslash_h \quad \overline{inf \vdash inf}\ axiom}{(np, ((np \backslash_0 S) /_4 inf, inf)^4)^0 \vdash S}\ /_h \quad \overline{np \vdash np}\ axiom}{(np, ((np \backslash_0 S) /_4 inf, (np, np \backslash_0 inf)^0)^4)^0 \vdash S}\ \backslash_h}{(np, (np, ((np \backslash_0 S) /_4 inf, np \backslash_0 inf)^4)^0)^0 \vdash S}\ MC_{0,4}$$

Observe, however, that we can now also derive '(dat) Marie wil Jan plagen' by simply omitting the last rule of the proof. In the next section, we will see how additional tools allow us to require that the verbs in the verb cluster are adjacent.

5.2 Unary Connectives

In this section we will look at two different ways of adding unary connectives to L and NL. We will start by looking at unary connectives which are inspired by those of linear logic, but where in linear logic they are used to add the structural rules of weakening and contraction in a controlled way, here they are used to add commutativity in a controlled way. After that, we will introduce a second way of looking at the unary connectives which are based on unary residuation. Finally, we will look at the new possibilities for structural rules these new connectives give us.

5.2.1 The Unary Connectives of Linear Logic

The unary connectives or exponentials were first introduced by Girard (1987) for linear logic. They allow a controlled use of the structural rules of weakening and contraction. Inspired by this example, extensions of the Lambek calculus using the exponentials to control access to the structural rule of commutativity were soon introduced, for example in (Morrill et al, 1990; Barry et al, 1991).

The logical rules for introducing a unary connective ! permitting permutation in L are shown in Figure 5.9.

The notation $!\Gamma$ in the rule $!_i$ means: every formula in the antecedent Γ has ! as its principal connective. The intuition behind this rule is that if we can derive C using only assumptions which have a special property (in this case, being able to move around) then we can also derive $!C$. The rule $!_h$ simply states that if we can derive something using a formula A as a hypothesis, then we can derive it using a formula $!A$ as a hypothesis as well.

We have two rules for permutation, one for movement to the left, the other for movement to the right, though we could consider these separately, as done in (Morrill et al, 1990). The other rules are unchanged from those of the Lambek calculus.

Example 5.4 (Medial extraction). The permute modality allows us to treat some phenomena which are difficult to handle in the Lambek calculus. An example is medial extraction. By assigning the word 'which' the types $(n \setminus n) / (np \setminus S)$ and $(n \setminus n) / (S / np)$ we can derive examples 5.4, where the extracted np is the leftmost constituent of the subordinate clause, and 5.5, where it is the rightmost constituent. However, adding a sentential modifier to 5.5 as in sentence 5.6 will make the resulting sentence underivable unless we add another type for 'which' to the lexicon.

(5.4) book which $[\,]_{np}$ fell.
(5.5) book which John read $[\,]_{np}$.
(5.6) book which John read $[\,]_{np}$ yesterday.

$$\frac{\Gamma, B, \Gamma' \vdash C \quad \Delta \vdash A}{\Gamma, \Delta, A \backslash B, \Gamma' \vdash C} \backslash_h \qquad \frac{A, \Gamma \vdash C}{\Gamma \vdash A \backslash C} \backslash_i \quad \Gamma \neq \varepsilon$$

$$\frac{\Gamma, B, \Gamma' \vdash C \quad \Delta \vdash A}{\Gamma, B / A, \Delta, \Gamma' \vdash C} /_h \qquad \frac{\Gamma, A \vdash C}{\Gamma \vdash C / A} /_i \quad \Gamma \neq \varepsilon$$

$$\frac{\Gamma, A, B, \Gamma' \vdash C}{\Gamma, A \bullet B, \Gamma' \vdash C} \bullet_h \qquad \frac{\Delta \vdash A \quad \Gamma \vdash B}{\Delta, \Gamma \vdash A \bullet B} \bullet_i$$

$$\frac{\Gamma, A, \Gamma' \vdash C}{\Gamma, !A, \Gamma' \vdash C} !_h \qquad \frac{!\Gamma \vdash C}{!\Gamma \vdash !C} !_i$$

$$\frac{\Gamma, !A, B, \Gamma' \vdash C}{\Gamma, B, !A, \Gamma' \vdash C} !_p \qquad \frac{\Gamma, B, !A, \Gamma' \vdash C}{\Gamma, !A, B, \Gamma' \vdash C} !_p$$

$$\frac{\Gamma \vdash A \quad \Delta_1, A, \Delta_2 \vdash B}{\Delta_1, \Gamma, \Delta_2 \vdash B} \mathit{cut} \qquad \frac{}{A \vdash A} \mathit{axiom}$$

Fig. 5.9. Logical rules for L!, L with the linear logic modality "!"

By refining the type for 'which' to $(n \backslash n) / (S / !np)$, as in the lexicon below — where the $S / !np$ argument intuitively means "a sentence missing a noun phrase anywhere" — we can derive all of the examples above.

Word	Type(s)
book	n
which	$(n \backslash n) / (S / !np)$
John	np
Mary	np
fell	$np \backslash S$
read	$(np \backslash S) / np$
gave	$((np \backslash S) / np) / np$
yesterday	$S \backslash S$

We give only the derivation of sentence 5.6. The other derivations are easy exercises.

$$
\begin{prooftree}
\AxiomC{}
\RightLabel{$axiom$}
\UnaryInfC{$n \vdash n$}
\AxiomC{}
\RightLabel{$axiom$}
\UnaryInfC{$n \vdash n$}
\RightLabel{$\backslash_h$}
\BinaryInfC{$n, n \backslash n \vdash n$}
\AxiomC{}
\RightLabel{$axiom$}
\UnaryInfC{$np \vdash np$}
\AxiomC{}
\RightLabel{$axiom$}
\UnaryInfC{$S \vdash S$}
\RightLabel{$\backslash_h$}
\BinaryInfC{$np, np \backslash S \vdash S$}
\AxiomC{}
\RightLabel{$axiom$}
\UnaryInfC{$np \vdash np$}
\RightLabel{$/_h$}
\BinaryInfC{$np, (np \backslash S) / np, np \vdash S$}
\AxiomC{}
\RightLabel{$axiom$}
\UnaryInfC{$S \vdash S$}
\RightLabel{$\backslash_h$}
\BinaryInfC{$np, (np \backslash S) / np, np, S \backslash S \vdash S$}
\RightLabel{$!_h$}
\UnaryInfC{$np, (np \backslash S) / np, !np, S \backslash S \vdash S$}
\RightLabel{$!_p$}
\UnaryInfC{$np, (np \backslash S) / np, S \backslash S, !np \vdash S$}
\RightLabel{$/_i$}
\UnaryInfC{$np, (np \backslash S) / np, S \backslash S \vdash S / !np$}
\RightLabel{$/_h$}
\BinaryInfC{$n, (n \backslash n) / (S / !np), np, (np \backslash S) / np, S \backslash S \vdash n$}
\end{prooftree}
$$

5.2.2 Unary Residuation

Moortgat (1995); Kurtonina and Moortgat (1997) propose a different way of introducing unary operators to the Lambek calculus. Their proposal uses a pair of unary connectives '$\Diamond$' and '$\Box$' which have their own structural punctuation. [1] The logical rules either remove or introduce this structural punctuation, similar to the logical rules for the binary connectives.

For all j in the set of unary modes J and all i in the set of binary modes I, the formulae for NL$\Diamond$ are the following.

$$\mathsf{Lp} \quad ::= \quad \mathsf{P} \quad | \quad \Box_j \mathsf{Lp} \quad | \quad \Diamond_j \mathsf{Lp} \quad | \quad (\mathsf{Lp} \backslash_i \mathsf{Lp}) \quad | \quad (\mathsf{Lp} /_i \mathsf{Lp}) \quad | \quad (\mathsf{Lp} \bullet_i \mathsf{Lp})$$

The definition of antecedent term now has a unary punctuation, written as $\langle . \rangle^j$ — the visual similarity with the $\Diamond$ should help distinguish them from the round brackets which correspond to $\bullet$.

Definition 5.5. *The* antecedent terms *of the multimodal Lambek calculus with unary residuated connectives are defined as follows.* Lp *is a formula, i is a binary mode in I, j is a unary mode in J.*

$$\mathscr{A} \quad ::= \quad \mathsf{Lp} \quad | \quad \langle \mathscr{A} \rangle^j \quad | \quad (\mathscr{A}, \mathscr{A})^i$$

With the extensions to the definitions of antecedent terms and formulae in place, Figure 5.10 presents the sequent calculus for NL$\Diamond$.

As a useful intuition, we can see $\Diamond A$ (and the structural punctuation $\langle . \rangle$) as an abbreviation for $A \bullet d$ and $\Box A$ as an abbreviation for A / d for some atomic formula d (to simplify notation, we will often drop the indices and write $\Box$ and $\Diamond$ instead of $\Box_i$ and $\Diamond_i$). For readers familiar with (minimal) tense logic (Prior, 1967), we can also see $\Diamond$ as the future possibility operator F ('it will be the case, at some future time') and $\Box$ as the past necessity operator H ('it has always been the case') (this is

[1] In the literature, the connective $\Box A$ is often written as $\Box^{\downarrow} A$.

$$\frac{\Gamma[B] \vdash C \quad \Delta \vdash A}{\Gamma[(\Delta, A \backslash_i B)^i] \vdash C} \backslash_h \qquad \frac{(A, \Gamma)^i \vdash C}{\Gamma \vdash A \backslash_i C} \backslash_i$$

$$\frac{\Gamma[B] \vdash C \quad \Delta \vdash A}{\Gamma[(B /_i A, \Delta)^i] \vdash C} /_h \qquad \frac{(\Gamma, A)^i \vdash C}{\Gamma \vdash C /_i A} /_i$$

$$\frac{\Gamma[(A, B)^i] \vdash C}{\Gamma[A \bullet_i B] \vdash C} \bullet_h \qquad \frac{\Delta \vdash A \quad \Gamma \vdash B}{(\Delta, \Gamma)^i \vdash A \bullet_i B} \bullet_i$$

$$\frac{\Gamma[A] \vdash C}{\Gamma[\langle \Box_j A \rangle^j] \vdash C} \Box_h \qquad \frac{\langle \Gamma \rangle^j \vdash C}{\Gamma \vdash \Box_j C} \Box_i$$

$$\frac{\Gamma[\langle A \rangle^j] \vdash C}{\Gamma[\Diamond_j A] \vdash C} \Diamond_h \qquad \frac{\Gamma \vdash C}{\langle \Gamma \rangle^j \vdash \Diamond_j C} \Diamond_i$$

$$\frac{\Gamma \vdash A \quad \Delta[A] \vdash B}{\Delta[\Gamma] \vdash B} \mathit{cut} \qquad \frac{}{A \vdash A} \mathit{axiom}$$

Fig. 5.10. Sequent calculus for NL$\Diamond$

of course *not* the intended interpretation, though, as we will see in Section 5.5, the semantics of the connectives is essentially the same as those of temporal logic; see Kurtonina (1995, 1998) for more on the relation between categorial grammar and tense logic).

Looking back at the rules, it is easy to verify that we can still restrict the axiom rule to the atomic formulas, as indicated by the following proposition.

Proposition 5.6. *Every axiom $A \vdash A$ can be derived from axioms $p \vdash p$, with p being a primitive type (and the proof does not use the cut rule).*

In addition, there are already some interesting derivability patterns available in the system without structural rules. For example we have

$$\Diamond\Box A \vdash A \vdash \Box\Diamond A$$

with the derivations shown below.

$$\dfrac{\dfrac{\overline{A \vdash A}\ axiom}{\langle \Box A \rangle \vdash A}\ \Box_h}{\Diamond \Box A \vdash A}\ \Diamond_h \qquad \dfrac{\dfrac{\overline{A \vdash A}\ axiom}{\langle A \rangle \vdash \Diamond A}\ \Diamond_i}{A \vdash \Box \Diamond A}\ \Box_i$$

Neither of these derivations should be surprising given the two intuitions we have indicated above: if we see $\Diamond A$ as $A \bullet d$ and $\Box A$ as A / b then $\Diamond \Box A$ corresponds to $(A / d) \bullet d$ and we have $(A / d) \bullet d \vdash A$. Similarly, we can see $A \vdash \Box \Diamond A$ as $A \vdash (A \bullet d) / d$. On the other hand, the temporal interpretation gives as that $\Diamond \Box A$ corresponds to FHA, ie. at some point in the future it will be the case that A has always been true in the past, but if this is true, then A is true now as well. Now suppose A is true 'now'. Then, at any point in the past, it has been the case that there was a point in the future — namely 'now' — such that A held, or A implies HFA[2], which corresponds to $A \vdash \Box \Diamond A$. Though they should be used with care, both intuitions can be valuable.

So the unary modes permit some simple derivability patterns, of which we have seen two so far. However, the relations between different unary prefixes don't end there. Figure 5.11 shows the relations between unary prefixes with equal numbers of diamonds and boxes up to length four (the formulae $\Diamond \Box \Diamond \Box A$ and $\Box \Diamond \Box \Diamond A$ are not shown in the figure; these formulae are the topic of Exercise 5.12). We can increase the complexity of the graph by augmenting the length of the prefix (with each increase augmenting the number of non-equivalent prefixes), but for most applications, the graph of Figure 5.11 suffices.

An important linguistic application of the unary modalities is their use as linguistic *features* (Heylen, 1999), but they have also been used for negative polarity items (Bernardi, 2002), quantifier scope (Bernardi and Moot, 2003) and clitics (Moot and Retoré, 2006).

Example 5.7 (English case). In English, the pronouns 'he' and 'she' can only occur in subject position, whereas the pronouns 'him' and 'her' can only occur in non-subject positions. Proper nouns, like 'Mary' can occur in either position. The examples below illustrate this.

(5.7) he likes Mary.
(5.8) $*$ her likes Mary.
(5.9) $*$ Mary likes he.
(5.10) Mary likes her.

The following lexicon distinguishes between English pronouns in nominative (mode n) and accusative case (mode a).

Word	Type(s)
he	$\Box_n \Diamond_n np$
her	$\Box_a \Diamond_a np$
Mary	np
likes	$(\Box_n \Diamond_n np \setminus S) / \Box_a \Diamond_a np$

[2] $A \to HFA$ is one of the axioms of minimal tense logic. $FHA \to A$ is easily derived from the minimal tense logic axiom $A \to GPA$.

The verb selects a subject in nominative case and an object in accusative case. Note how the assignment of $\Box_j \Diamond_j np$ to both arguments of the verb permits a proper noun like 'Mary', which retains its simple np type, to play the role of subject as well as object by means of the derivability relation $np \vdash \Box_j \Diamond_j np$.

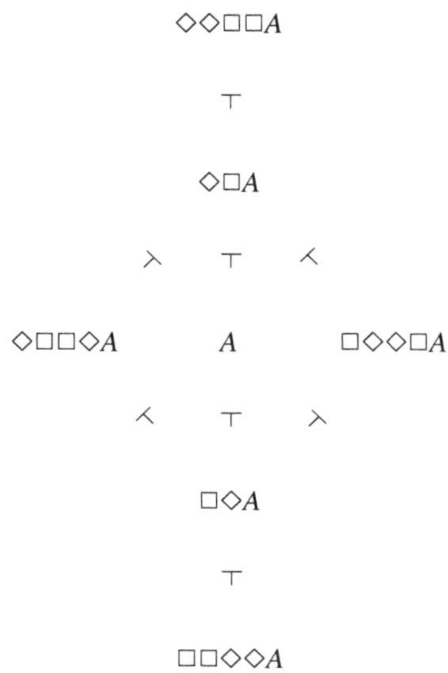

Fig. 5.11. Derivability relations between different formulae using the unary logical rules

5.2.3 Structural Rules

The unary connectives extend the possibilities for the structural rules in several ways. First of all, we can introduce structural rules which change the properties of the unary connectives (the names T and 4 for these rules correspond to the axioms of the same name in modal logic; it is easy to verify that structural rule T for unary mode j allows us to derive $\Box_j A \vdash A$ and that structural rule 4 allows us to derive $\Box_j A \vdash \Box_j \Box_j A$).

$$\frac{\Gamma[\langle\Delta\rangle^j] \vdash C}{\Gamma[\Delta] \vdash C}\,T \qquad \frac{\Gamma[\langle\Delta\rangle^j] \vdash C}{\Gamma[\langle\langle\Delta\rangle^j\rangle^j] \vdash C}\,4$$

Secondly, we can have interaction rules which tell us how the unary structural connectives distribute over the binary ones, either moving to the left, the right or both branches (the K rule corresponds to the axiom of the same name in modal logic, $K1$ and $K2$ modify K by moving the unary structural connective j to only one of the two daughters of a binary structural connective i).

$$\frac{\Gamma[(\langle\Delta_1\rangle^j, \langle\Delta_2\rangle^j)^i] \vdash C}{\Gamma[\langle(\Delta_1, \Delta_2)^i\rangle^j] \vdash C} K$$

$$\frac{\Gamma[(\langle\Delta_1\rangle^j, \Delta_2)^i] \vdash C}{\Gamma[\langle(\Delta_1, \Delta_2)^i\rangle^j] \vdash C} K1 \quad \frac{\Gamma[(\Delta_1, \langle\Delta_2\rangle^j)^i] \vdash C}{\Gamma[\langle(\Delta_1, \Delta_2)^i\rangle^j] \vdash C} K2$$

Modal logicians may be surprised that K is not a valid principle in the base logic. The reason is of course that — as we have seen in the previous chapter — the binary connectives are modal operators themselves. It is easily verified that all proof attempts of K fail in NL$\diamond$ without structural rules. If we want K to be valid, we need to add it explicitly in the form of a structural rule.

Note that adding the K, T and 4 structural rules together will give us the logical rules for the exponentials of linear logic and that they allow us to derive the characteristic axioms of the modal logic S4: that is, for all formulae A we can derive $\Box A \vdash A$, $\Box A \vdash \Box\Box A$ and $\Box(A \setminus B) \vdash \Box A \setminus \Box B$.

In practice, the T and 4 structural rules are never used in multimodal categorial grammars. Moortgat (1997, definition 4.13) (Moortgat, 2011, definition 2.4.10) remarks that we can simulate $\Box_{t4}A$ (where $t4$ is a mode having access to T and 4 but not K) by $\diamond_0\Box_0 A$, where 0 is a mode to which neither T nor 4 applies. This translation makes use of the derivability of $\diamond_0\Box_0 A \vdash A$ to replace $\Box_{tu}A \vdash A$ (Exercise 5.13 at the end of this chapter asks you to demonstrate that this translation behaves correctly in other respects as well).

To finish the discussion about the T and 4 structural rules, the difference between external and internal modes should again be emphasized. Remember that when we want to decide whether or not a list of formulae $A_1, \dots, A_n$ derives a formula C, the real question is rather: what antecedent term Γ, which contains a subset $E \subseteq I$ of the total modes I and which has $A_1, \dots, A_n$ as its yield, allows us to derive $\Gamma \vdash C$. We need to decide for the unary modes as well if we allow them to appear in Γ, selecting a subset E_1 of J to be external modes. To illustrate why this is important, look at the following simply example: if we want to decide whether $A \vdash \diamond_0 A$ is derivable for $\diamond_0$ without any structural rules, then there exists an antecedent term $\langle A\rangle^0$, such that $\langle A\rangle^0 \vdash \diamond_0 A$ is derivable. Or, in other words, if unary mode 0 is external (that is, $0 \in E_1$), then — since we allow ourselves to add unary parentheses 0 whenever needed — T becomes derivable even without any structural rules.

Example 5.8 (Dutch verb clusters II)
The unary modalities give us the tools we need to refine our analysis of Dutch verb clusters to require that the verbs in the cluster occur adjacently. In addition to the two binary modes 4 for head adjunction and 0 for phrasal adjunction, we use two unary modes 0 for phrasal heads and 1 for lexical heads. The revised lexicon is shown below.

Word	Type(s)	Translation
Marie	np	*Mary*
Jan	np	*John*
wil	$\Box_1((np\backslash_0 S)/_4 inf)$	*wants to*
plagen	$\Box_1(np\backslash_0 inf)$	*tease*

Note how the only difference between the previous lexicon is that the two verbs now have $\Box_1$ as their main connective to indicate they are lexical heads.

The additional structural rules we need are the following.

$$\frac{\Gamma[(\Delta_1,\langle\Delta_2\rangle^0)^0]\vdash C}{\Gamma[\langle(\Delta_1,\Delta_2)^0\rangle^0]\vdash C}\,K2 \qquad \frac{\Gamma[(\langle\Delta_1\rangle^1,\langle\Delta_2\rangle^1)^4]\vdash C}{\Gamma[\langle(\Delta_1,\Delta_2)^4\rangle^1]\vdash C}\,K \qquad \frac{\Gamma[\langle\Delta\rangle^1]\vdash C}{\Gamma[\langle\Delta\rangle^0]\vdash C}\,I$$

The $K2$ rule searches the phrasal head. Given that Dutch subordinate clauses are verb-final, it searches the head on the right branch. The I rule says that every lexical head is a phrasal head as well. Finally, the K rule states that we can combine two adjacent lexical heads together into a complex head. The K rule will force the verbs to appear in a single cluster and the $K2$ rule will require that it occurs clause-finally.

We can see these structural rules in action in the proof below.

$$\begin{array}{cl}
\vdots & \\
(np,(np,((np\backslash_0 S)/_4 inf, np\backslash_0 inf)^4)^0)^0\vdash S & \\
\hline
(np,(np,((np\backslash_0 S)/_4 inf,\langle\Box_1(np\backslash_0 inf))\rangle^1)^4)^0)^0\vdash S & \Box_h \\
\hline
(np,(np,(\langle\Box_1((np\backslash_0 S)/_4 inf))\rangle^1,\langle\Box_1(np\backslash_0 inf))\rangle^1)^4)^0)^0\vdash S & \Box_h \\
\hline
(np,(np,\langle(\Box_1((np\backslash_0 S)/_4 inf),\Box_1(np\backslash_0 inf))^4\rangle^1)^0)^0\vdash S & K \\
\hline
(np,(np,\langle(\Box_1((np\backslash_0 S)/_4 inf),\Box_1(np\backslash_0 inf))^4)\rangle^0)^0)^0\vdash S & I \\
\hline
(np,\langle(np,(\Box_1((np\backslash_0 S)/_4 inf),\Box_1(np\backslash_0 inf))^4))^0\rangle^0)^0\vdash S & K2 \\
\hline
\langle(np,(np,(\Box_1((np\backslash_0 S)/_4 inf),\Box_1(np\backslash_0 inf))^4)^0)^0\rangle^0\vdash S & K2 \\
\hline
(np,(np,(\Box_1((np\backslash_0 S)/_4 inf),\Box_1(np\backslash_0 inf))^4)^0)^0\vdash \Box_0 S & \Box_i
\end{array}$$

We can continue the proof above as shown in Example 5.3 and derive the correct word order as before.

Licensing and Restricting Structural Rules with Unary Modes

Another application of the unary modalities is in the form of modally licensed structural rules. This is the application we have seen for the linear logic unary connectives and it applies equally well here. Figure 5.12 shows the structural rules for mixed associativity and mixed commutativity we have seen in Figure 5.5, but this time the structural rules are licensed by a unary mode $\langle . \rangle^1$. Figure 5.13 shows the structural rules in the form of trees.

If we translate $(\Gamma,\Delta)^1$ as $(\Gamma,\langle\Delta\rangle^1)^0$, the two packages of structural rules from Figure 5.5 and Figure 5.12 clearly behave the same. The structural rules which use the unary mode permit us to model the same "right branch" extraction we have seen in Section 5.1.1, but it is easy to add an additional structural rule of modally licensed commutativity (shown at the bottom of Figure 5.12) which moves the unary mode from a left branch to a right branch and the combination of these three rules therefore allows extraction from anywhere inside a context of binary modes 0 (what would be the problem of adding a version of *Com* to the rules of Figures 5.5 and 5.6?).

So the modally licensed structural rules can either implement right branch extraction, using the top row of Figure 5.12 (the structural rules $MA\diamond_1$ and $MC\diamond_1$) or extraction from anywhere inside an antecedent term built from binary mode 0, using all three rules of the figure ($MA\diamond_1$, $MC\diamond_1$ and $Com\diamond_1$).

$$\frac{\Gamma[(\Delta_1,(\Delta_2,\langle\Delta_3\rangle^1)^0)^0]\vdash C}{\Gamma[((\Delta_1,\Delta_2)^0,\langle\Delta_3\rangle^1)^0]\vdash C}\,MA\diamond_1 \qquad \frac{\Gamma[((\Delta_1,\langle\Delta_3\rangle^1)^0,\Delta_2)^0]\vdash C}{\Gamma[((\Delta_1,\Delta_2)^0,\langle\Delta_3\rangle^1)^0]\vdash C}\,MC\diamond_1$$

$$\frac{\Gamma[(\langle\Delta_2\rangle^1,\Delta_1)^0]\vdash C}{\Gamma[(\Delta_1,\langle\Delta_2\rangle^1)^0]\vdash C}\,Com\diamond_1$$

Fig. 5.12. Mixed associativity and mixed commutativity licensed by $\diamond_1$ (top row) and commutativity for $\diamond_1$ (bottom row)

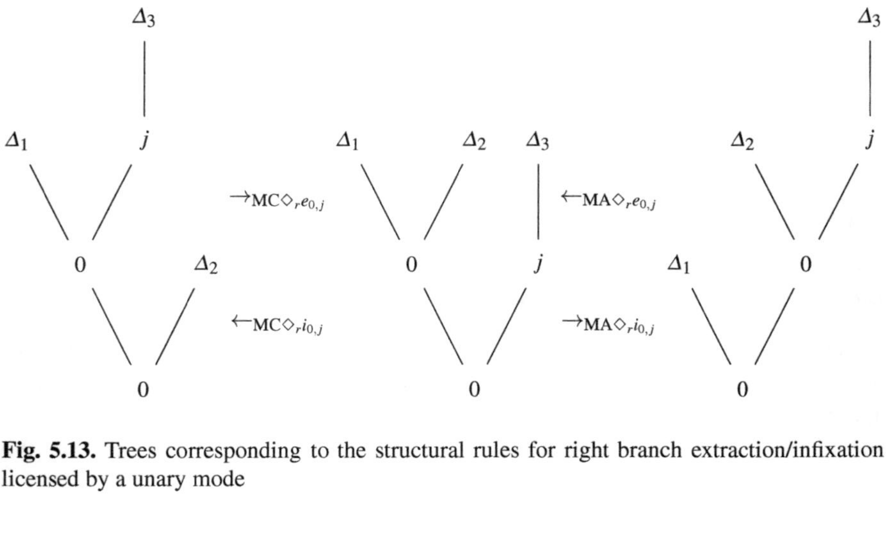

Fig. 5.13. Trees corresponding to the structural rules for right branch extraction/infixation licensed by a unary mode

The structural rules for left branch extraction and infixation are symmetric; they correspond to replacing $(\Gamma,\Delta)^j$ by $(\langle\Gamma\rangle^j,\Delta)^0$ in Figures 5.6 and 5.8 or to taking the left-to-right mirror image of the trees in Figure 5.13.

Example 5.9 (Medial extraction II). With all this in place, we can revisit Example 5.4, which treated medial extraction in English.

Replacing $!np$ by $\Diamond_1\Box_1 np$ as discussed above gives us the following lexicon.

Word	Type(s)
book	n
which	$(n\backslash_0 n)/_0(S/_0\Diamond_1\Box_1 np)$
John	np
fell	$np\backslash_0 S$
read	$(np\backslash_0 S)/_0 np$
yesterday	$S\backslash_0 S$

This lexicon, together with the structural rules of Figure 5.12 allows us to derive the medial extraction cases of Example 5.4. The structural rules allow us to extract from right branches only and is thus more restricted than full commutativity. However, as Moortgat (1999) argues, this is actually a point in favor of the current analysis and the differences between left branch and right branch extraction explain some differences in extraction between English and Dutch. For example, given that a preposition is assigned the formula pp/np, right branch extraction (as used in Moortgat's analysis of English) allows us to extract noun phrases from prepositional phrases, whereas left branch extraction (used in Moortgat's analysis of Dutch) does not.

The derivation below shows how "which John read yesterday" is of type $n\backslash_0 n$. The derivation starts with the hypothesis rule for $/$, which requires us to prove that "John read yesterday" is of type $S/_0\Diamond_1\Box_1 np$. We introduce $\Diamond_1\Box_1 np$ on the right branch using the $/_i$ rule, then introduce the unary brackets by means of the $\Diamond_h$ rule. This produces the correct configuration to apply the structural rules; we move the np past the adverb and to the object position of the verb using the *MC* and *MA* rules respectively. Finally, we eliminate the structural bracketing and finish the proof using the $\backslash_h$ and $/_h$ rules (not displayed in the proof, but an easy exercise).

$$\dfrac{\dfrac{}{n\backslash_0 n\vdash n\backslash_0 n}\,axiom \qquad \dfrac{\dfrac{\dfrac{\dfrac{\dfrac{\dfrac{\vdots}{((np,((np\backslash S)/np,np)^0)^0,S\backslash_0 S)^0\vdash S\vdash S}}{((np,((np\backslash_0 S)/_0 np,\langle\Box_1 np\rangle^1)^0)^0,S\backslash_0 S)^0\vdash S\vdash S}\,\Box_h}{(((np,(np\backslash_0 S)/_0 np)^0,\langle\Box_1 np\rangle^1)^0,S\backslash_0 S)^0\vdash S}\,MA\Diamond}{(((np,(np\backslash_0 S)/_0 np)^0,S\backslash_0 S)^0,\langle\Box_1 np\rangle^1)^0\vdash S}\,MC\Diamond}{(((np,(np\backslash_0 S)/_0 np)^0,S\backslash_0 S)^0,\Diamond_1\Box_1 np)^0\vdash S}\,\Diamond_h}{((np,(np\backslash_0 S)/_0 np)^0,S\backslash_0 S)^0\vdash S/_0\Diamond_1\Box_1 np}\,/_i}{(((n\backslash_0 n)/_0(S/_0\Diamond_1\Box_1 np),(np,(np\backslash_0 S)/_0 np)^0)^0,S\backslash_0 S)^0\vdash n\backslash_0 n}\,/_h$$

In addition to using the unary modes to *license* structural rules, we can also use them to impose *constraints* on the application of structural rules. So, for example, while the sequent

$$(A /_a B, B /_a C) \vdash A /_a C$$

is derivable assuming associativity for mode a, replacing the succedent formula $A /_a C$ by $\Box_0 A /_a C$ makes the sequent underivable, as shown by the following partial proof.

$$\dfrac{\dfrac{\dfrac{???}{(\langle (A /_a B, B /_a C)^a \rangle^0, C)^a \vdash A} /_i}{\langle (A /_a B, B /_a C)^a \rangle^0 \vdash A /_a C} \Box_i}{(A /_a B, B /_a C)^a \vdash \Box_0 (A /_a C)}$$

The $\Box_i$ rule puts a pair of unary brackets around the antecedent term. Applying the $/_i$ rule afterwards adds a C formula to the antecedent. Crucially, the unary brackets block the associativity rule (which would be applicable in its absence) thereby making the sequent underivable.

Kurtonina and Moortgat (1997) show that this type of result can be fully generalized: from any of the base logics (NL, L, NLP and LP) we can:

- recover a weaker logic by using a systematic translation which adds unary modalities to *constrain* structural rule applications,
- recover a stronger logic by using a systematic translation which adds unary modalities to *license* the application of structural rules normally unavailable in this logic.

Exercise 5.6 gives an example of how this type of result can be used to implement some well-known constraints from linguistics.

5.2.4 The General Form of Structural Rules

After these examples, it is time to define formally what are the structural rules which are allowed in the multimodal Lambek calculus.

Before we give the definition, we first present a slight generalization of context. A context, according to Definition 4.4, is an antecedent term with a single hole. We generalize this notion of context (as an antecedent term with a single hole) to a context with n holes with the additional requirement that the only leaves of the context are holes and that these holes occur in the indicated left-to-right order.

Definition 5.10. *A* generalized context $\Xi[]$ *is defined as follows.*

$$\mathscr{C} \quad ::= \quad [] \quad | \quad \langle \mathscr{C} \rangle^j \quad | \quad (\mathscr{C}, \mathscr{C})^i$$

The arity *of a generalized context is equal to the number of holes it contains.*

For a context $\Xi[]$ of arity n, we denote the simultaneous substitution *of n antecedent terms $\Gamma_1, \ldots, \Gamma_n$ in left-to-right order by $\Xi[\Gamma_1, \ldots, \Gamma_n]$. Since this fills all holes of $\Xi[]$ by antecedent terms, the result of this simultaneous substitution is an antecedent term.*

Now, given the definitions above and given two contexts $\Xi[]$ and $\Xi'[]$, where $\Xi[]$ is a non-empty context of arity n, $\Xi'[]$ is a context of the same arity n, and given a permutation π of n, we define the general structural rule schema as follows.

$$\frac{\Gamma[\Xi[\Delta_1, \ldots, \Delta_n]] \vdash C}{\Gamma[\Xi'[\Delta_{\pi_1}, \ldots, \Delta_{\pi_n}]] \vdash C} SR$$

Example 5.11. To illustrate the above definitions, let's look at an instantiation of this schema.

Define the two generalized contexts as $\Xi[] = ([], ([], [])^0)^4$, which has arity 3, and $\Xi'[] = ([], ([], [])^4)^0$ which has the same arity as required and the permutation π which assigns $\pi_1 = 2, \pi_2 = 1, \pi_3 = 3$. Performing simultaneous substitution for both $\Xi[\Delta_1, \Delta_2, \Delta_3]$ and $\Xi'[\Delta_{\pi_1}, \Delta_{\pi_2}, \Delta_{\pi_3}] = \Xi'[\Delta_2, \Delta_1, \Delta_3]$ gives us the following structural rule.

$$\frac{\Gamma[(\Delta_1, (\Delta_2, \Delta_3)^0)^4] \vdash C}{\Gamma[(\Delta_2, (\Delta_1, \Delta_3)^4)^0] \vdash C} MC_{r,4}$$

It is often convenient to restrict the size of $\Xi'[]$ to be equal to or strictly less than the size of $\Xi[]$. We will call a set of structural rules satisfying this condition *non-increasing*. Since both contexts have the same number of leaves, both have the same number of binary brackets by definition. So requiring the size of the contexts to be non-increasing amounts to saying that the number of unary branches cannot increase. As shown in (Moot, 2002), this requirement makes the calculus decidable: it is PSPACE complete and generates exactly the context-sensitive languages. Finding interesting classes of structural rules generating more restricted classes of languages — such as the so-called *mildly context-sensitive languages* — and with better computational complexity remains an interesting open question (though see Moot, 2008a,b, for some partial answers).

To conclude, the general form of structural rules is defined by replacing a context with n leaves by another context with a permutation of these leaves. These conditions on the structural rules prohibit the copying or deletion of material. Though some authors have added rules which do not respect these conditions to the Lambek calculus (notably Jäger, 2001, 2005, who adds a restricted form of contraction to the Lambek calculus to treat anaphora) we will consider only the logics without. Doing so has several advantages: first of all, as anyone who has seen a cut elimination proof of classical or intuitionistic logic knows, it keeps the cut elimination proof simple, secondly, it allows us to stay within the multiplicative fragment of linear logic (Danos and Regnier, 1989), a fragment which, as we will see in Chapters 6 and 7, has a particularly elegant proof theory.

5.2.5 Cut Elimination

To show that the new additions to the sequent calculus, the structural rules and the unary connectives, still give a calculus which satisfies cut elimination, we verify the new cases, following (Moortgat, 1996b).

The cut elimination theorem should by now be familiar (if not refer to Section 2.7 for L and Section 4.2.3 for NL). We look at proofs where all axioms are atomic and we look at the cut rule which is of the lowest depth, which will be of the following schematic form.

$$\dfrac{\dfrac{\vdots\,\gamma}{\Gamma \vdash D}\,R^a \qquad \dfrac{\vdots\,\delta}{\Delta[D] \vdash C}\,R^f}{\Delta[\Gamma] \vdash C}\,cut\ d$$

Remember that the structural rules only reorder the formulae and that for each formula in the conclusion of a structural rule we can find exactly one formula in the premise of this structural rule which corresponds to it (and vice versa). This means that the multiset of formulae does not change when we apply a structural rule and therefore that we do not have to consider the cases where a formula is introduced by a structural rule (which would correspond to a structural rule which operates like the structural rule of weakening) or cases where a formula is copied by a structural rule (which would correspond to a structural rule like contraction). Though it is unproblematic to handle contraction and or weakening for a cut elimination proof, their absence makes the structural rules easy to handle in the cut elimination proof.

1. If at least one of the rules is an axiom, we can remove the cut as before.

$$\dfrac{A \vdash A \qquad \begin{array}{c}\vdots\,\delta\\ \Gamma[D] \vdash C\end{array}}{\Gamma[D] \vdash C}\,cut \qquad \text{reduces to} \qquad \begin{array}{c}\vdots\,\delta\\ \Gamma[D] \vdash C\end{array}$$

2. If R^a does not produce the cut formula, we proceed by case analysis as before. In addition to rules $\backslash_h$, $/_h$ and $\bullet_h$ which we have seen before, rules $\Diamond_h$, $\Box_h$ and one of the structural rules *SR* can apply.

2 **R^a does not create D, the cut formula**		
R^a	Before reduction	After reduction
SR	$\dfrac{\dfrac{\overset{\vdots\,\delta}{\Delta \vdash D}}{\Delta' \vdash D}\,SR \qquad \overset{\vdots\,\gamma}{\Gamma[D] \vdash C}}{\Gamma[\Delta'] \vdash C}\,cut\,d$	$\dfrac{\dfrac{\overset{\vdots\,\delta}{\Delta \vdash D} \quad \overset{\vdots\,\gamma}{\Gamma[D] \vdash C}}{\Gamma[\Delta] \vdash C}\,cut\,d}{\Gamma[\Delta'] \vdash C}\,SR$
$\Diamond_h$	$\dfrac{\dfrac{\overset{\vdots\,\delta}{\Delta[\langle A\rangle^j] \vdash D}}{\Delta[\Diamond_j A] \vdash D}\,\Diamond_h \qquad \overset{\vdots\,\gamma}{\Gamma[D] \vdash C}}{\Gamma[\Delta[\Diamond_j A]] \vdash C}\,cut\ d$	$\dfrac{\dfrac{\overset{\vdots\,\delta}{\Delta[\langle A\rangle^j] \vdash D} \quad \overset{\vdots\,\gamma}{\Gamma[D] \vdash C}}{\Gamma[\Delta[\langle A\rangle^j]] \vdash D}\,cut\ d}{\Gamma[\Delta[\Diamond_j A]] \vdash C}\,\Diamond_h$
$\Box_h$	$\dfrac{\dfrac{\overset{\vdots\,\delta}{\Delta[A] \vdash D}}{\Delta[\langle\Box_j A\rangle^j] \vdash D}\,\Box_h \qquad \overset{\vdots}{\Gamma[D] \vdash C]}}{\Gamma[\Delta[\langle\Box_j A\rangle^j]] \vdash C}\,cut\,d$	$\dfrac{\dfrac{\overset{\vdots\,\delta}{\Delta[A] \vdash D} \quad \overset{\vdots}{\Gamma[D] \vdash C]}}{\Gamma[\Delta[A]] \vdash C}\,cut\,d}{\Gamma[\Delta[\langle\Box_j A\rangle^j]] \vdash C}\,\Box_h$

As indicated above, the structural rules keep the formulas in the antecedent (seen as a multiset) constant and therefore do not require special attention.

Note that if we want to be really precise, we have to verify that the previous cases are unaffected by the use of modes. However, since all these rules have the same general form of where R^a, which before reduction is applied in a context $\Delta[]$, is after reduction applied in a larger context $\Gamma[\Delta[]]$ this condition is trivially satisfied.

3. If R^f does not product the cut formula, we check all new rules.

   For the structural rule, we know that if Γ contains a distinguished subformula D before application of the structural rule, then we can find its image D after application of the structural rule by the conditions on the structural rules.

   For the $\Diamond_h$ and the $\Box_h$ rule we will again denote by $\Gamma^{\{D:=\Delta\}}[]$ the context $\Gamma[]$ with the cut formula D replaced by Δ.

3 R^f does not create D, the cut formula		
R^f	Before reduction	After reduction
SR	$\dfrac{\overset{\vdots\,\delta}{\Delta \vdash D} \quad \dfrac{\overset{\vdots\,\gamma}{\Gamma[D] \vdash C}}{\Gamma'[D] \vdash C}\,SR}{\Gamma'[\Delta]}\,cutd$	$\dfrac{\dfrac{\overset{\vdots\,\delta}{\Delta \vdash D} \quad \overset{\vdots\,\gamma}{\Gamma[D] \vdash C}}{\Gamma[\Delta] \vdash C}\,cutd}{\Gamma'[\Delta]}\,SR$
$\Diamond_h$	$\dfrac{\overset{\vdots\,\delta}{\Delta \vdash D} \quad \dfrac{\overset{\vdots\,\gamma}{\Gamma[\langle A\rangle^j] \vdash C}}{\Gamma[\Diamond_j A] \vdash C}\,\Diamond_h}{\Gamma^{\{D:=\Delta\}}[\Delta[\Diamond_j A]] \vdash C}\,cut\ d$	$\dfrac{\dfrac{\overset{\vdots\,\delta}{\Delta \vdash D} \quad \overset{\vdots\,\gamma}{\Gamma[\langle A\rangle^j] \vdash C}}{\Gamma^{\{D:=\Delta\}}[\langle A\rangle^j] \vdash C}\,cut\ d}{\Gamma^{\{D:=\Delta\}}[\Diamond_j A] \vdash C}\,\Diamond_h$
$\Box_h$	$\dfrac{\overset{\vdots\,\delta}{\Delta \vdash D} \quad \dfrac{\overset{\vdots\,\gamma}{\Gamma[A] \vdash C}}{\Gamma[\langle \Diamond_j A\rangle^j] \vdash C}\,\Box_h}{\Gamma^{\{D:=\Delta\}}[\Delta[\langle \Box_j A\rangle^j]] \vdash C}\,cut\ d$	$\dfrac{\dfrac{\overset{\vdots\,\delta}{\Delta \vdash D} \quad \overset{\vdots\,\gamma}{\Gamma[A] \vdash C}}{\Gamma^{\{D:=\Delta\}}[A] \vdash C}\,cut\ d}{\Gamma^{\{D:=\Delta\}}[\langle \Box_j A\rangle^j] \vdash C}\,\Box_h$
$\Diamond_i$	$\dfrac{\overset{\vdots\,\delta}{\Delta \vdash D} \quad \dfrac{\overset{\vdots\,\gamma}{\Gamma \vdash C}}{\langle \Gamma[D]\rangle^j \vdash \Diamond_j C}\,\Diamond_i}{\langle \Gamma[\Delta]\rangle^j \vdash \Diamond_j C}\,cutd$	$\dfrac{\dfrac{\overset{\vdots\,\delta}{\Delta \vdash D} \quad \overset{\vdots\,\gamma}{\Gamma[D] \vdash C}}{\Gamma[\Delta] \vdash C}\,cutd}{\langle \Gamma[\Delta]\rangle^j \vdash \Diamond_j C}\,\Diamond_i$
$\Box_i$	$\dfrac{\overset{\vdots\,\delta}{\Delta \vdash D} \quad \dfrac{\overset{\vdots\,\gamma}{\langle \Gamma[D]\rangle^j \vdash C}}{\Gamma[D] \vdash \Box_j C}\,\Box_i}{\Gamma[\Delta] \vdash \Box_j C}\,cutd$	$\dfrac{\dfrac{\overset{\vdots\,\delta}{\Delta \vdash D} \quad \overset{\vdots\,\gamma}{\langle \Gamma[D]\rangle^j \vdash C}}{\langle \Gamma[\Delta]\rangle^j \vdash C}\,cutd}{\Gamma[\Delta] \vdash \Box_j C}\,\Diamond_i$

Again we should verify that the binary connectives can still be converted as before with the addition of the mode information, but this is easily done.

4. Finally, we verify that when both R^a and R^f create the cut formula we can replace the cut by one or two cuts of lesser degree. We verify all cases except $/$, which is symmetric to $\backslash$.

	4 Both R^a and R^f create the cut-formula	
	Before reduction	After reduction
$\bullet$	$\dfrac{\dfrac{\overset{\vdots\,\delta}{\Delta \vdash A} \quad \overset{\vdots\,\theta}{\Theta \vdash B}}{(\Delta,\Theta)^i \vdash A \bullet_i B}\bullet_i \quad \dfrac{\overset{\vdots\,\gamma}{\Gamma[(A,B)^i] \vdash C}}{\Gamma[(A \bullet_i B)] \vdash C}\bullet_h}{\Gamma[(\Delta,\Theta)^i] \vdash C}\,\mathit{cut}\ d$	$\dfrac{\overset{\vdots\,\delta}{\Delta \vdash A} \quad \dfrac{\overset{\vdots\,\theta}{\Theta \vdash B} \quad \overset{\vdots\,\gamma}{\Gamma[(A,B)^i] \vdash C}}{\Gamma[(A,\Theta)^i] \vdash C}\,\mathit{cut} < d}{\Gamma[(\Delta,\Theta)^i] \vdash C}\,\mathit{cut} < d$
$\backslash$	$\dfrac{\dfrac{\overset{\vdots\,\delta}{(A,\Delta)^i \vdash B}}{\Delta \vdash A \backslash B}\backslash_i \quad \dfrac{\overset{\vdots\,\gamma}{\Gamma[B] \vdash C} \quad \overset{\vdots\,\theta}{\Theta \vdash A}}{\Gamma[(\Theta, A \backslash_i B)^i] \vdash C}\backslash_h}{\Gamma[(\Theta,\Delta)^i] \vdash C}\,\mathit{cut}\ d$	$\dfrac{\dfrac{\overset{\vdots\,\theta}{\Theta \vdash A} \quad \overset{\vdots\,\delta}{(A,\Delta)^i \vdash B}}{(\Theta,\Delta) \vdash B}\,\mathit{cut} < d \quad \overset{\vdots\,\gamma}{\Gamma[B] \vdash C}}{\Gamma[(\Theta,\Delta)^i] \vdash C}\,\mathit{cut} < d$
$\Diamond$	$\dfrac{\dfrac{\overset{\vdots\,\delta}{\Delta \vdash A}}{\langle \Delta \rangle^j \vdash \Diamond_j A}\Diamond_i \quad \dfrac{\overset{\vdots\,\gamma}{\Gamma[\langle A \rangle^j] \vdash C}}{\Gamma[\Diamond_j A] \vdash C}\Diamond_h}{\Gamma[\langle \Delta \rangle^j] \vdash C}\,\mathit{cut}\ d$	$\dfrac{\overset{\vdots\,\delta}{\Delta \vdash A} \quad \overset{\vdots\,\gamma}{\Gamma[\langle A \rangle^j] \vdash C}}{\Gamma[\langle \Delta \rangle^j] \vdash C}\,\mathit{cut} < d$
$\Box$	$\dfrac{\dfrac{\overset{\vdots\,\delta}{\langle \Delta \rangle^i \vdash B}}{\Delta \vdash \Box B}\Box_i \quad \dfrac{\overset{\vdots\,\gamma}{\Gamma[B] \vdash C}}{\Gamma[(\langle \Box B \rangle^j)] \vdash C}\Box_h}{\Gamma[(\langle \Delta \rangle^j)] \vdash C}\,\mathit{cut}\ d$	$\dfrac{\overset{\vdots\,\delta}{\langle \Delta \rangle^j \vdash B} \quad \overset{\vdots\,\gamma}{\Gamma[B] \vdash C}}{\Gamma[(\langle \Delta \rangle^j)] \vdash C}\,\mathit{cut} < d$

This completes our case analysis.

5.3 Natural Deduction

The multimodal Lambek calculus has a natural deduction calculus as well. Figure 5.14 shows the rules. The structural rules for natural deduction are the same as those of the sequent calculus.

Because of the similarity between the $\bullet_e$ rule and the $\Diamond_e$ rule, it is slightly less natural for the unary connectives. For example, normalization requires commutative conversions like we need for the Lambek calculus with product (Section 2.6.3).

Though it might be argued that linguistic applications of • in the Lambek calculus are rather limited, the $\diamond$ connective plays a vital role in the applications we have seen, which makes the inconvenience of its natural deduction rules a more serious problem. We will see a solution which combines the good properties of natural deduction and sequent calculus in Chapter 7.

$$\frac{\Gamma \vdash A \quad \Delta \vdash A \backslash_i B}{(\Gamma,\Delta)^i \vdash B} \backslash_e \qquad \frac{(A,\Gamma)^i \vdash C}{\Gamma \vdash A \backslash_i C} \backslash_i$$

$$\frac{\Delta \vdash B /_i A \quad \Gamma \vdash A}{(\Delta,\Gamma)^i \vdash B} /_e \qquad \frac{(\Gamma,A)^i \vdash C}{\Gamma \vdash C /_i A} /_i$$

$$\frac{\Delta \vdash A \bullet_i B \quad \Gamma[(A,B)^i] \vdash C}{\Gamma[\Delta] \vdash C} \bullet_e \qquad \frac{\Delta \vdash A \quad \Gamma \vdash B}{(\Delta,\Gamma)^i \vdash A \bullet_i B} \bullet_i$$

$$\frac{\Gamma \vdash \Box_j B}{\langle \Gamma \rangle^j \vdash B} \Box_e \qquad \frac{\langle \Gamma \rangle^j \vdash C}{\Gamma \vdash \Box_j C} \Box_i$$

$$\frac{\Delta \vdash \diamond_j A \quad \Gamma[\langle A \rangle^j] \vdash C}{\Gamma[\Delta] \vdash C} \diamond_e \qquad \frac{\Gamma \vdash A}{\langle \Gamma \rangle^j \vdash \diamond_j A} \diamond_i$$

$$\frac{}{A \vdash A} \textit{axiom}$$

Fig. 5.14. Natural deduction rules for the multimodal Lambek calculus

Example 5.12. The sequent proof shown in Example 5.9 (on page 169) corresponds to the natural deduction proof shown below. Note that in order to save some space, we have replaced the formulas in the antecedent by the first letter of the corresponding word, as indicated by the table below

$$\begin{aligned}
&\mathit{Lex}(\text{which}) &&= w = (n \backslash n) / (S / \diamond_1 \Box_1 np)\\
&\mathit{Lex}(\text{John}) &&= j = np\\
&\mathit{Lex}(\text{read}) &&= r = (np \backslash_0 S) /_0 np\\
&\mathit{Lex}(\text{yesterday}) &&= y = S /_0 S
\end{aligned}$$

These are of course the same lexical entries as those in Example 5.9.

$$
\dfrac{w \vdash (n \backslash_0 n) /_0 (S /_0 \Diamond_1 \Box_1 np) \qquad \dfrac{\dfrac{\Diamond_1 \Box_1 np \vdash \Diamond_1 \Box_1 np \qquad \dfrac{\dfrac{\dfrac{\dfrac{j \vdash np \qquad \dfrac{r \vdash (np \backslash_0 S) /_0 np \qquad \dfrac{\Box_1 np \vdash \Box_1 np}{\langle \Box_1 np \rangle^1 \vdash np}\;\Box_e}{(r, \langle \Box_1 np \rangle^1)^0 \vdash np \backslash_0 S}\;/_e}{(j, (r, \langle \Box_1 np \rangle^1)^0)^0 \vdash S}\;\backslash_e \qquad y \vdash S \backslash_0 S}{((j, (r, \langle \Box_1 np \rangle^1)^0)^0, y)^0 \vdash S}\;\backslash_e}{(((j, r)^0, \langle \Box_1 np \rangle^1)^0, y)^0 \vdash S}\;MA\Diamond}{(((j, r)^0, y)^0, \langle \Box_1 np \rangle^1)^0 \vdash S}\;MC\Diamond}{(((j, r)^0, y)^0, \Diamond_1 \Box_1 np)^0 \vdash S}\;\Diamond_e}{((j, r)^0, y)^0 \vdash S /_0 \Diamond_1 \Box_1 np}\;/_i}{(w, ((j, r)^0, y)^0)^0 \vdash n \backslash n}
$$

To start the proof, we use the $\Box_1 np \vdash \Box_1 np$ axiom followed by the $\Box_e$ rule, which gives us the sequent $\langle \Box_1 np \rangle^1 \vdash np$, which we can use as a normal np to derive "John read $\langle \Box_1 np \rangle^1$ yesterday" as being of type S, after which we can apply the structural rules like we did in the sequent calculus derivation to derive "John read yesterday" of type $S /_0 \Diamond_1 \Box_1 np$. The only slight complication is the $\Diamond_e$ rule.

Equivalence between the natural deduction calculus and the sequent calculus is again easily established. We only show the unary cases. As for the binary rules, both calculi share the introduction rules, so we only need to show that the elimination rules are equivalent to the hypothesis rules.

Natural Deduction to Sequent Calculus

Replace:	with:
$\dfrac{\Gamma \vdash \Box C}{\langle \Gamma \rangle^j \vdash C}\;\Box_e$	$\dfrac{\Gamma \vdash \Box C \qquad \dfrac{\dfrac{}{C \vdash C}\;ax}{\langle \Box_j C \rangle^j \vdash C}\;\Box_h}{\langle \Gamma \rangle^j \vdash C}\;cut$
$\dfrac{\Delta \vdash \Diamond_j A \quad \Gamma[\langle A \rangle^j] \vdash C}{\Gamma[\Delta] \vdash C}\;\Diamond_e$	$\dfrac{\Delta \vdash \Diamond_j A \qquad \dfrac{\Gamma[\langle A \rangle^j] \vdash C}{\Gamma[\Diamond_j A] \vdash C}\;\Diamond_h}{\Gamma[\Delta] \vdash C}\;cut$

Sequent Calculus to Natural Deduction

If the proof ends with a $\Box_h$ rule:

$$
\dfrac{\begin{matrix} \vdots\,\gamma \\ \Gamma[A] \vdash C \end{matrix}}{\Gamma[\langle \Box_j A \rangle^j] \vdash C}\;\Box_h
$$

then by induction hypothesis we have a natural deduction proof, γ^* of $\Gamma[A] \vdash C$ which we extend as follows.

Replace:	with:
$\dfrac{}{A \vdash A}\,axiom$ $\vdots\,\gamma^*$ $\Gamma[A] \vdash C$	$\dfrac{\dfrac{}{\Box_j A \vdash \Box_j A}\,axiom}{\langle \Box_j A \rangle \vdash A}\,\Box_e$ $\vdots\,\gamma^*$ $\Gamma[\langle \Box_j A \rangle^j] \vdash C$

If the proof ends with a $\Diamond_h$ rule:

$$\dfrac{\begin{array}{c}\vdots\,\gamma\\ \Gamma[\langle A \rangle^j] \vdash C\end{array}}{\Gamma[\Diamond_j A] \vdash C}\,\Diamond_h$$

then by induction hypothesis we have a natural deduction proof, γ^* of $\Gamma[\langle A \rangle^j] \vdash C$ which we extend as follows.

$$\dfrac{\dfrac{}{\Diamond_j A \vdash \Diamond_j A}\,axiom \qquad \begin{array}{c}\vdots\,\gamma^*\\ \Gamma[\langle A \rangle^j] \vdash C\end{array}}{\Gamma[\Diamond_j A] \vdash C}\,\Diamond_e$$

5.4 Axiomatic Presentation

As a prelude to our section on model theory we will present an axiomatic presentation for the unary connectives in the style of the axiomatic presentation of NL which we saw in Section 4.4.

The axiomatic presentation is a simple extension of the axiomatic presentation of NL.

We add two new axioms and two new rules. The new calculus is displayed in Figure 5.15. There are two new axioms, (*Unit*) and (*Co-unit*), which are the unary equivalents of application and co-applications and two new monotonicity rules (one for each of the new connectives).

We need to add a new case to the translation from antecedent terms to formulae and verify this does not affect the proof of the Substitution Lemma, Lemma 4.12. Both are trivial.

Definition 5.13. *We define the function* $\|.\|^\bullet$ *from NL*$\Diamond$ *antecedent terms to NL*$\Diamond$ *formulae as follows.*

$\|\mathsf{Lp}\|^\bullet = \mathsf{Lp}$
$\|\langle \mathscr{A} \rangle\|^\bullet = \Diamond \|\mathscr{A}\|^\bullet$
$\|(\mathscr{A}, \mathscr{A})\|^\bullet = \|\mathscr{A}\|^\bullet \bullet \|\mathscr{A}\|^\bullet$

Axioms

$$\frac{}{A \vdash A}\,(Id)$$

$$\frac{}{\Diamond\Box A \vdash A}\,(Unit) \qquad \frac{}{A \vdash \Box\Diamond A}\,(Co\text{-}unit)$$

$$\frac{}{A \bullet (A \backslash B) \vdash B}\,(Appl\backslash) \qquad \frac{}{(B / A) \bullet A \vdash B}\,(Appl/)$$

$$\frac{}{A \vdash B \backslash (B \bullet A)}\,(Co\text{-}appl\backslash) \qquad \frac{}{A \vdash (A \bullet B) / B}\,(Co\text{-}appl/)$$

Rules

$$\frac{A \vdash B}{\Diamond A \vdash \Diamond B}\,(Mon\Diamond) \qquad \frac{A \vdash B}{\Box A \vdash \Box B}\,(Mon\Box)$$

$$\frac{A \vdash B \quad C \vdash D}{A \bullet C \vdash B \bullet D}\,(Mon\bullet) \qquad \frac{A \vdash B \quad C \vdash D}{B \backslash C \vdash A \backslash D}\,(Mon\backslash) \qquad \frac{A \vdash B \quad C \vdash D}{C / B \vdash D / A}\,(Mon/)$$

$$\frac{A \vdash B \quad B \vdash C}{A \vdash C}\,(Trans)$$

Fig. 5.15. Axiomatic or combinator presentation of NL$\Diamond$

Lemma 5.14. *If* $\|\Gamma[B]\|^{\bullet} \vdash C$ *and* $A \vdash B$ *then* $\|\Gamma[A]\|^{\bullet} \vdash C$

Proof. The inductive case for $\Diamond$ is a direct consequence of rule $(Mon\Diamond)$. $\Box$

Lemma 5.15. *The axiomatic presentation of NL*$\Diamond$ *is equivalent to the sequent calculus presentation of NL*$\Diamond$.

Proof. The proof for the binary connectives is unchanged from the proof of Lemma 4.13. We verify only the new cases.

The translation from axiomatic proofs to sequent proofs is trivial. The two new axioms are derived as follows

$$\frac{\dfrac{\dfrac{}{A \vdash A}\,axiom}{\langle \Box A \rangle \vdash A}\,\Box_h}{\Diamond\Box A \vdash A}\,\Diamond_h \qquad \frac{\dfrac{\dfrac{}{A \vdash A}\,axiom}{\langle A \rangle \vdash \Diamond A}\,\Diamond_i}{A \vdash \Box\Diamond A}\,\Box_i$$

and the two new rules

$$\frac{\dfrac{\dfrac{\vdots}{A \vdash A}}{\langle A \rangle \vdash \Diamond A}\,\Diamond_i}{\Diamond A \vdash \Diamond A}\,\Diamond_h \qquad \frac{\dfrac{\dfrac{\vdots}{A \vdash A}}{\langle \Box A \rangle \vdash A}\,\Box_h}{\Box A \vdash \Box A}\,\Box_i$$

For the translation from sequent proofs to axiomatic proofs, the rule $\Diamond_h$ is the identity function and the rule for $\Diamond_i$ corresponds directly to monotonicity rule ($Mon\Diamond$) in the axiomatic calculus.

The rule $\Box_h$ is an application of the Substitution Lemma with the axiom $\Diamond\Box A \vdash A$ (*Unit*). This leaves only the rule $\Box_i$, for which we have to prove that given an axiomatic proof of $\Diamond A \vdash C$ we can produce an axiomatic proof of $A \vdash \Box C$.

The following proof does exactly that.

$$\dfrac{\dfrac{}{A \vdash \Box\Diamond A}\,(Co\text{-}unit) \qquad \dfrac{\begin{matrix}\vdots\\ \Diamond A \vdash C\end{matrix}}{\Box\Diamond A \vdash \Box C}\,(Mon\Box)}{A \vdash \Box C}\,(Trans)$$

This completes our analysis of the new cases. □

5.5 Model Theory

As we have seen in Section 4.5, NL has Kripke models which use a ternary accessibility relation R^3 (which models the relation of two resources with their composition) for which we can establish soundness and completeness by means of a canonical model construction. As was first shown in Kurtonina and Moortgat (1997), this construction extends quite naturally to the unary connectives $\Diamond$ and $\Box$, which are interpreted by means of a binary accessibility relation R^2.

The interpretation of the complex formulae is given below.

$$\begin{array}{lll}
\mathcal{M},a \models A \bullet B & \Longleftrightarrow & \exists b \exists c (R^3abc \;\&\; \mathcal{M},b \models A \;\&\; \mathcal{M},c \models B)\\
\mathcal{M},a \models A \backslash B & \Longleftrightarrow & \forall b \forall c (R^3cba \;\&\; \mathcal{M},b \models A \Rightarrow \mathcal{M},c \models B)\\
\mathcal{M},a \models A / B & \Longleftrightarrow & \forall b \forall c (R^3cab \;\&\; \mathcal{M},b \models A \Rightarrow \mathcal{M},c \models B)\\
\mathcal{M},a \models \Diamond A & \Longleftrightarrow & \exists b (R^2ab \;\&\; \mathcal{M},b \models A)\\
\mathcal{M},a \models \Box A & \Longleftrightarrow & \forall b (R^2ba \Rightarrow \mathcal{M},b \models A)
\end{array}$$

Only the cases for $\Diamond$ and $\Box$ are new. Note that the interpretation of the unary connectives corresponds to the future possibility and past necessity operations if we interpret R^2ab as a occurs before b (even though this is not the intended interpretation in the current context).

Lemma 5.16 (Soundness). *If $A \vdash B$ is derivable then for all models M and worlds a, $M,a \models (A \vdash B)$*

Proof. We show that whenever there is an axiomatic proof p of $A \vdash B$, then $M,a \models A$ implies $M,a \models B$. This is again a trivial verification, which we prove by induction on the depth d of p.

If $d = 0$ then the proof consists of just an axiom. We verify only the new cases (*Unit*) and (*Co-unit*).

($Unit$) Suppose $M,b \models \Diamond\Box A$. We need to show that $M,b \models A$. Spelling out the definition of $M,b \models \Diamond\Box A$ gives us $\exists c(R^2bc \,\&\, M,c \models \Box A)$, which according to the interpretation of $\Box A$ gives $\exists c(R^2bc \,\&\, \forall d(R^2dc \Rightarrow M,d \models A))$. Let c be an arbitrary word and choose world b for d. This means that R^2bc and $(R^2bc \Rightarrow M,b \models A)$. Combining these two gives us $M,b \models A$.

(Co-$unit$) Suppose $M,b \models A$. We need to show that $M,b \models \Box\Diamond A$. Writing out the evaluation of $M,b \models \Box\Diamond A$ gives us $\forall c(R^2cb \Rightarrow M,c \models \Diamond A)$, which we can spell out further by evaluating $\Diamond A$, which produces $\forall c(R^2cb \Rightarrow \exists d(R^2cd \,\&\, M,d \models A))$. In other words, we have to show that in any world c such that R^2cb there exists a world d such that R^2cd and A is true at world d. Now take an arbitrary world c such that R^2cb and take b as value for d. World b satisfies R^2cb and satisfies $M,b \models A$ by our initial hypothesis. Therefore, any world such that $M,b \models A$ also satisfies $M,b \models \Box\Diamond A$.

If $d > 0$ we verify the new rules ($Mon\Diamond$) and ($Mon\Box$). We know by induction hypothesis that for all M and all c, if $M,c \models A$ then $M,c \models B$.

($Mon\Diamond$) We need to show that for all M and d, if $M,d \models \Diamond A$ then $M,d \models \Diamond B$. Suppose $M,d \models \Diamond A$. This means that there exists an e such that R^2de and $M,e \models A$. By induction hypothesis $M,e \models A$ implies $M,e \models B$, and given that there exists an e such that R^2de and $M,e \models B$, we can conclude $M,d \models \Diamond B$.

($Mon\Box$) We need to show that for all M and d, if $M,d \models \Box A$ then $M,d \models \Box B$. Suppose $M,d \models \Box A$. This means that for all e such that R^2ed we have $M,e \models A$. By applying the induction hypothesis, we can conclude that for all e such that R^2ed, $M,e \models B$, which means that $M,d \models \Box B$ as required. □

For the completeness proof, we extend the canonical model construction for NL to NL$\Diamond$. A canonical model for NL$\Diamond$ is a canonical model for NL with the addition of a binary accessibility relation R^2. The interpretation clause for R^2 links R^2 to $\Diamond$, just as the clause for R^3 links it to $\bullet$.

Definition 5.17. *A* canonical model $\mathcal{M} = \langle W, R^2, R^3 V\rangle$ *for NL$\Diamond$ is defined as follows:*

- *W is the set of all formulae,*
- *R^3abc iff $a \vdash b \bullet c$, and*
- *R^2ab iff $a \vdash \Diamond b$, and*
- *$a \models p$ iff $a \in V(p)$.*

Lemma 5.18. *The* Truth Lemma, $\mathcal{M}, a \models A$ *iff* $a \vdash A$

The proof of the Truth Lemma is a simple extension of the Truth Lemma for NL which we proved as Lemma 4.20. We treat only the new cases.

Proof

$\Longleftarrow$

Given that the canonical model is a model, soundness is a direct consequence of Lemma 5.16.

$\Longrightarrow$

For the completeness part we show that if $\mathcal{M}, a \models A$ then $a \vdash A$, which we prove by induction on A.

- $\Diamond A$ Suppose $\mathcal{M}, b \models \Diamond A$. We need to show that $b \vdash \Diamond A$. Given that $\mathcal{M}, b \models \Diamond A$, the definition of $\mathcal{M}, b \models \Diamond A$ gives us that there exists a c such that R^2bc and $\mathcal{M}, c \models A$. We can apply the induction hypothesis to $\mathcal{M}, c \models A$ to obtain an axiomatic proof $c \vdash A$, which we can extend by rule $(Mon\Diamond)$ to a proof of $\Diamond c \vdash \Diamond A$. The definition of R^2bc in the canonical model gives us $b \vdash \Diamond c$, which we can combine with the proof of $\Diamond c \vdash \Diamond A$ using the transitivity rule (*Trans*) to obtain a proof of $b \vdash \Diamond A$ as required.
- $\Box A$ Suppose $\mathcal{M}, b \models \Box A$. This means that for all c such that R^2cb we have $\mathcal{M}, c \models A$. Now, take $c = \Diamond b$. Given that $c \vdash \Diamond b$ we know by the definition of R^2 in the canonical model that R^2cb and therefore $\mathcal{M}, c \models A$ or, equivalently, $\mathcal{M}, \Diamond b \models A$. Given that A is a subformula of $\Box A$ we can apply the induction hypothesis to obtain an axiomatic proof of $\Diamond b \vdash A$. We can transform this proof into a proof of $b \vdash \Box A$ as follows.

$$\dfrac{\dfrac{}{b \vdash \Box\Diamond b}\,(\textit{Co-unit}) \qquad \dfrac{\begin{array}{c}\vdots\\ \Diamond b \vdash A\end{array}}{\Box\Diamond b \vdash \Box A}\,(\textit{Mon}\Box)}{b \vdash \Box A}\,(\textit{Trans})$$

This completes our analysis of the new cases and our soundness proof. □

Theorem 5.19. *NL◇ is sound and complete with respect to all models.*

Proof. We proved soundness in Lemma 5.16. As before, completeness is a simple consequence of the Truth Lemma and is proved in exactly the same way. We need to show that if a for every model M, $M \models (A \vdash B)$ then $A \vdash B$ is derivable. If this holds for every model, then it holds in particular for the canonical model $\mathcal{M}$, which by the definition of $\mathcal{M} \models (A \vdash B)$ means that for all worlds a, if $\mathcal{M}, a \models A$ then $\mathcal{M}, a \models B$. If this is true for all a, the it is true for A as well. Since by the Truth Lemma we have that $\mathcal{M}, A \models A$ (since $A \vdash A$ is derivable), we can therefore conclude $\mathcal{M}, A \models B$ which by application of the Truth Lemma means that $A \vdash B$ is derivable. □

5.5.1 Completeness for Weak Sahlqvist Postulates

We have given a soundness and completeness result for NL$\Diamond$ and general Kripke frames. In Section 4.5, we have seen that it is possible to add frame constraints corresponding to associativity and to commutativity. In this chapter, we will discuss the results of Kurtonina (1995, 1998) who extends modal logic's Sahlqvist-van Benthem strategy to give a general translation of a class of postulates into frame constraints in such a way that NL with the addition of one of these postulates is sound and complete with respect to a canonical model which satisfies the corresponding frame constraint.

Definition 5.20. *A* canonical model

$$\mathcal{M} = \langle W, \bigcup_{j \in J} R^2_i, \bigcup_{i \in I} R^3_i, V \rangle$$

for the multimodal version of NL$\Diamond$ is defined as follows:

- *W is the set of all formulae,*
- *For each $i \in I$, $R^3_i abc$ iff $a \vdash b \bullet_i c$, and*
- *For each $j \in J$, $R^2_j ab$ iff $a \vdash \Diamond_j b$, and*
- *$a \models p$ iff $a \in V(p)$.*

$$\begin{array}{lll}
\mathcal{M}, a \models A \bullet_i B & \Longleftrightarrow & \exists b \exists c (R^3_i abc \;\&\; \mathcal{M}, b \models A \;\&\; \mathcal{M}, c \models B) \\
\mathcal{M}, a \models A \backslash_i B & \Longleftrightarrow & \forall b \forall c (R^3_i cba \;\&\; \mathcal{M}, b \models A \;\Rightarrow\; \mathcal{M}, c \models B) \\
\mathcal{M}, a \models A /_i B & \Longleftrightarrow & \forall b \forall c (R^3_i cab \;\&\; \mathcal{M}, b \models A \;\Rightarrow\; \mathcal{M}, c \models B) \\
\mathcal{M}, a \models \Diamond_j A & \Longleftrightarrow & \exists b (R^2_j ab \;\&\; \mathcal{M}, b \models A) \\
\mathcal{M}, a \models \Box_j A & \Longleftrightarrow & \forall b (R^2_j ba \;\Rightarrow\; \mathcal{M}, b \models A)
\end{array}$$

Definition 5.21. *A* weak Sahlqvist axiom *is a statement of the form $A \vdash B$ such that*

- *A is a formula using only the connectives $\bullet$ and $\Diamond$,*
- *none of the atomic subformulae of A occurs more than once,*
- *B is a formula using only the connectives $\bullet$ and $\Diamond$ and which contains at least one occurrence of either $\bullet$ or $\Diamond$.*
- *all atomic subformulae of B are atomic subformulae of A.*

The definition of weak Sahlqvist axiom allows more structural rules than the class of structural rules we defined in Section 5.2.4. Since we required the context $\Xi[]$ to be non-empty, they are a proper subset of the weak Sahlqvist rules (and therefore benefit from the soundness and completeness result below). In particular, the definition above allows,

- multiple occurrences of atomic subformulae in B,
- atomic subformulae of A which are not atomic subformulae of B,

thereby permitting axioms like the following

$p \vdash p \bullet_{co} p$
$p \bullet_{wk} q \vdash p$

which would correspond to the following structural rules.

$$\frac{\Gamma[(\Delta,\Delta)^{co}] \vdash C}{\Gamma[\Delta] \vdash C} Co \qquad \frac{\Gamma[\Delta_1] \vdash C}{\Gamma[(\Delta_1,\Delta_2)^{wk}] \vdash C} Wk$$

Readers familiar with sequent calculus for classical or intuitionistic logic will immediately recognize these structural rules as versions of contraction and weakening.

Kurtonina's result is therefore stronger than we need in the current context. The main theorem, which we only state without proof (detailed proofs can be found in Kurtonina, 1995, 1998) is the following.

Theorem 5.22. *If $A \vdash B$ is a weak Sahlqvist axiom then:*

- *there is a first-order frame constraint FO which corresponds to $A \vdash B$, such that for any frame F, F satisfies FO if and only if $F \models (A \vdash B)$*
- *$\mathcal{M}_{FO}$ is a canonical model over frames satisfying FO.*
- *$NL\diamond + A \vdash B$ is sound and complete with respect to this canonical model $\mathcal{M}_{FO}$*

5.6 Concluding Remarks

We have seen the multimodal Lambek calculus and shown how it provides us with a flexible logic which is able to treat a wide range of linguistic phenomena. The class of structural rules allowed is rather large, which is both an advantage and a drawback: it is an advantage, since it gives us a lot of expressive power and gives us several options for giving an analysis of linguistic facts. It is a drawback since this makes it hard to choose *how* to model certain facts and since it gives us a rather high computational complexity — though it could be argued that PSPACE complete for non-increasing structural rules is not a lot worse than NP complete for the structural rules of associativity only.

So the big open question for multimodal categorial grammars is to find the "right" combination of structural rules both from the point of view of being able to make the relevant linguistic generalizations and from the point of view of computational complexity.

Vermaat (2005) makes a good case for using the mixed associativity and mixed commutativity structural rules of Figure 5.12 as the *universal* set of structural rules and gives a cross-linguistic account of *wh* extraction using this package of structural rules (Example 5.9 gives a very simple example of *wh* extraction). For the moment, we leave it as an important open question whether or not other structural rules are needed in our universal grammar.

Exercises for Chapter 5

Exercise 5.1. Use the multimodal calculus with modes n, a, nc and c and the structural rules shown in Figure 5.2 on page 152 to verify that $(A /_a B, B /_a C)^a \vdash A /_a C$ is derivable but $(A /_n B, B /_n C)^n \vdash A /_n C$ is not.

Similarly, verify that $A /_c B \vdash B \backslash_c A$ is derivable, but $A /_a B \vdash B \backslash_a A$ is not.

Exercise 5.2. Return to Exercise 4.8 on page 145 and give a multimodal lexicon (using the structural rules of Figure 5.2 on page 152) which handles all and only the correct sentences, using only one formula assignment per word.

Motivate your choice of modes for each lexical entry.

Exercise 5.3. The placement of adverbs is often rather free. Look at the following sequence of sentences (take from Dowty, 1997).

(5.11) Supposedly someone has been eating the cake.
(5.12) Someone supposedly has been eating the cake.
(5.13) Someone has supposedly been eating the cake.
(5.14) Someone has been supposedly eating the cake.
(5.15) Someone has been eating the cake, supposedly.

The adverb “supposedly” can occur at most places in the sentence (a similar observation is made by Morrill, 1994).

Give lexical entries for all words in the sentence; in particular, give a single lexical entry to “supposedly” which generates all sentences above. What are the structural rules required to assure this?

More difficult: how can we assure we generate *only* the correct permutations of “supposedly”?

Exercise 5.4. Pied-piping is the linguistic phenomenon where a relativizer — such as “which”, which we’ve seen in the previous exercise and in Examples 5.4 and 5.9, for example — can allow additional material to be fronted. Morrill (1994) gives the following example of the phenomenon.

(5.16) the contract which John talked about $[]_{np}$
(5.17) the contract about which John talked $[]_{pp}$

Sentence 5.16 gives the configuration without pied-piping. “which” selects for a sentence missing an np to its right. This is the “basic” type for a relativizer, of which we have seen examples before.

The difference between Sentence 5.16 and Sentence 5.17 is that the preposition “about” has moved to the position before the relativizer “which”.

Morrill (1994), using the wrapping structural rules we have seen in Figure 5.3 on page 153, proposes the following lexicon for Sentence 5.17

Word	Type(s)
the	$np\,/_a\,n$
contract	n
about	$pp\,/_a\,np$
which	$(pp\,/_w\,np)\,\backslash_w\,((n\,\backslash_a\,n)\,/_a\,(S\,/_a\,pp))$
John	np
talked	$(np\,\backslash_a\,S)\,/_a\,pp$

- using the lexicon above and the structural rules of Figure 5.3, show that "about which John talked" is an expression of type $n\,\backslash_a\,n$.

Exercise 5.5. In Combinatory Categorial Grammars (Steedman, 1997), the following rule of "mixed composition" is assumed.

$$A\,/\,B, C\,\backslash\,B \vdash C\,\backslash\,A$$

- Show that this sequent is underivable in the Lambek calculus.
- Show that given the structural rules of Figure 5.6, the following sequent *is* derivable in a multimodal Lambek calculus (interestingly, modern CCG analyses have adopted the ideas of the multimodal Lambek calculus and "index" their categories in a way very similar to the multimodal sequent shown below).

$$(A\,/\,B_0, C\,\backslash_2\,B)^0 \vdash C\,\backslash_2\,A$$

Exercise 5.6. One of the observations of Ross (1967) is that extraction, as we have seen it in Example 5.9 on page 169, is subject to certain constraints.

Kurtonina and Moortgat (1997) give the following examples as motivation for their use of unary connectives to constrain derivability (in his section 7.8, Morrill (1994) gives a similar analysis). They correspond to the constraints of Ross: we cannot extract from inside a coordination (as shown by example sentences 1 and 2) unless the same element is extracted from all conjuncts (sentence 3). These constraints are called the "coordinate structure constraint" and its exception is called the "across-the-board constraint".

(5.18) $*$the book that Melville wrote Moby Dick and John read $[]_{np}$.
(5.19) $*$the book that Melville wrote $[]_{np}$ and John read Moby Dick.
(5.20) the book that Melville wrote $[]_{np}$ and John read $[]_{np}$.

As we have seen in Section 4.2.2 (and before that in Exercise 4), in categorial grammars we assign words like "and" the formula $(X\,\backslash_0\,X)\,/_0\,X$, where X is instantiated by another formula.

- Show that given the analysis of extraction of Example 5.9, there are instantiations of X which make all three example sentences above derivable.
- Show that if we follow Kurtonina and Moortgat (1997) and change the type of the coordinator to $(X \backslash_0 \Box_0 X) /_0 X$, then sentences 1 and 2 above become underivable, whereas sentence 3 will stay derivable.

Exercise 5.7. Give an alternative solution for treating medial extraction (as discussed in Example 5.4 on page 160 using the structural rules of mixed associativity and mixed commutativity between a mode 0 one of the modes 1,2,3,4 discussed in Section 5.1.1. Use the lexicon below, adding only the formula for "which".

Word	Type(s)
book	n
which	???
John	np
fell	$np \backslash_0 S$
read	$(np \backslash_0 S) /_0 np$
yesterday	$S \backslash_0 S$

Take special care of

- which combination of mixed associativity and mixed commutativity you need (look at Figures 5.5 and 5.6 (page 157),
- the inclusion between modes to assure derivability.

Exercise 5.8. Example 5.3 provides a first step towards the treatment of Dutch verb clusters. While talking about the left branch extraction and infixation structural rules of Figure 5.6 on page 157, the possibility of using a deterministic version of the infixation rules is briefly mentioned. One way of implementing this idea is shown below.

$$\frac{\Gamma[(\Delta_1, (\Delta_2, \Delta_3)^l)^4] \vdash C}{\Gamma[((\Delta_1, \Delta_2)^4, \Delta_3)^l] \vdash C} MA_{l,4} \qquad \frac{\Gamma[(\Delta_1, (\Delta_2, \Delta_3)^r)^4] \vdash C}{\Gamma[(\Delta_2, (\Delta_1, \Delta_3)^4)^r] \vdash C} MC_{r,4}$$

Modify the Dutch lexicon from Example 5.3, repeated below

Word	Type(s)	Translation
Marie	np	*Mary*
Jan	np	*John*
wil	$(np \backslash_0 S) /_4 inf$	*wants*
plagen	$np \backslash_0 inf$	*to tease*

in such a way that "(dat) Jan Marie wil plagen" is derivable using these new structural rules. Are any other word orders derivable using your type assignments?

Exercise 5.9. Revisiting the Dutch verb clusters as they have been treated in Example 5.8, show that, using the structural rules of Example 5.8 and the following extended lexicon

Word	Type(s)	Translation
ik	np	*I*
Marie	np	*Mary*
Henk	np	*Henk*
de	np / n	*the*
nijlpaarden	n	*hippopotami*
zag	$\Box_1((np \backslash_0 (np \backslash_0 S)) /_4 inf)$	*saw*
helpen	$\Box_1((np \backslash_0 inf) /_4 inf)$	*to help*
voeren	$\Box_1(np \backslash_0 inf)$	*to feed*

derives the famous "hippopotami" sentences shown below.

(5.21) *(dat) ik Marie de nijlpaarden zag voeren.*
(that) I Mary the hippopotami saw feed.

'(that) I saw Mary feed the hippopotami.'

(5.22) *(dat) ik Henk Marie de nijlpaarden zag helpen voeren.*
(that) I Henk Mary the hippopotami saw help feed.

'(that) I saw Henk help Mary feed the hippopotami.'

Make sure in your analyses that "Henk" is the object of "see", "Marie" is the object of "help" and "the hippopotami" is the object of "feed".

Exercise 5.10. Prove Proposition 5.6 on page 163.

Exercise 5.11. Give derivations for the remaining related pairs in Figure 5.11 on page 165.

Exercise 5.12. The formulae $\Diamond\Box\Diamond\Box A$ and $\Box\Diamond\Box\Diamond A$ are not portrayed in the Figure 5.11 on page 165. Show that these formulae are equivalent to $\Diamond\Box A$ and $\Box\Diamond A$ respectively by proving the more general results of $\Diamond\Box\Diamond A$ being equivalent to $\Diamond A$ and of $\Box\Diamond\Box A$ being equivalent to $\Box A$.

Exercise 5.13. Inspired by a remark from Moortgat (2011, definition 2.4.10), we propose the following translation of multimodal formulae formed using the single unary mode $t4$, to which the postulates T and 4 of modal logic apply, to multimodal formulae formed using the single unary mode 0 to which neither T nor 4 applies.

$$
\begin{aligned}
\|p\| &= p \\
\|A \bullet_i B\| &= \|A\| \bullet_i \|B\| \\
\|A \backslash_i B\| &= \|A\| \backslash_i \|B\| \\
\|A /_i B\| &= \|A\| /_i \|B\| \\
\|\Box_{t4} A\| &= \Diamond_0 \Box_0 \|A\| \\
\|\Diamond_{t4} A| &= \Box_0 \Diamond_0 \|A\|
\end{aligned}
$$

Remember that the T and 4 postulates for $\Box_{t4}$ and $\Diamond_{t4}$ are the following.

$$
\begin{aligned}
&\Box_{t4} A \vdash A \\
&\Box_{t4} A \vdash \Box_{t4} \Box_{t4} A \\
&A \vdash \Diamond_{t4} A \\
&\Diamond_{t4} \Diamond_{t4} A \vdash \Diamond_{t4} A
\end{aligned}
$$

- Prove that the translations of the above sequents using the function $\|.\|$ defined in this exercise are all theorems of $NL\Diamond$.
- Look back to Figure 5.11 on page 165. Where this is possible, use the inverse of $\|.\|$ to translate the formulae in the figure to formulae using $\Box_{t4}$ and $\Diamond_{t4}$. Do the derivability relations between the formula correspond to the expected derivability relations among formulae using the T and 4 modal postulates? Justify your answer.
- Where would the formulae $\Box_{t4} \Diamond_{t4} \Box_{t4} A$ and $\Diamond_{t4} \Box_{t4} \Diamond_{t4} A$ "fit" in Figure 5.11? That is, from which formulae in the figure are these two formulae derivable and which formulae in the figure are derivable from these two formulae? Give proofs of all relations that you can find.
- Look at the following seven formulae:

$$
\begin{aligned}
&A, \\
&\Diamond_{t4} A, \Box_{t4} A, \\
&\Box_{t4} \Diamond_{t4}, \Diamond_{t4} \Box_{t4} A, \\
&\Diamond_{t4} \Box_{t4} \Diamond_{t4} A, \Box_{t4} \Diamond_{t4} \Box_{t4} A
\end{aligned}
$$

  Show that for each formula B which consists of a formula A prefixed with any number of $\Diamond_{t4}$ and $\Box_{t4}$ is equivalent to one of these seven formulae.
  Hint: show that prefixing either $\Diamond_{t4}$ or $\Box_{t4}$ to one of the formulae produces a formula equivalent to another of these formulae.

Exercise 5.14. For Example 5.7 on page 164, show how — with the lexicon given in the example — sentences 5.7 and 5.10 are derivable but sentences 5.8 and 5.10 are not.

Exercise 5.15. How can we treat the difference between nominative and accusative case in English using only a single unary modality? You might wish to look back to Figure 5.11 to find appropriate formulae for nominative and accusative noun phrases as well as for a type which can serve as either.

References

Barry, G., Hepple, M., Leslie, N., Morrill, G.: Proof figures and structural operators for categorial grammar. In: Proceedings of EACL 1995, Berlin, pp. 198–203 (1991)

Bernardi, R.: Reasoning with polarity in categorial type logic. PhD thesis, Utrecht Institute of Linguistics OTS, Utrecht University (2002)

Bernardi, R., Moot, R.: Generalized quantifiers in declarative and interrogative sentences. Logic Journal of the IGPL 11(4), 419–434 (2003)

Chomsky, N.: The minimalist program. MIT Press, Cambridge (1995)

Danos, V., Regnier, L.: The structure of multiplicatives. Archive for Mathematical Logic 28, 181–203 (1989)

Dowty, D.: Non constituent coordination, wrapping and multimodal categorial grammars: Syntactic form as logical form. Tech. rep., Ohio State University (1997); expanded draft of an August 1996 paper from the International Congress of Logic, Methodology and Philosophy of Science, downloaded from the author's website (March 17, 2011)

Girard, J.Y.: Linear logic. Theoretical Computer Science 50(1), 1–102 (1987)

de Groote, P.: Partially commutative linear logic: sequent calculus and phase semantics. In: Abrusci, V.M., Casadio, C. (eds.) Third Roma Workshop: Proofs and Linguistics Categories – Applications of Logic to the Analysis and Implementation of Natural Language, pp. 199–208. CLUEB, Bologna (1996)

Hepple, M.: A general framework for hybrid substructural categorial logics. Tech. rep., Institute for Research in Cognitive Science, University of Pennsylvania (1993)

Heylen, D.: Types and sorts – resource logic for feature checking. PhD thesis, Universiteit Utrecht (1999)

Jäger, G.: Anaphora and Quantification in Categorial Grammar. In: Moortgat, M. (ed.) LACL 1998. LNCS (LNAI), vol. 2014, pp. 70–90. Springer, Heidelberg (2001)

Jäger, G.: Anaphora and Type Logical Grammar, Trends in Logic – Studia Logic Library, vol. 24. Springer (2005)

Kurtonina, N.: Frames and labels. A modal analysis of categorial inference. PhD thesis, OTS Utrecht, ILLC Amsterdam (1995)

Kurtonina, N.: Categorial inference and modal logic. Journal of Logic, Language and Information 7(4), 399–411 (1998)

Kurtonina, N., Moortgat, M.: Structural control. In: Blackburn, P., de Rijke, M. (eds.) Specifying Syntactic Structures, pp. 75–113. CSLI, Stanford (1997)

Moortgat, M.: Multimodal linguistic inference. Bulletin of the IPGL Special Issue (1995) Kempson, R. (ed.)

Moortgat, M.: In situ binding: A modal analysis. In: Dekker, P., Stokhof, M. (eds.) Proceedings 10th Amsterdam Colloquium, pp. 539–549. ILLC, Amsterdam (1996a)

Moortgat, M.: Multimodal linguistic inference. Journal of Logic, Language and Information 5(3-4), 349–385 (1996b)

Moortgat, M.: Categorial type logics. In: van Benthem, J., ter Meulen, A. (eds.) Handbook of Logic and Language, ch. 2, pp. 93–177. North-Holland Elsevier, Amsterdam (1997)

Moortgat, M.: Constants of grammatical reasoning. In: Bouma, G., Hinrichs, E., Kruijff, G.J., Oehrle, R.T. (eds.) Constraints and Resources in Natural Language Syntax and Semantics, pp. 195–219. CSLI, Stanford (1999)

Moortgat, M.: Categorial type logics. In: van Benthem, J., ter Meulen, A. (eds.) Handbook of Logic and Language, 2nd edn., ch. 2, pp. 95–179. North-Holland Elsevier, Amsterdam (2011)

Moortgat, M., Morrill, G.: Heads and phrases: Type calculus for dependency and constituent structure. Tech. rep., Research Institute for Language and Speech (OTS), Utrecht (1991)

Moortgat, M., Oehrle, R.T.: Logical parameters and linguistic variation. In: Fifth European Summer School in Logic, Language and Information, Lisbon. Lecture notes on categorial grammar (1993)

Moortgat, M., Oehrle, R.T.: Adjacency, dependency and order. In: Proceedings 9th Amsterdam Colloquium, pp. 447–466 (1994)

Moot, R.: Proof nets for linguistic analysis. PhD thesis, Utrecht Institute of Linguistics OTS, Utrecht University (2002)

Moot, R.: Lambek grammars and hyperedge replacement grammars. Tech. rep., LaBRI, CNRS (2008a)

Moot, R.: Lambek grammars, tree adjoining grammars and hyperedge replacement grammars. In: Gardent, C., Sarkar, A. (eds.) Proceedings of TAG+9, The Ninth International Workshop on Tree Adjoining Grammars and Related Formalisms, pp. 65–72 (2008)

Moot, R., Retoré, C.: Les indices pronominaux du français dans les grammaires catégorielles. Lingvisticae Investigationes 29(1), 137–146 (2006)

Morrill, G.: Type Logical Grammar. Kluwer Academic Publishers, Dordrecht (1994)

Morrill, G.: Discontinuity in categorial grammar. Linguistics and Philosophy 18(2), 175–219 (1995)

Morrill, G., Leslie, N., Hepple, M., Barry, G.: Categorial deductions and structural operations. In: Barry, G., Morrill, G. (eds.) Studies in Categorial Grammar, Edinburgh Working Papers in Cognitive Science, vol. 5, pp. 1–21. Centre for Cognitive Science (1990)

Oehrle, R., Zhang, S.: Lambek calculus and preposing of embedded subjects. Chicago Linguistics Society 25 (1989)

Prior, A.: Past, Present and Future. Oxford University Press (1967)

Ross, J.R.: Constraints on variables in syntax. PhD thesis, Massachusets Institute of Technology (1967)

Shieber, S.: Evidence against the context-freeness of natural language. Linguistics & Philosophy 8, 333–343 (1985)

Steedman, M.: Surface structure and interpretation. Linguistic Inquiry Monographs, vol. 30. MIT Press, Cambridge (1997)

Vermaat, W.: The minimalist move operation in a deductive perspective. Journal of Language and Computation 2(1), 69–85 (2004); Special Issue on Resource Logics and Minimalist Grammars

Vermaat, W.: The logic of variation. A cross-linguistic account of wh-question formation. PhD thesis, Utrecht Institute of Linguistics OTS, Utrecht University (2005)

6

Lambek Calculus and Linear Logic: Proof Nets as Parse Structures

Summary. This chapter, a large part of which is a translation of (Retoré, 1996), deals with the connection between Lambek categorial grammar and linear logic, the main objective being the presentation of proof nets which are excellent parse structures, because they identify linguistically equivalent analyses of a given sentence.

This graphical notation for proofs that are parse structures in categorial grammar is a not a mere variation for convenience. On a technical ground, it avoids the so-called spurious ambiguity problem of categorial grammars (the fact that we can find many different proofs/parse structures for what corresponds to a single analysis or lambda term). Conceptually, this proof syntax is a justification of the use of the expression *parsing as deduction* often associated with categorial grammar. Indeed proof nets only distinguish between proofs which correspond to different syntactic analyses.

We first give a rather complete presentation of the correspondence between the Lambek calculus and variants of multiplicative linear logic, since the Lambek calculus can be defined as non-commutative intuitionistic multiplicative linear logic without empty antecedents.

Next we define proof nets and establish their correspondence with the more traditional sequent calculus, present parsing as proof net construction and present some recent descriptions of non commutative proof nets.

As an evidence of their linguistic relevance, we explain how they provide a formal account of some performance questions, like the complexity of the processing of several intricate syntactic constructs, like center embedded relatives, garden path phenomena and preferred readings.

6.1 The Formula Language of Categorial Grammar and of Linear Logic

6.1.1 The Formula Language of Multiplicative Linear Logic

Let us recall the language of the Lambek calculus:

$$\mathsf{Lp} \quad ::= \quad \mathsf{P} \quad | \quad (\mathsf{Lp} \bullet \mathsf{Lp}) \quad | \quad (\mathsf{Lp} / \mathsf{Lp}) \quad | \quad (\mathsf{Lp} \setminus \mathsf{Lp})$$

As we have seen in the previous chapters $\backslash$ and $/$ are implications, and the product $\bullet$ is a conjunction. All these connectives are linear logic connectives, but are rather denoted by: $\circ\!-, -\!\circ, \otimes$ in the linear logic community.

Lambek calculus	$\backslash$	$/$	$\bullet$
Linear logic	$-\!\circ$	$\circ\!-$	$\otimes$

Multiplicative linear logic is a classical calculus which extends the Lambek calculus by a negation denoted by $(\ldots)^{\perp}$ (*the orthogonal of* ...) together with the symmetries induced by a classical negation: the familiar De Morgan identities of classical logic.

To be precise, Multiplicative Linear Logic extends the Lambek calculus without the non empty antecedent requirement, and allows for permutation (hypotheses can be permuted). In order to have a single involutive negation and two distinct implications $\circ\!-$ and $-\!\circ$, one must restrict the allowed permutations to *cyclic* permutations. In the absence of any form of permutation, there have to be two negations (Abrusci, 1991, 1995).

Because of the De Morgan identities, there will be a disjunction $\wp$ (*par*, standing for *in parallel with*) corresponding to the conjunction $\otimes$. As we are especially interested in having a non commutative conjunction, the disjunction, by duality, will be non commutative as well.

Such a disjunction and a classical negation allow the implication $A \backslash B$ to be defined as $A^{\perp} \wp B$ and the implication B / A to be defined as $B \wp A^{\perp}$ — just like it is possible to define $A \Rightarrow B$ as $\neg A \vee B$ in classical logic. Notice that the non commutativity of the disjunction is necessary if we want to be able to distinguish between these two implications.

In the Lambek calculus, one has the following equivalence: $(C / B) / A \equiv C / (A \bullet B)$: indeed $(C / B) / A$ is a formula which requires an A and then a B to obtain C, and $C / (A \bullet B)$ is a formula which requires an A *followed by a* B, to obtain a C. The formula $(C / B) / A$ can be written as $C \wp B^{\perp} \wp A^{\perp}$ using the (associative) disjunction and the formula $C / (A \bullet B)$ as $C \wp (A \otimes B)^{\perp}$. Therefore if there is a classical extension of the Lambek calculus then negation has to swap the components of a disjunction (and of a conjunction, by duality).

Linear logic, when seen as a classical extension of the Lambek calculus, has the following language:

$$\mathsf{Li}_+ \quad ::= \quad \mathsf{P} \mid \mathsf{Li}_+^{\perp} \mid (\mathsf{Li}_+ \wp \mathsf{Li}_+) \mid (\mathsf{Li}_+ \otimes \mathsf{Li}_+) \mid (\mathsf{Li}_+ \backslash \mathsf{Li}_+) \mid (\mathsf{Li}_+ / \mathsf{Li}_+)$$

and enjoys the elimination of double negation and the De Morgan identities, as shown below.

$$(A^{\perp})^{\perp} \equiv A \qquad (A \wp B)^{\perp} \equiv B^{\perp} \otimes A^{\perp} \qquad (A \otimes B)^{\perp} \equiv B^{\perp} \wp A^{\perp}$$

6.1.2 Reduced Linear Language (Negative Normal Form)

For every formula X in Li_+ there exists a unique equivalent formula $+X$ such that negation only applies to propositional variables, and its only connectives are conjunction and disjunction. In some books, the analogous of $+X$ for classical logic is called its negative normal form. The formula $+X$ is obtained by replacing its implication by its definition as a disjunction, and then applying De Morgan identities as rewriting rules from left to right, and, finally by cancelling double negations. Notice that, unlike disjunctive normal form and conjunctive normal form, this form does not require distributivity of $\wp$ w.r.t. $\otimes$ or $\otimes$ w.r.t. $\wp$ — these distributivity identities do not hold in linear logic.[1]

So every formula in Li_+ is equivalent to a formula $+X$ in Li, where Li is:

$$\mathsf{Li} ::= \mathsf{N} \mid \mathsf{Li} \wp \mathsf{Li} \mid \mathsf{Li} \otimes \mathsf{Li} \qquad \text{where } \mathsf{N} = \mathsf{P} \cup \mathsf{P}^{\perp} \text{ is the set of } \textit{atoms}.$$

Observe that if $F \in \mathsf{Li}$ then $+F = F$.

Let us denote by $-F$ the unique formula in Li equivalent to $(F)^{\perp} \in \mathsf{Li}_+$ — $-F = +(F^{\perp})$. Given $+F$, $-F$ is obtained by replacing every propositional variable in $+F$ with its negation, every conjunction by a disjunction, every disjunction by a conjunction, and finally by *reversing the left to right order of the result.*

Given $F = (\alpha^{\perp} \wp \beta) \otimes \gamma^{\perp}$ one first obtains $F' = (\alpha \otimes \beta^{\perp}) \wp \gamma$, which yields $F^{\perp} \equiv -F = \gamma \wp (\beta^{\perp} \otimes \alpha)$ by rewriting F' from right to left.

6.1.3 Relating Categories and Linear Logic Formulae: Polarities

Since Lp is a sublanguage of Li_+, for every formula L in Lp there exists a unique formula $+L$ in Li which is equivalent to L and a unique formula $-L$ which is equivalent to $L^{\perp}$. These two maps from Lp to Li can be inductively defined as follows:

L	$\alpha \in \mathsf{P}$	$L = M \bullet N$	$L = M \setminus N$	$L = N / M$
$+L$	α	$+M \otimes +N$	$-M \wp +N$	$+N \wp -M$
$-L$	$\alpha^{\perp}$	$-N \wp -M$	$-N \otimes +M$	$+M \otimes -N$

Example 6.1

L	$+L$	$-L$	
np	np	$np^{\perp}$	noun phrase
np / n	$np^{\perp} \wp n$	$n^{\perp} \otimes np$	determiner
n	n	$n^{\perp}$	common noun
$n \setminus n$	$n^{\perp} \wp n$	$n^{\perp} \otimes n$	right adjective
$(n \setminus n) / (n \setminus n)$	$(n^{\perp} \wp n) \wp (n^{\perp} \otimes n)$	$(n^{\perp} \wp n) \otimes (n^{\perp} \otimes n)$	left modifier for right adjectives
$\beta \setminus ((\alpha / \beta) \setminus \alpha)$	$\beta^{\perp} \wp ((\beta \otimes \alpha^{\perp}) \wp \alpha)$	$(\alpha^{\perp} \otimes (\alpha \wp \beta^{\perp})) \otimes \beta)$	type raising

[1] Though the classical distributivities, such as the equivalences $A \wedge (B \vee C) \leftrightarrow (A \wedge B) \vee (A \wedge C)$ and $A \vee (B \wedge C) \leftrightarrow (A \vee B) \wedge (A \vee C)$ which are required for disjunctive and conjunctive normal form do not hold between tensor and par, we do have some weaker *implications*, eg. $A \wedge (B \vee C) \rightarrow (A \wedge B) \vee C$, or written using tensor and par: $A \otimes (B \wp C) \vdash (A \otimes B) \wp C$.

Let us consider the following sets of formulae, which enable us to recognize, among linear formulae the ones which are Lambek formulae or the negation of Lambek formulae.

$$\begin{array}{ll}\mathsf{Li}^{\circ} = \{F \in \mathsf{Li} / \exists L \in \mathsf{Lp} \quad +L = F\} & : \text{positive linear formulae} \\ \mathsf{Li}^{\bullet} = \{F \in \mathsf{Li} / \exists L \in \mathsf{Lp} \quad -L = F\} & : \text{negative linear formulae} \\ \mathsf{Li}^{\circ} \cup \mathsf{Li}^{\bullet} \quad : & \text{intuitionistic or polarized linear formulae}\end{array}$$

We then have:

$$F \in \mathsf{Li}^{\bullet} \Leftrightarrow -F \in \mathsf{Li}^{\circ} \qquad \text{and} \qquad F \in \mathsf{Li}^{\circ} \Leftrightarrow -F \in \mathsf{Li}^{\bullet}$$
$$\mathsf{Li}^{\circ} \cup \mathsf{Li}^{\bullet} \neq \mathsf{Li} \text{ — for instance } \alpha \wp \beta \notin \mathsf{Li}^{\circ} \cup \mathsf{Li}^{\bullet}$$
$$\mathsf{Li}^{\bullet} \cap \mathsf{Li}^{\circ} = \emptyset \text{ — because of the following proposition:}$$

Proposition 6.2. *The sets of formulae* Li° *and* $\mathsf{Li}^{\bullet}$ *are inductively defined by:*

$$\left[\begin{array}{l}\mathsf{Li}^{\circ} ::= \mathsf{P} \mid (\mathsf{Li}^{\circ} \otimes \mathsf{Li}^{\circ}) \mid (\mathsf{Li}^{\bullet} \wp \mathsf{Li}^{\circ}) \mid (\mathsf{Li}^{\circ} \wp \mathsf{Li}^{\bullet}) \\ \mathsf{Li}^{\bullet} ::= \mathsf{P}^{\perp} \mid (\mathsf{Li}^{\bullet} \wp \mathsf{Li}^{\bullet}) \mid (\mathsf{Li}^{\circ} \otimes \mathsf{Li}^{\bullet}) \mid (\mathsf{Li}^{\bullet} \otimes \mathsf{Li}^{\circ})\end{array}\right.$$

The maps $+$ *and* $-$ *are bijections from* Lp *to* Li° *and* $\mathsf{Li}^{\bullet}$ *respectively.*

If $(\ldots)^{\circ}_{\mathsf{Lp}}$ denotes the inverse bijection of $+$, from Li° to Lp and if $(\ldots)^{\bullet}_{\mathsf{Lp}}$ denotes the inverse bijection of $-$ from $\mathsf{Li}^{\bullet}$ to Lp. Then these two maps are inductively defined as follows:

$F \in \mathsf{Li}^{\circ}$	$\alpha \in \mathsf{P}$	$(G \in \mathsf{Li}^{\circ}) \otimes (H \in \mathsf{Li}^{\circ})$	$(G \in \mathsf{Li}^{\bullet}) \wp (H \in \mathsf{Li}^{\circ})$	$(G \in \mathsf{Li}^{\circ}) \wp (H \in \mathsf{Li}^{\bullet})$
F°_{Lp}	α	$G^{\circ}_{\mathsf{Lp}} \otimes H^{\circ}_{\mathsf{Lp}}$	$G^{\bullet}_{\mathsf{Lp}} \setminus H^{\circ}_{\mathsf{Lp}}$	$G^{\circ}_{\mathsf{Lp}} / H^{\bullet}_{\mathsf{Lp}}$

$F \in \mathsf{Li}^{\bullet}$	$\alpha^{\perp} \in \mathsf{P}^{\perp}$	$(G \in \mathsf{Li}^{\bullet}) \wp (H \in \mathsf{Li}^{\bullet})$	$(G \in \mathsf{Li}^{\circ}) \otimes (H \in \mathsf{Li}^{\bullet})$	$(G \in \mathsf{Li}^{\bullet}) \otimes (H \in \mathsf{Li}^{\circ})$
$F^{\bullet}_{\mathsf{Lp}}$	α	$H^{\bullet}_{\mathsf{Lp}} \otimes G^{\bullet}_{\mathsf{Lp}}$	$H^{\bullet}_{\mathsf{Lp}} / G^{\circ}_{\mathsf{Lp}}$	$H^{\circ}_{\mathsf{Lp}} \setminus G^{\bullet}_{\mathsf{Lp}}$

The inductive definition of Li° and $\mathsf{Li}^{\bullet}$ yields an easy decision procedure to check whether a formula F is in Li° or $\mathsf{Li}^{\bullet}$ — if so, all subformulae of F are in Li° or in $\mathsf{Li}^{\bullet}$: replace every propositional variable with $\circ$ and every negation of a propositional variable with $\bullet$ and compute using $\wp$ and $\otimes$ as the following operations on $\star, \circ, \bullet$:

$\wp$	$\star$	$\circ$	$\bullet$
$\star$	$\star$	$\star$	$\star$
$\circ$	$\star$	$\star$	$\circ$
$\bullet$	$\star$	$\circ$	$\bullet$

$\otimes$	$\star$	$\circ$	$\bullet$
$\star$	$\star$	$\star$	$\star$
$\circ$	$\star$	$\circ$	$\bullet$
$\bullet$	$\star$	$\bullet$	$\star$

The result of this simple computation is used as follows:

- $\star$ whenever the formula is neither in Li° nor in $\mathsf{Li}^{\bullet}$
- $\circ$ whenever the formula is in Li°
- $\bullet$ whenever the formula is in $\mathsf{Li}^{\bullet}$

Example 6.3

F	computation	conclusion	F°_{Lp}	$F^{\bullet}_{\mathsf{Lp}}$
$(\alpha^{\perp} \wp \beta) \wp \alpha$	$(\bullet \wp \circ) \wp \circ = \circ \wp \circ = \star$	$F \notin \mathsf{Li}^{\circ} \cup \mathsf{Li}^{\bullet}$	undefined	undefined
$(\alpha^{\perp} \wp \beta) \wp \alpha^{\perp}$	$(\bullet \wp \circ) \wp \bullet = \circ \wp \bullet = \circ$	$F \in \mathsf{Li}^{\circ}$	$(\alpha \setminus \beta) / \alpha$	undefined
$(\alpha^{\perp} \wp \beta) \otimes \alpha^{\perp}$	$(\bullet \wp \circ) \otimes \bullet = \circ \otimes \bullet = \bullet$	$F \in \mathsf{Li}^{\bullet}$	undefined	$\alpha / (\alpha \setminus \beta)$

6.2 Two Sided Calculi

Here is the two sided linear calculus MLL_{+} for all connectives of the language Li_{+}. In the Section 6.3.1, we shall see how it embeds the Lambek calculus.

Exchange	$\dfrac{\Gamma, A, B, \Delta \vdash \Psi}{\Gamma, B, A, \Delta \vdash \Psi}\,(\mathbf{x})_h$		$\dfrac{\Theta \vdash \Gamma, A, B, \Delta}{\Theta \vdash \Gamma, B, A, \Delta}\,(\mathbf{x})_i$
Axiom	$\dfrac{}{A \vdash A}\,ax$		$A \in \mathsf{Li}_{+}$
Logical rules	$\dfrac{\Gamma \vdash A, \Delta}{A^{\perp}, \Gamma \vdash \Delta}\,\perp_h$	Negation	$\dfrac{A, \Gamma \vdash \Delta}{\Gamma \vdash A^{\perp}, \Delta}\,\perp_i$
	$\dfrac{\Gamma, A \vdash \Theta \qquad B, \Gamma' \vdash \Theta'}{\Gamma, A \wp B, \Gamma' \vdash \Theta, \Theta'}\,\wp_h$	Disjunction	$\dfrac{\Theta \vdash \Gamma, A, B, \Delta}{\Theta \vdash \Gamma, A \wp B, \Delta}\,\wp_h$
	$\dfrac{\Gamma, A, B, \Delta \vdash \Psi}{\Gamma, A \otimes B, \Delta \vdash \Psi}\,\otimes_h$	Conjunction	$\dfrac{\Theta \vdash \Phi, A \qquad \Theta' \vdash B, \Phi'}{\Theta, \Theta' \vdash \Phi, A \otimes B, \Phi'}\,\otimes_i$
	$\dfrac{\Gamma \vdash \Phi, A \qquad \Gamma', B, \Delta' \vdash \Psi'}{\Gamma', \Gamma, A \setminus B, \Delta' \vdash \Phi, \Psi'}\,\setminus_h$	Implications	$\dfrac{A, \Gamma \vdash C, \Phi}{\Gamma \vdash A \setminus C, \Phi}\,\setminus_i$
	$\dfrac{\Gamma \vdash \Phi, A \qquad \Gamma', B, \Delta' \vdash \Psi'}{\Gamma', B / A, \Gamma, \Delta' \vdash \Phi, \Psi'}\,/_h$		$\dfrac{\Gamma, A \vdash \Phi, C}{\Gamma \vdash \Phi, C / A}\,/_i$

6.2.1 Properties of the Linear Two Sided Sequent Calculus

Cut Elimination

We left out the cut rule on purpose. There are two ways to formulate the cut rule in a classical calculus:

$$\dfrac{\Theta \vdash \Phi, A \qquad A, \Theta' \vdash \Psi'}{\Theta, \Theta' \vdash \Phi, \Psi'}\ cut \qquad\qquad \dfrac{\Theta \vdash \Phi, A \qquad \Theta' \vdash A^{\perp}, \Phi'}{\Theta, \Theta' \vdash \Phi, \Phi'}\ cut$$

As in the Lambek calculus, this rule is redundant, and the proof is more or less the same. As a consequence, the subformula property also holds for this calculus.

De Morgan Identities and Double Negation Elimination

As we claimed before, these identities hold for linear logic. For instance:

$$\dfrac{\dfrac{A \vdash A}{A^{\perp}, A \vdash}\ \perp_i}{A \vdash (A^{\perp})^{\perp}}\ \perp_i \qquad\qquad \dfrac{\dfrac{A \vdash A}{\vdash A^{\perp}, A}\ \perp_i}{(A^{\perp})^{\perp} \vdash A}\ \perp_i$$

Restriction to Atomic Axioms

As for the Lambek calculus, an easy induction on A, shows that every axiom $A \vdash A$ can be derived from axioms $\alpha \vdash \alpha$, where α is a propositional variable, without using the exchange rule. For instance let us show that $A \vdash A$ with $A = \alpha \wp \beta^{\perp}$ can be derived from the axioms $\alpha \vdash \alpha$ and $\beta \vdash \beta$:

$$\dfrac{\dfrac{\dfrac{}{\alpha \vdash \alpha}\ ax \qquad \dfrac{\dfrac{\dfrac{}{\beta \vdash \beta}\ ax}{\beta, \beta^{\perp} \vdash}\ \perp_h}{\beta^{\perp} \vdash \beta^{\perp}}\ \perp_i}{\alpha \wp \beta^{\perp} \vdash \alpha, \beta^{\perp}}\ \wp_h}{\alpha \wp \beta^{\perp} \vdash \alpha \wp \beta^{\perp}}\ \wp_i$$

Equality of the Two Implications

In this calculus, the implication $A \setminus B$ can be viewed as a shorthand for $A^{\perp} \wp B$, while $A \,/\, B$ is a shorthand for $B \wp A^{\perp}$. Indeed the rules for the implications can be derived when implications are defined this way. Furthermore, in the presence of a full exchange rule, one has: $A \setminus B \equiv B \,/\, A$.

$$\dfrac{\dfrac{\Gamma \vdash A}{\Gamma, A^{\perp} \vdash}\,{\perp_h} \qquad \Delta', B, \Gamma' \vdash \Theta'}{\Delta', \Gamma, A^{\perp} \wp B, \Gamma' \vdash \Theta'}\,\wp_h \qquad\qquad \dfrac{\dfrac{\Gamma, A \vdash B}{\Gamma \vdash A^{\perp}, B}\,{\perp_h}}{\Gamma \vdash A^{\perp} \wp B}\,\wp_i$$

$$\dfrac{\dfrac{\dfrac{\dfrac{\dfrac{}{B \vdash B}\,ax \qquad \dfrac{}{A^{\perp} \vdash A^{\perp}}\,ax}{B \wp A^{\perp} \vdash B, A^{\perp}}\,\wp_h}{B \wp A^{\perp} \vdash A^{\perp}, B}\,(\mathbf{x})_i}{B \wp A^{\perp} \vdash A^{\perp} \wp B}\,\wp_i}{B / A \vdash A \backslash B}\,\equiv \qquad\qquad \dfrac{\dfrac{\dfrac{\dfrac{\dfrac{}{A^{\perp} \vdash A^{\perp}}\,ax \qquad \dfrac{}{B \vdash B}\,ax}{A^{\perp} \wp B \vdash A^{\perp}, B}\,\wp_h}{A^{\perp} \wp B \vdash B, A^{\perp}}\,(\mathbf{x})_i}{A^{\perp} \wp B \vdash B \wp A^{\perp}}\,\wp_i}{A \backslash B \vdash B / A}\,\equiv$$

Negation and Symmetrical Rules

If one considers formulae up to De Morgan identities, then right rules are enough.

For instance the rule $\wp_h$ can be simulated by the rule $\otimes_i$ as shown in the following derivation.

$$\dfrac{\dfrac{\dfrac{\dfrac{\Gamma, A \vdash \Theta}{\Gamma \vdash A^{\perp}, \Theta}\,{\perp_i} \qquad \dfrac{B, \Gamma' \vdash \Theta'}{\Gamma' \vdash B^{\perp}, \Theta'}\,{\perp_i}}{\Gamma, \Gamma' \vdash \Theta, A^{\perp} \otimes B^{\perp}, \Theta'}\,\otimes_i}{[A \wp B \equiv (A^{\perp} \otimes B^{\perp})^{\perp}], \Gamma', \Gamma \vdash \Theta', \Theta}\,(\mathbf{x})_h \text{ and } {\perp_h}}{\Gamma, [B \wp A \equiv (A^{\perp} \otimes B^{\perp})^{\perp}], \Gamma' \vdash \Theta', \Theta}\,(\mathbf{x})_h$$

In order to avoid the exchange rule, one has to consider a more subtle sequent calculus like the one of (Abrusci, 1991, p. 1415) but identifying the two negations — this actually forces a restricted form of the exchange rule known as cyclic exchange, that we shall present later on.

6.2.2 The Intuitionistic Two Sided Calculus LP_ε

The calculus LP_ε, that is Lambek calculus with permutation and empty antecedents is exactly intuitionistic multiplicative linear logic. This calculus is obtained from MLL_+ by forcing sequents to always have exactly one formula on the right hand side.

By inspection of the rules, it is clear that restricting the right hand side of the sequent to one formula means that we can no longer formulate the rules for negation. Therefore the natural language for LP_ε is Lp. The rules are obtained from the ones of MLL_+ in Section 6.2, by replacing the sequences of formulae denoted by Φ and Φ' by the empty sequence, and the sequences of formulae denoted by Ψ and Ψ' by a single formula F or F'. This yields the following rules:

Exchange	$\dfrac{\Gamma,A,B,\Delta \vdash F}{\Gamma,B,A,\Delta \vdash F}\,(\mathbf{x})_h$	
Axiom	$\dfrac{}{A \vdash A}\,ax \qquad A \in \mathsf{Lp}$	
Logical rules	$\dfrac{\Gamma,A,B,\Delta \vdash F}{\Gamma,A\otimes B,\Delta \vdash F}\,\otimes_h$	Conjunction	$\dfrac{\Theta \vdash A \qquad \Theta' \vdash B}{\Theta,\Theta' \vdash A\otimes B}\,\otimes_h$
	$\dfrac{\Gamma \vdash A \qquad \Gamma',B,\Delta' \vdash F'}{\Gamma',\Gamma,A\backslash B,\Delta' \vdash F'}\,\backslash_h$ $\dfrac{\Gamma \vdash A \qquad \Gamma',B,\Delta' \vdash F'}{\Gamma',B/A,\Gamma,\Delta' \vdash F'}\,/_h$	Implications	$\dfrac{A,\Gamma \vdash C}{\Gamma \vdash A\backslash C}\,\backslash_i$ $\dfrac{\Gamma,A \vdash C}{\Gamma \vdash C/A}\,/_i$

This calculus LP_ε and its variants are studied in a slightly different perspective in (van Benthem, 1991), and is also the basis of works on the semantics of LFG in a series of articles like (Dalrymple et al, 1995).

This calculus allows for several variants according to the presence or absence of the exchange rule, or the allowance or prohibition of sequents with an empty antecedent, that is: the sequence of formulae Π is not empty when the rule $\backslash_i$ or $/_i$ is applied or, equivalently, every sequent in a proof has a non empty antecedent.

This last restriction is harmless from a logical viewpoint, i.e. preserves cut-elimination, but is essential for a grammatical use of the Lambek calculus, as we have seen in Section 2.5. Let us give another example of an incorrect analysis due to empty antecedents:

Example 6.4. Look at the following small lexicon.

Word	Type(s)	Translation
exemple	n	*example*
simple	$n\backslash n$	*simple*
très	$(n\backslash n)/(n\backslash n)$	*very*
un	np/n	*a*

$$\dfrac{\dfrac{\dfrac{}{n \vdash n}\,ax}{\boxed{\varepsilon} \vdash n/n}\,/_i \qquad \dfrac{\dfrac{\dfrac{}{n \vdash n}\,ax \quad \dfrac{}{np \vdash np}\,ax}{np/n,n \vdash np}\,/_h \qquad \dfrac{}{n \vdash n}\,ax}{np/n,n,n\backslash n \vdash np}\,\backslash_h}{np/n,\quad n,\quad (n\backslash n)/(n\backslash n) \vdash np}\,/_h$$

`un exemple très`

6.2.3 Proofs as Parse Structures: Too Many of Them

When we look at parsing a Lambek grammar, then, given that the Lambek calculus is a logic, a parse for a Lambek grammar is a proof in the Lambek calculus. However, sequent calculus proof search, which is one way of implementing parsing for the Lambek calculus is problematic: it is easy to find several proofs which should correspond to the same parse structure, but which nevertheless are distinct. For instance, with the previous lexicon, the following sequent calculus proofs are different

Example 6.5

$$\dfrac{\dfrac{\dfrac{\dfrac{}{n \vdash n}\,ax \quad \dfrac{}{n \vdash n}\,ax}{n, n\backslash n \vdash n}\,\backslash_h}{n\backslash n \vdash n\backslash n}\,\backslash_i \qquad \dfrac{\dfrac{\dfrac{}{n \vdash n}\,ax \quad \dfrac{}{n \vdash n}\,ax}{n, n\backslash n \vdash n}\,\backslash_h \quad \dfrac{}{np \vdash np}\,ax}{np\,/\,n, n, n\backslash n \vdash np}\,/_h}{\underset{\texttt{un}}{np\,/\,n,} \quad \underset{\texttt{exemple}}{n,} \quad \underset{\texttt{très}}{(n\backslash n)\,/\,(n\backslash n),} \quad \underset{\texttt{simple}}{n\backslash n} \quad \vdash np}\,/_h$$

Example 6.6

$$\dfrac{\dfrac{\dfrac{\dfrac{}{n \vdash n}\,ax \quad \dfrac{}{n \vdash n}\,ax}{n, n\backslash n \vdash n}\,\backslash_h}{n\backslash n \vdash n\backslash n}\,\backslash_i \qquad \dfrac{\dfrac{\dfrac{}{n \vdash n}\,ax \quad \dfrac{}{np \vdash np}\,ax}{np/n, n \vdash np}\,/_h \quad \dfrac{}{n \vdash n}\,ax}{np\,/\,n, n, n\backslash n \vdash np}\,\backslash_h}{\underset{\texttt{un}}{np\,/\,n,} \quad \underset{\texttt{exemple}}{n,} \quad \underset{\texttt{très}}{(n\backslash n)\,/\,(n\backslash n),} \quad \underset{\texttt{simple}}{n\backslash n} \quad \vdash np}\,/_h$$

In the two proofs above the order of the $\backslash_h$ and $/_h$ rules is reversed, but both rules have the same formula occurrences as their active and main formulae; the only way the two proofs differ is in the way the context variables of the rules are instantiated.

This problem, that there can be many proofs of what we would want to be the same *parse* is sometimes called the *spurious ambiguity problem*. Natural deduction is a bit better in this respect, though, as we have seen in Section 2.6.3 problems of multiple derivations corresponding to a single parse exist for the product formula. One of the main objective of this chapter is to find a notion of proof that yields one proof per parse structure; this is a key motivation for proof nets, to be introduced in Section 6.4: proof nets will solve the problem of multiple equivalent proofs which exists for the sequent calculus and, unlike natural deduction, will treat the product formulae as easily as the other connectives.

6.3 A One Sided Calculus for Linear Logic: MLL

As we have seen in the paragraph 6.1.2 for every formula X of Li_+ there exists a unique formula $+X$ of Li which is equivalent to it by De Morgan identities, and as explained in paragraph 6.2.1, right rules can be simulated by left rules. Therefore, if one considers formulae up to De Morgan identities then the following one sided sequent calculus, defined as follows, is enough:

Exchange	$\dfrac{\vdash \Gamma, A}{\vdash A, \Gamma}$ (cx)	$\dfrac{\vdash \Gamma, A, B}{\vdash \Gamma, B, A}$ (tx)
Axiom	$\dfrac{}{\vdash \alpha, \alpha^{\perp}}\, ax$	$\alpha \in \mathsf{P}$
Logical rules	$\dfrac{\vdash \Gamma, A, B, \Delta}{\vdash \Gamma, A \wp B, \Delta}\, \wp$	$\dfrac{\vdash \Gamma, A \qquad \vdash B, \Gamma'}{\vdash \Gamma, A \otimes B, \Gamma'}\, \otimes$

The exchange rule $(\mathbf{x})_h$ of MLL_+ has been split into two rules (tx) (transposition exchange) and (cx) (cyclic exchange). Therefore $(\mathbf{x})_h$ is derivable but, this formulation allows to consider the calculus NC-MLL of (Yetter, 1990), which only has the (cx) exchange, but not the (tx) exchange.

The simple calculus MLL whose language is Li, proves exactly the same sequents as the bigger two sided calculus MLL_+ :

Proposition 6.7. *Let $A_1, \ldots, A_n, B_1, \ldots, B_p$ be formulae in Li_+; then one has:*

$$(A_1, \ldots, A_n \vdash_{\mathsf{MLL}_+} B_1, \ldots, B_p) \Leftrightarrow (\vdash_{\mathsf{MLL}} -A_n, \ldots, -A_1, +B_1, \ldots, +B_p)$$

For the converse implication, notice that given a formula $F \in \mathsf{Li}$ there usually exist several formulae $X \in \mathsf{Li}_+$ such that $+X = F$ or $-X = F$.

6.3.1 Variants

We are about to introduce several variants of MLL according to the following restrictions:

INTUI **intuitionistic calculi**
in two sided presentation: one formula in the right hand side of every sequent
in one sided presentation: only polarized formulae (formulae of $\mathsf{Li}^{\circ} \cup \mathsf{Li}^{\bullet}$)[2]

NC **non commutative calculi**
in two sided presentation: no exchange at all
in one sided presentation: cyclic exchange (cx) only (no transposition exchange (tx))

[2] Note that by Proposition 6.8 of the next section, we do not have to require explicitly that there is only one formula in Li°.

ε-FREE **no empty antecedent**
in two sided presentation: no empty antecedent, at least one formula on the left hand side of every sequent
in one sided presentation: at least two formulae in every sequent

The names for these calculi somehow differ in the categorial tradition and in the linear logic community, for instance, calculi without empty antecedents are never considered in linear logic and, though classical calculi are sometimes discussed in the categorial tradition (see, for example, Lambek, 1993; de Groote and Lamarche, 2002), there are, to the best of our knowledge, no linguistic applications of formulas *not* in $\mathsf{Li}^{\circ} \cup \mathsf{Li}^{\bullet}$. For linear calculi, the restriction which corresponds to forbidding empty antecedents will be denoted by $(\cdots)^{*}$. Conversely, for categorial grammar and Lambek calculus, allowing for empty antecedents will be denoted by $(\cdots)^{\varepsilon}$. The non-commutative restriction of a linear calculus will be denoted by a prefix NC, and the commutative extension of a Lambek style calculus will be denoted by a suffix P

Because of these two communities, we have two names for the intuitionistic calculi, and we hope it will not confuse the reader. Table 6.1 lists all the different systems, together with their different names and the restrictions which apply to them. Figure 6.1 portrays the relations between the logics by means of a commutative diagram. All these restrictions will appear again for describing the proof nets corresponding to each calculus.

Although this might be surprising we are able to provide a one sided formulation for intuitionistic calculi. So we will use the *linear name* $\cdots$ *MLL for one sided calculi* and the *categorial name L*$\cdots$ *for two sided calculi.*

Table 6.1. The different logical systems and their properties

INTUI	NC	ε-FREE	Linear name	Categorial name
yes	yes	yes	NC-IMLL*	L
yes	yes	no	NC-IMLL	L_{ε}
yes	no	yes	IMLL*	LP
yes	no	no	IMLL	$\mathsf{LP}_{\varepsilon}$
no	yes	yes	NC-MLL*	
no	yes	no	NC-MLL	
no	no	yes	MLL*	
no	no	no	MLL	
			one sided	two sided

6.3.2 The Intuitionistic Restriction in One Sided Calculi

The two sided intuitionistic calculus $\mathsf{LP}_{\varepsilon}$ is a proper restriction of its classical counterpart MLL. For instance, if we look at the formula $F = (\beta \wp \alpha) \wp (\alpha^{\perp} \otimes \beta^{\perp})$ one

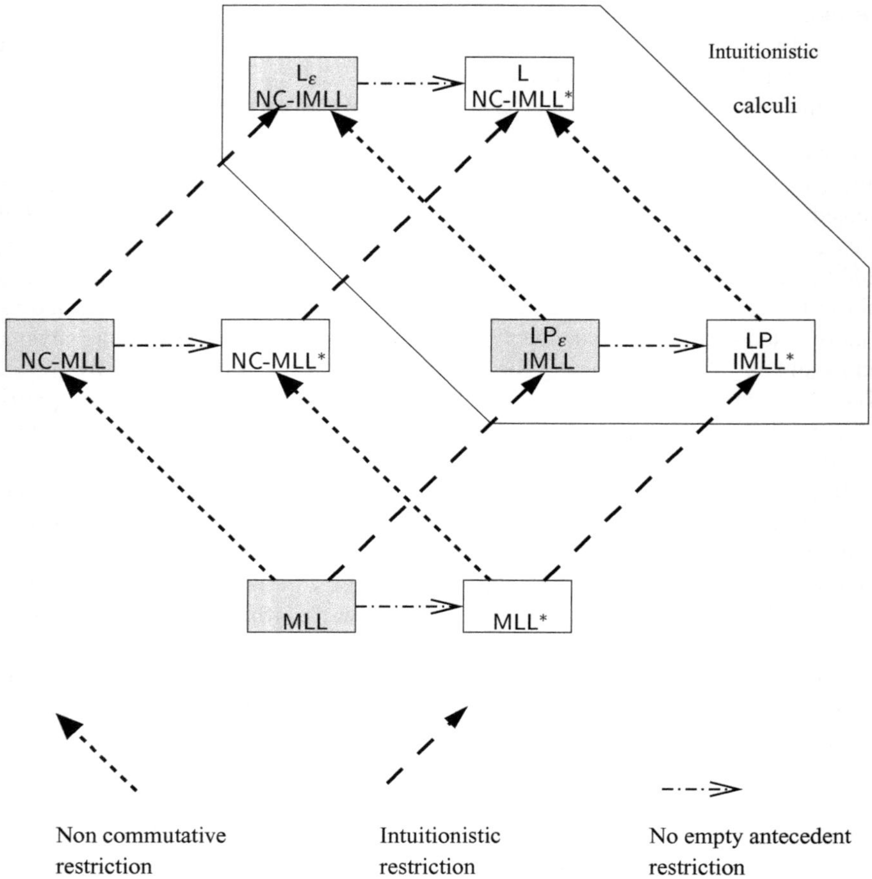

Fig. 6.1. Commutative diagram listing the relations between the different logics

has $\vdash_{\mathsf{MLL}} F$ but there is no formula G equivalent to F such that $\vdash_{\mathsf{LP}_\varepsilon} G$. Actually, this restriction is only a restriction of the *language*, which we have already studied in Section 6.1.3. Indeed, it is only because there is no formula in Lp equivalent to F, i.e. because $F \notin \mathsf{Li}^\bullet \cup \mathsf{Li}^\circ$ that F is not a theorem of IMLL. More precisely we have the following result.

Proposition 6.8. *If* $\forall i \in [1,n]\ A_i \in \mathsf{Li}^\bullet \cup \mathsf{Li}^\circ$ *then*

$$(\vdash_{MLL} A_1, \ldots, A_n) \Leftrightarrow (\vdash_{IMLL} A_1, \ldots, A_n)$$

and whenever these properties hold, then exactly one formula of the sequent is in Li°*, all others being in* $\mathsf{Li}^\bullet$*. This also holds for the variants NC-MLL and NC-IMLL.*

Proof. Easy induction on the proofs. □

Proposition 6.8 was first studied by van de Wiele in the typed case and then taken up by Bellin and Scott (1994) and by Danos and Regnier (Danos, 1990; Regnier, 1992) in the untyped case. This property has lead Lamarche to an interesting theory of intuitionistic proof nets (Lamarche, 1994) which is orthogonal to our presentation.

From the previous proposition we easily deduce the correspondence between one sided intuitionistic calculi and the two sided intuitionistic calculi:

Proposition 6.9. *If* $\vdash_{\mathsf{MLL}} F_1,\ldots,F_n$, *with* $\forall i \in [1,n]\ F_i \in \mathsf{Li}^\bullet \cup \mathsf{Li}^\circ$, *then:*

- *there exists a unique index* $i_0 \in [1,n]$ *such that* $F_{i_0} \in \mathsf{Li}^\circ$ *and for every other index* $i \in [1,n]$ *we have* $F_i \in \mathsf{Li}^\bullet$ *because of the Proposition 6.8*
- *because of Section 6.1.3, every formula* $F_i^\perp$ *with* $i \neq i_0$ *is equivalent to a unique formula* $(F_i)^\bullet_{\mathsf{Lp}} \in \mathsf{Lp}$, *while* F_{i_0} *is equivalent to a unique formula* $(F_{i_0})^\circ_{\mathsf{Lp}}$
- $(F_{i_0-1})^\bullet_{\mathsf{Lp}}, (F_{i_0-2})^\bullet_{\mathsf{Lp}}, \ldots, (F_1)^\bullet_{\mathsf{Lp}}, (F_n)^\bullet_{\mathsf{Lp}}, \ldots, (F_{i_0+1})^\bullet_{L} \vdash_{\mathsf{LP}_\varepsilon} (F_{i_0})^\circ_{\mathsf{Lp}}$

Conversely, $(X_1,\ldots,X_n \vdash_{\mathsf{LP}_\varepsilon} Y) \Rightarrow (\vdash_{\mathsf{MLL},\mathsf{IMLL}} -X_n,\ldots,-X_1,+Y)$.

If one replaces MLL *with* NC-MLL *(resp.* NC-MLL*) *and* LP_ε *with* L_ε *(resp.* L*) the result also holds (As announced in the commutative diagram of Figure 6.1, the restrictions* INTUI, NC *and* ε-FREE *commute).*

For these non commutative variants NC-MLL, NC-MLL*, L_ε *and* L, *with a restricted exchange rule, one has to abide by the order between formulae: this order is reversed when formulae move from one side of the sequent's turnstile to the other.*

Proof. The "conversely" is obvious.

The direct implication is shown by induction on the proof. For the proof to work in the non commutative case, the rule (tx) is only used for the translation of the $(\mathbf{x})_h$ rule of IMLL. Here is, for instance, the translation of the $/_h$.

Assume that the sequences of formulae involved in $/_h$ are $\Gamma = G_1,\ldots,G_n$, $\Gamma' = G'_1,\ldots,G'_k$, $\Delta' = D'_1,\ldots,D'_l$. Here is the NC-MLL proof which simulates the rule $/_h$ of L_ε — remember that $+A \otimes -B = -(A \setminus B)$ (c.f. Section 6.1.3):

$$\dfrac{\dfrac{\dfrac{\vdash -G_n,\ldots,-G_1,+A \qquad \dfrac{\vdash -D'_l,\ldots,-D'_1,-B,-G'_k,\ldots,-G'_1,+C'}{\vdash -B,-G'_k,\ldots,-G'_1,+C',-D'_l,\ldots,-D'_1}\,l(EC)}{\vdash -G_n,\ldots,-G_1,+A\otimes -B,-G'_k,\ldots,-G'_1,+C',-D'_l,\ldots,-D'_1}\,\otimes}{\vdash -D'_l,\ldots,-D'_1,-G_n,\ldots,-G_1,+A\otimes -B,-G'_k,\ldots,-G'_1,+C'}\,l(EC)}{\vdash -D'_l,\ldots,-D'_1,-G_n,\ldots,-G_1,-(A\setminus B),-G'_k,\ldots,-G'_1,+C'}\,=$$

□

Let us provide the NC-MLL translation of the proofs or parse structures given in examples 6.5 and 6.6 :

Example 6.10

$$
\dfrac{
\dfrac{
\dfrac{
\dfrac{
\dfrac{}{\vdash n,n^{\perp}}\,ax \quad \dfrac{}{\vdash n,n^{\perp}}\,ax
}{\vdash n,n^{\perp}\otimes n,n^{\perp}}\,\otimes
}{\vdash n^{\perp}\otimes n,n^{\perp},n}\,CX
}{\vdash n^{\perp}\otimes n,n^{\perp}\wp n}\,\wp
\quad
\dfrac{
\dfrac{
\dfrac{}{\vdash n,n^{\perp}}\,ax \quad \dfrac{}{\vdash n,n^{\perp}}\,ax
}{\vdash n,n^{\perp}\otimes n,n^{\perp}}\,\otimes
}{\vdash n^{\perp}\otimes n,n^{\perp},n}\,CX
}{\vdash n^{\perp}\otimes n,(n^{\perp}\wp n)\otimes(n^{\perp}\otimes n),n^{\perp},n}\,\otimes
\quad
\dfrac{}{\vdash np^{\perp},np}\,ax
}{
\begin{array}{cccc}
\vdash n^{\perp}\otimes n, & (n^{\perp}\wp n)\otimes(n^{\perp}\otimes n), & n^{\perp}, & n\otimes np^{\perp},\ np\\
\texttt{simple} & \texttt{très} & \texttt{exemple} & \texttt{un}\\
n\backslash n & (n\backslash n)/(n\backslash n) & n & np/n
\end{array}
}\,\otimes
$$

Example 6.11

$$
\dfrac{
\dfrac{
\dfrac{
\dfrac{
\dfrac{}{\vdash n,n^{\perp}}\,ax \quad \dfrac{}{\vdash n,n^{\perp}}\,ax
}{\vdash n,n^{\perp}\otimes n,n^{\perp}}\,\otimes
}{\vdash n^{\perp}\otimes n,n^{\perp},n}\,CX
}{\vdash n^{\perp}\otimes n,n^{\perp}\wp n}\,\wp
\quad
\dfrac{
\dfrac{
\dfrac{
\dfrac{
\dfrac{}{\vdash n,n^{\perp}}\,ax \quad \dfrac{}{\vdash np,np^{\perp}}\,ax
}{\vdash n^{\perp},n\otimes np^{\perp},np}\,\otimes
}{\vdash n\otimes np^{\perp},np,n^{\perp}}\,CX
\quad
\dfrac{}{\vdash n,n^{\perp}}\,ax
}{\vdash n\otimes np^{\perp},np,n^{\perp}\otimes n,n^{\perp}}\,\otimes
}{\vdash n^{\perp}\otimes n,n^{\perp},n\otimes np^{\perp},np}\,2\times CX
}{
\begin{array}{cccc}
\vdash n^{\perp}\otimes n, & (n^{\perp}\wp n)\otimes(n^{\perp}\otimes n), & n^{\perp}, & n\otimes np^{\perp},\ np\\
\texttt{simple} & \texttt{très} & \texttt{exemple} & \texttt{un}\\
n\backslash n & (n\backslash n)/(n\backslash n) & n & np/n
\end{array}
}\,\otimes
$$

6.4 Proof Nets: Concise and Expressive Proofs

We now turn our attention to proof nets; they are for linear logic what natural deductions (or typed lambda terms) are for intuitionistic logic, in the sense that the contexts are not copied at each step of the proof.

From a logical viewpoint, they are much more compact than sequent calculus proofs: well-formedness is a global condition but easy (and fast) to verify, and cut-elimination is a local and efficient process. But their main advantage is that they are a better representation of proofs. Indeed, many sequent calculus proofs which only differ in the order of application of the rules convert to the same proof net. For example, the two proofs given in the Examples 6.10 and 6.11 will yield the same proof net. It should be noticed that when these proofs are viewed as a representation of syntactic analyses in the Lambek calculus (they correspond to the parses of Examples 6.5 and 6.6. in a Lambek grammar), they both describe the same linguistic analysis, so it is really a good feature of proof nets that we are able to describe this analysis by a single object.

6.4.1 Proof Nets for MLL

R&B Graphs

A *matching* in a graph is a subset of the set of edges such that no two edges of the matching are adjacent. The matching is said to be *perfect* whenever each vertex of the graph is incident to an edge of the matching – because it is a matching, each vertex is incident to exactly one edge of the matching.

Definition 6.12 (R&B **graphs).** *A R&B graph is an edge colored graph, whose edges either are of color B (blue or bold), or R (red or regular), such that the B edges define a perfect matching of the graph.*

B edges correspond to formulae and R edges to connectives. The recognition, among all such graphs, of the ones which are proofs, will involve the notion of alternate elementary path.

Definition 6.13 (æ **paths and cycles).** *An æ path in a R&B graph is an alternating elementary path, that is a path the edges of which are alternatively in B and in R which does not use twice the same edge — as B edges are a matching, this is equivalent to the property that the path does not contain the same vertex twice (except, possibly the first and last vertices that might be the same). More precisely, an æ path is a finite sequence of edges $(a_i)_{i\in[1,n]}$ such that:*

$$i \neq j \Longrightarrow a_i \neq a_j \qquad \#(a_i \cap a_{i+1}) = 1$$
$$a_i \in B \Longrightarrow a_{i+1} \in R \qquad a_i \in R \Longrightarrow a_{i+1} \in B$$

An æ cycle is an æ path of even length, whose end vertices are equal.

Prenets

Definition 6.14 (Prenets or proof structures, links). *Prenets are R&B graphs built from basic R&B graphs called links, which are shown in Figure 6.2 (where α denotes an atomic formula) in such a way that each formula is the conclusion of exactly one link and the premise of at most one link. Formulae that are not the premise of a link are called conclusions of the prenet.*

Definition 6.15 (R&B **subformula tree).** *Given a formula C, its* R&B subformula tree $T(C)$ *is a R&B graph defined inductively as follows.*

- *If $C = \alpha$ is a propositional variable then $T(C)$ is:* ○ α
- *given $T(A)$ and $T(B)$, $T(A \otimes B)$ and $T(A \wp B)$ are defined as follows.*

$T(A \otimes B)$:

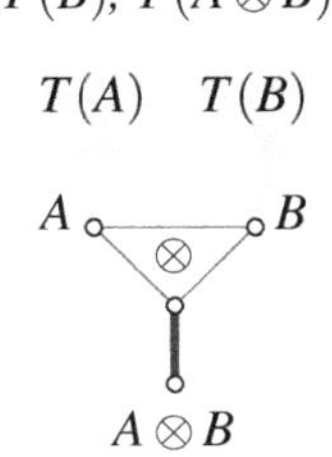

$T(A \wp B)$:

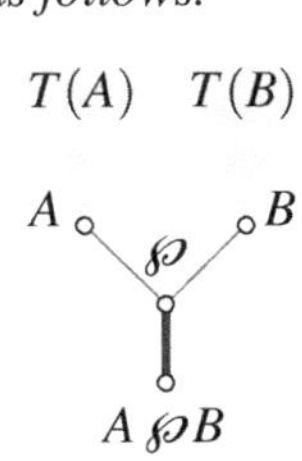

Links			
Name	Graph	Premises	Conclusions
Axiom		none	α and $\alpha^{\perp}$
Times	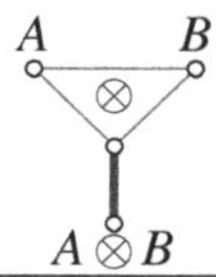	A and B	$A \otimes B$
Par	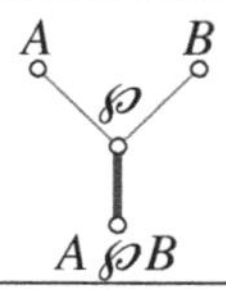	A and B	$A \wp B$

Fig. 6.2. Links for constructing prenets

Beware that the R&B subformula tree of a formula C is not, from a graph theoretical point of view, a tree: indeed, every *Times* link contains a cycle. We nevertheless chose this name because it is very similar to the subformula tree, and because of the fact that w.r.t. the æ paths, the only paths we shall use, the R&B subformulae trees are acyclic.

The vertices corresponding to propositional variables in a subformula tree will be called leaves of the subformula tree.

Definition 6.16 (prenet with conclusions Γ). *Given a sequence of formulae Γ, a prenet Π with conclusions Γ consists of:*

- *the R&B subformula trees of the formulae in Γ*
- *a set of B edges joining dual leaves, called axioms, such that each leaf is incident to exactly one axiom.*

The structure of a prenet is the following.

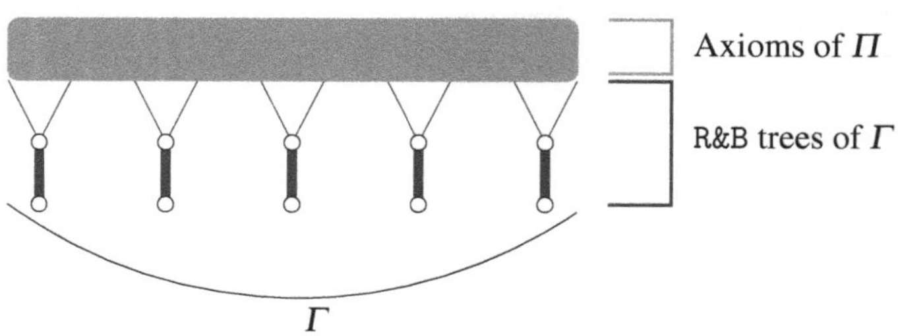

Notice that the order between formulae of Γ or their subformula trees is not part of the structure, but because of the labeling of the vertices, R&B subformula trees make a distinction between their right and left subtrees.

The examples below — Example 6.17 to 6.23 — give some examples of prenets. Note that not all of these prenets correspond to sequent proofs: we will see how to distinguish the correct prenets, the *proof nets*, from the other prenets below.

Example 6.17

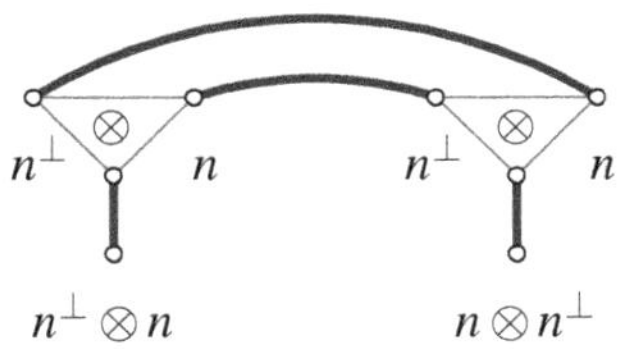

Example 6.18

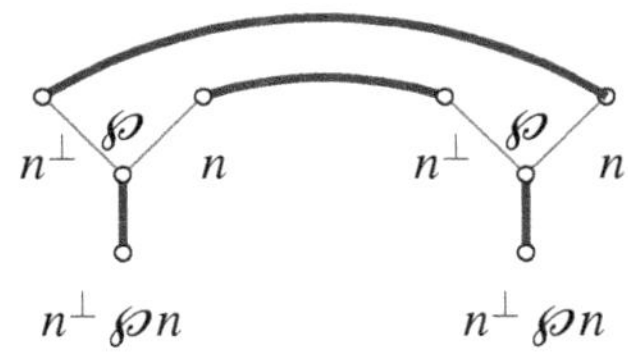

Example 6.19

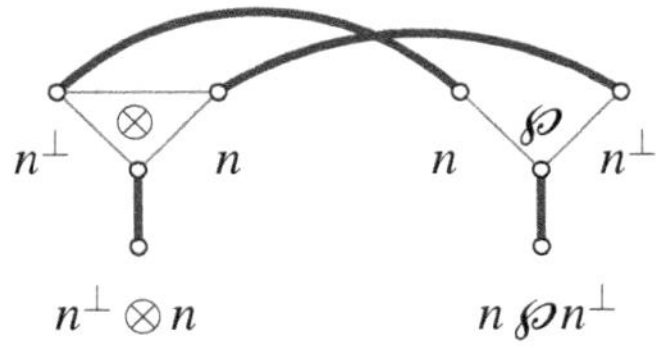

Example 6.20

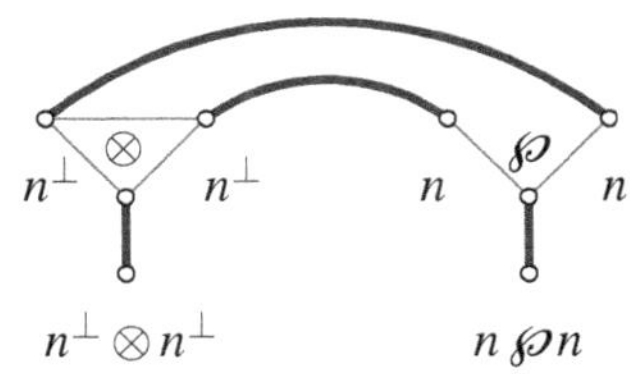

Example 6.21

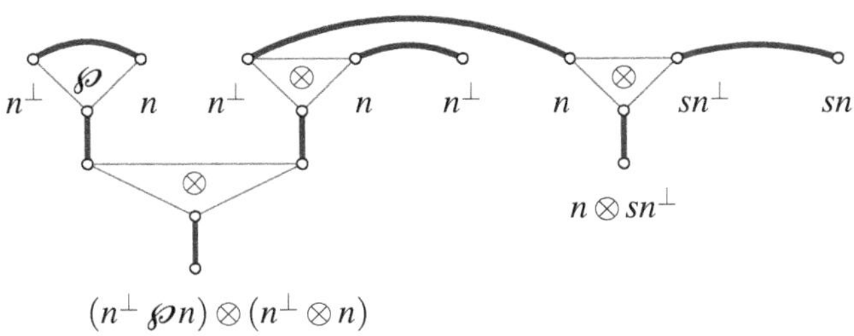

Example 6.22

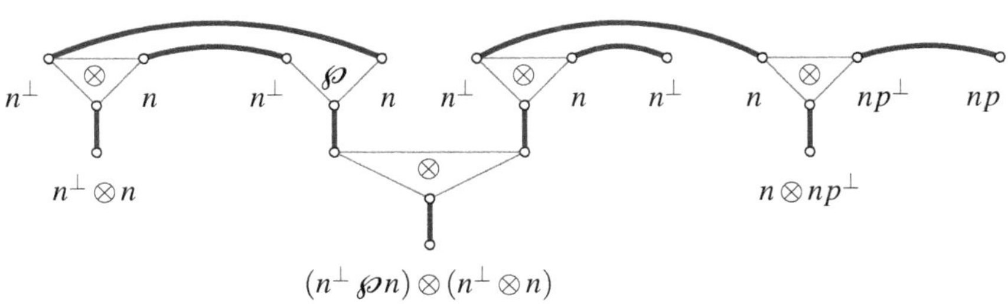

Example 6.23

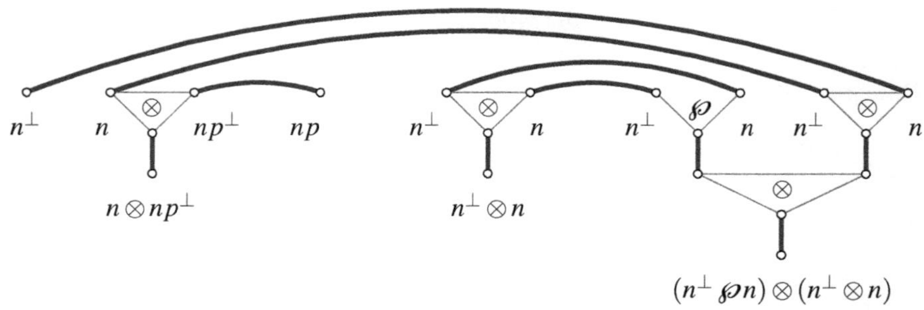

Proof Nets

Definition 6.24 (proof net). *A* proof net *is a prenet satisfying the following properties:*

ØÆ *there is no æ cycle.*
SAT *there exists an æ path between any two vertices.*

To facilitate a comparison with the well-known presentation of proof nets according to (Danos and Regnier, 1989; Girard, 1995), we will introduce the Danos-Regnier correctness condition, which is stated using correction graphs of prenets, defined as follows.

Definition 6.25 (correction graph). *From a prenet we obtain a* correction graph *by rewriting the logical links as follows.*

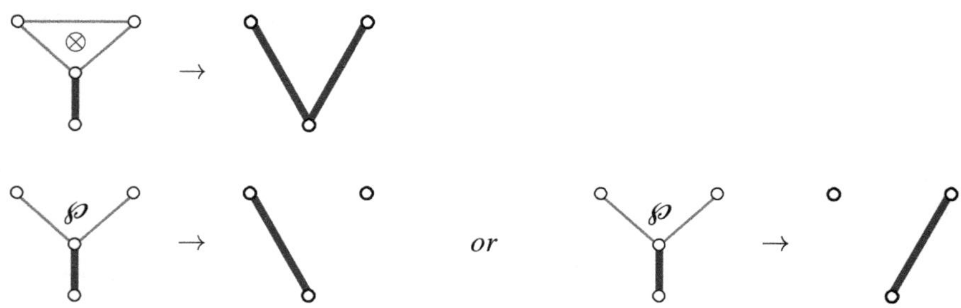

Note that there are two ways of rewriting the par links, which means that for a prenet with p par links there are 2^p *correction graphs. In addition, correction graphs only have a single type of edges (all edges are B edges) so correction graphs really* are *graphs (ie. a set of vertices and a set of edges connecting these vertices).*

Definition 6.26 (Danos and Regnier (1989)). *A prenet is a* proof net *iff all its correction graphs are acyclic and connected.*

Compared to the Danos and Regnier presentation of proof nets, the property ØÆ corresponds to the acyclicity of all correction graphs and the property SAT to their connectedness (see Fleury and Retoré, 1994; Retoré, 1996). The advantage of the current representation of proof nets is that the correctness condition can be verified by inspection of only a single graph.

The following result of (Retoré, 1996; Retoré, 2003) shows that verifying the correctness of prenets is rather easy from an algorithmic point of view — recently some linear algorithms have been provided on the Danos-Regnier presentation of proof nets, and they certainly can be adapted to our formalism (Guerrini, 1999, 2011; Murawski and Ong, 2000).

Proposition 6.27. *Given a prenet with n vertices, their exists an algorithm which decides in n^2 steps whether the prenet is a proof net.*

Among the examples of prenets given above, only 6.19, 6.20, 6.21, 6.22 and 6.23 are proof nets. The prenet 6.17 contains an æ cycle, and the prenet 6.18 does not contain any æ path between the left most leaves $n^{\perp}$ and n.

6.4.2 Sequent Calculus and Proof Nets

The following proposition gives a precise account of the correspondence between proof nets and sequent calculus proofs, and its proofs shows how sequent calculus proofs are mapped onto proof nets. The converse correspondence relies on graph theoretical properties, and we refer the reader to (Retoré, 1996; Retoré, 2003).

Theorem 6.28. *Every sequent calculus proof in MLL of a sequent $\vdash A_1,\dots,A_n$ translates into a proof net with conclusions $A_1,\dots,A_n$. Conversely, every proof net with conclusions $A_1,\dots,A_n$ corresponds to at least one sequent calculus proof in MLL of $\vdash A_1,\dots,A_n$ in NC-MLL — every such proof is called a sequentialisation of the proof net.*

Proof. As said above, we limit ourselves to the first part of this statement.

The translation from sequent calculus proofs to proof nets is defined inductively. As the exchange rule has no effect on proof nets, since for the time being we have no order on the conclusions, we simply skip it. The effect of this rule would be to produce crossings of axiom links, but up to now this is not part of our description of a proof net. For instance, the Examples 6.22 and 6.23 shown above are considered to be the same proof net: the three rightmost conclusions of Example 6.22 ($n^{\perp}$, $n \otimes np^{\perp}$ and np) are the three leftmost conclusions of Example 6.23 but they are connected in exactly the same way both to each other and to the rest of the prenet.

Proof ∂ in MLL	Corresponding proof net $\partial^{\star}$
$\dfrac{}{\vdash \alpha^{\perp}, \alpha}\ ax$	
$\dfrac{\vdots\ \partial_1 \quad \vdash \Gamma, A, B}{\vdash \Gamma, A \wp B}\ \wp$	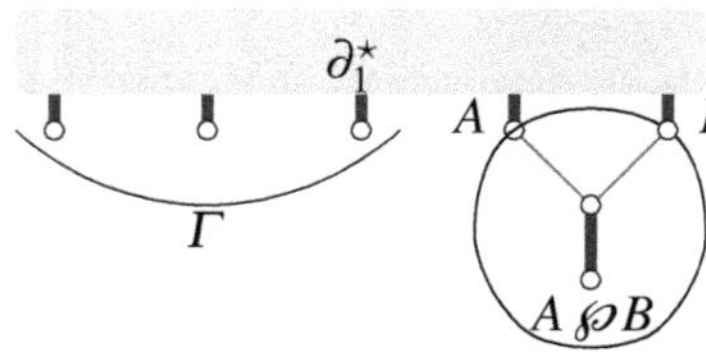
$\dfrac{\vdots\ \partial_1 \quad \vdash \Gamma, A \qquad \vdots\ \partial_2 \quad \vdash B, \Gamma'}{\vdash \Gamma, A \otimes B, \Gamma'}\ \otimes$	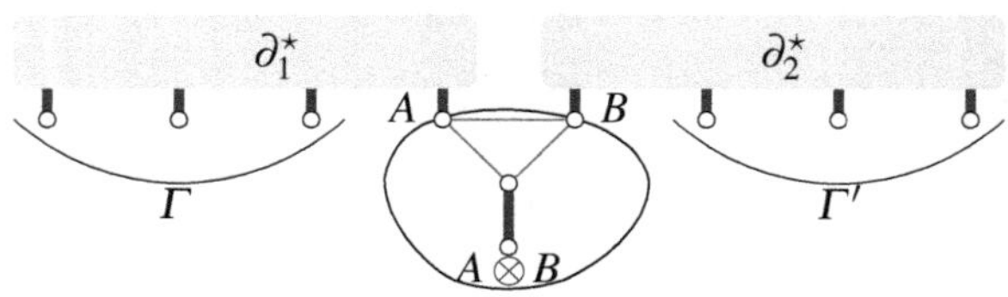

It is easily checked by induction that the prenet corresponding to a sequent calculus proof are proof nets: no æ cycle can appear during the construction, and the fact that there always exists an æ path between any two vertices is also preserved during the construction. □

Using this inductive definition, the proofs of Example 6.10 and 6.11, both yield the proof net of Example 6.22, so a single proof net corresponds to a single parse structure.

Rules and links are in a one-to-one correspondence (that is, *ax* with *Axiom*, $\wp$ with *Par* and $\otimes$ with *Tensor*), and the last logical rule in the sequent calculus proof correspond to a final link in the prenet — a link which is the root of one of the subformula trees — while the converse does not hold. We nevertheless have the following property, that will be useful later on:

Proposition 6.29. *Let Π be a proof net such that:*

- *all conclusions of Π are the conclusions of Times or Axioms links*
- *there is at least one Times link, that is Π is not a single Axiom*

then at least one of the final Times links is splitting*, that is each of the two premise B edges is a bridge — an edge the suppression of which increases the number of connected components.*

Proof. As we have a proof net, at least one sequent calculus proof translates into it. The final rule of the sequent calculus correspond to a final link, so is a *Times* link. From the translation given above, both the premise B edges of this link are bridges of the graph. □

Observe that not all final *Times* links are splitting. For instance in the example 6.22 the final *Times* $n^\perp \otimes n$ is not splitting, and can not be the translation of the final rule of a corresponding sequent calculus proof. The final *Times* links $(n^\perp \wp n) \otimes (n^\perp \otimes n)$ and $n \otimes np^\perp$ are splitting *Times* links, and this is supported by the sequentialisations given in examples 6.10 and 6.11.

We can generalize the notion of splitting *Times* link to a *hereditary* splitting *Times* link as follows (Retoré, 1993).

Proposition 6.30. *Let Π be a proof net and, as in Proposition 6.29, let all conclusions of Π be the conclusions of Times and Axiom links with the number of Times links being at least one. Π has a* hereditary splitting *Times link T; that is*

- *T is a splitting Times link, and therefore removing T from Π splits the proof net into two proof nets Π_1 and Π_2*
- *For each of the premises P_1 and P_2 of T, if P_i is the conclusion of a tensor link T_i then T_i is a hereditary splitting Times link in Π_i. Note that, since we know for both Π_1 and Π_2 that all conclusions are either the conclusions of Axiom or of Tensor links, it makes sense to talk about hereditary splitting Tensor links of these subnets.*

Proof. First, we remark that if one of the Π_i has a hereditary splitting *Times* link $T_i' \neq T_i$, then T_i' is a hereditary splitting *Times* link of Π. For suppose T_i' were not a hereditary splitting *Times* link of Π, this would mean that there would be a path from two of the "leaves" of the tensor tree with T_i' as its root passing through T which contradicts T being a splitting tensor link. The figure below illustrates the situation. Note that T is a splitting tensor but (because of T_i) not a hereditary splitting tensor although T_i' is a hereditary splitting tensor.

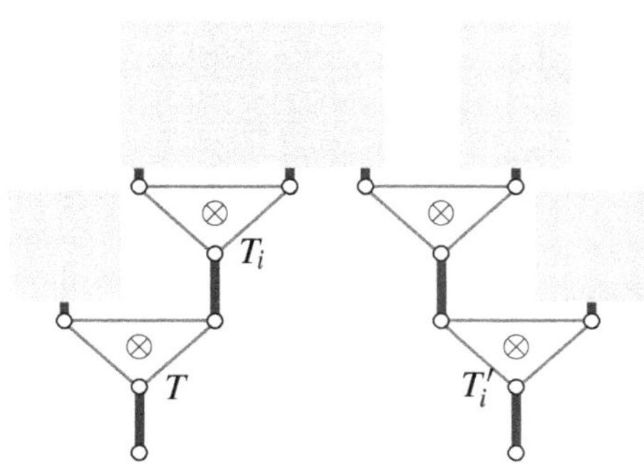

We proceed by induction on the number of tensor links in the proof net. Let T be a splitting tensor link of Π.

If none of the premises of T is the conclusion of a *Times* link, then T is hereditary splitting and we are done.

If one of the two premises of T, say P_1 is the conclusion of a *Times* link T_1 then, by induction hypothesis T_1 has a hereditary splitting *Times* link T_1'. If $T_1' \neq T_1$ then T_1' is a hereditary splitting *Times* link of Π according to the remark at the start of the proof. Otherwise T_1 is a hereditary splitting *Times* link in Π_1. We therefore look at the other premise P_2 of T. If it is not the conclusion of a tensor link, then we are done. However, if it is the conclusion of a *Times* link T_2 we proceed as before: we know by induction hypothesis that Π_2 has a hereditary splitting *Times* link T_2'. If T_2' is not equal to T_2 then T_2' is a hereditary splitting *Times* link of Π. But if T_2' is a hereditary splitting *Times* link of Π_2 then T is a hereditary splitting *Times* link of Π. □

A minimal representation of prenets and proof nets

To define a prenet or a proof net Π it is enough to give its conclusions and the pairs of propositional variables which are linked by an axiom link. These pairs can be depicted by a 2-permutation σ_Π — that is a permutation such that $\sigma_\Pi^2 = Id$ and $\forall x\ \sigma_\Pi(x) \neq x$ — defined on the set of occurrences of atoms in the sequence of conclusions. This representation will become necessary when we will deal with proof nets for the Lambek calculus, that are parse structures for Lambek categorial grammars.

Up to now, representing the conclusions by a graph is needed to check whether a prenet is a proof net (Girard, 1987; Danos and Regnier, 1989; Asperti, 1991; Asperti and Dore, 1994; Métayer, 1993). This graph can be minimized in more abstract representation (Retoré, 2003). There exists an alternative criterion relying on denotational semantics (Retoré, 1997) which does not need such a graph, but, unfortunately, checking the correctness becomes exponential.

Let us give the description of the examples 6.22 and 6.19 by means of 2-permutations.

Example 6.31

Proof Net Π	Example 6.22										Example 6.19			
Conclusions of Π	$n^\perp \otimes n$		$(n^\perp \wp n) \otimes (n^\perp \otimes n)$				$n^\perp$	$n \otimes np^\perp$		np	$n^\perp \otimes n$		$n \wp n^\perp$	
Atom occurrences x	$n_1^\perp$	n_2	$n_3^\perp$	n_4	$n_5^\perp$	n_6	$n_7^\perp$	n_8	$np_9^\perp$	np_{10}	$n_1^\perp$	n_2	n_3	$n_4^\perp$
$\sigma_\Pi(x)$	n_4	$n_3^\perp$	n_2	$n_1^\perp$	n_8	$n_7^\perp$	n_6	$n_5^\perp$	np_{10}	$np_9^\perp$	n_3	$n_4^\perp$	$n_1^\perp$	n_2

6.4.3 Intuitionistic Proof Nets

Definition 6.32. *An intuitionistic proof net with conclusions $F_1, \ldots, F_n$ is a proof net satisfying:*

INTUI: $\forall i \in [1,n]\ F_i \in \mathsf{Li}^\circ \cup \mathsf{Li}^\bullet$.

For instance the example 6.20 is not an intuitionistic proof net since $n \wp n \notin \mathsf{Li}^\bullet \cup \mathsf{Li}^\circ$.

Theorem 6.33. *Every sequent calculus proof* $A_1, \ldots, A_n \vdash B$ *in* IMLL *translates into an intuitionistic proof net with conclusions* $-A_n, \ldots, -A_1, +B$.

Conversely, let Π *a proof net with conclusions* $F_1, \ldots, F_n \in \mathsf{Li}$. *Then there exists a unique index* i_0 *in* $[1,n]$ *such that* $F_{i_0} \in \mathsf{Li}^\circ$ *and* $F_i \in \mathsf{Li}^\bullet$, *for* $i \neq i_0$, *and* Π *is the translation of a proof in* IMLL *of*

$$(F_{i_0-1})^\bullet_{\mathsf{Lp}}, (F_{i_0-2})^\bullet_{\mathsf{Lp}}, \ldots, (F_1)^\bullet_{\mathsf{Lp}}, (F_n)^\bullet_{\mathsf{Lp}}, \ldots, (F_{i_0+1})^\bullet_{\mathsf{Lp}} \vdash (X_{i_0})^\circ_{\mathsf{Lp}}$$

Proof. The first part is obvious.

For the converse, we first have to justify the existence of i_0. This existence is justified by Theorem 6.28 (it shows that Π is the translation of proof of MLL) and proposition 6.8 (which shows that a proof in MLL with all its conclusions in $\mathsf{Li}^\bullet \cup \mathsf{Li}^\circ$ has exactly one conclusion in Li° and all the others in $\mathsf{Li}^\bullet$). Once the existence of i_0 is established, the result follows from proposition 6.9, which shows that given a sequentialisation of Π in MLL, with conclusions $\vdash F_1, \ldots, F_n$ (with F_{i_0} in Li° and all the others in $\mathsf{Li}^\bullet$) corresponds to a proof in IMLL of

$$(F_{i_0-1})^\bullet_{\mathsf{Lp}}, (F_{i_0-2})^\bullet_{\mathsf{Lp}}, \ldots, (F_1)^\bullet_{\mathsf{Lp}}, (F_n)^\bullet_{\mathsf{Lp}}, \ldots, (F_{i_0+1})^\bullet_{\mathsf{Lp}} \vdash (X_{i_0})^\circ_{\mathsf{Lp}} \qquad \square$$

6.4.4 Cyclic Proof Nets

We now turn our attention towards proof nets for NC-MLL. These are proof nets which can be drawn in the plane without intersecting axioms, keeping the same design and up-down orientation for links. This condition is strictly stronger than being a planar graph (because we ask for the links to be drawn respecting left-right and up-down as shown in the figures). Consequently we shall present this condition without any reference to an embedding of the graph in the plane, but by means of a 2-permutation (bracketings from formal language theory would work just the same). This restriction, combined with the restriction for intuitionistic proof nets from the previous paragraph, will give us a characterization of proof nets for the Lambek calculus, and therefore give us a way to parse phrases and sentences with proof nets.

Cyclic Permutations and Compatibility of a 2-Permutation

A permutation ψ over a set E with n elements is said to be cyclic whenever:

$$\forall x, y \in E \quad \exists k \in [0, n-1] \quad y = \psi^k(x) \text{ (with } \psi^0(x) = x)$$

such a permutation ψ can be described by an expression:

$$\rhd\, x; \psi(x); \psi(\psi(x)); \cdots; \psi^{n-1}(x) \,\rhd$$

Given $x, y \in E$, and an index $k \in [0, n-1]$ such that $y = \psi^k(x)$, we write $[x, y]$ for the set $\{z \mid \exists j \in [0,k] \quad z = \psi^j(x)\}$; similarly $[x, y[$ is defined as $\{z \mid \exists j \in [0,k[\quad z = \psi^j(x)\}$ etc.

Given a set E endowed with a cyclic permutation ψ and a 2-permutation σ we can give an algebraic account of the following geometric fact: if we place the points of E on a circle following the cyclic order ψ, the chords joining x and $\sigma(x)$ *do not intersect* any other chord — in other words, σ is a correct bracketing, w.r.t. the cyclic order ψ over E.

Definition 6.34. *A 2-permutation σ of E is said to be* compatible *with a cyclic permutation ψ of E whenever $\forall x,y \in E\ x \in [y,\sigma(y)] \Rightarrow \sigma(x) \in [y,\sigma(y)]$.*

For instance the 2-permutation σ_Π of the example 6.31 $(n_1^\perp,n_3),(n_2,n_4^\perp)$ is not compatible with the cyclic permutation $\rhd n_1^\perp;n_2;n_3;n_4^\perp \rhd$. Indeed, $n_2 \in [n_1^\perp,\sigma_\Pi(n_1^\perp){=}n_3]$ while $\sigma_\Pi(n_2) = n_4^\perp \notin [n_1^\perp,n_3]$.

In the following definition the E_i's should be viewed as the conclusions of a proof net Π, endowed with the cyclic permutation Ψ_Π. The induced cyclic permutation is the cyclic permutation induced on the atoms — thus, viewing σ of the previous definition as the axioms of Π, we are able to express that axioms do not intersect.

Definition 6.35. *Let $\rhd E_1;\cdots;E_n \rhd$ be a cyclic permutation of $M = \{E_1,\ldots,E_n\}$ where each E_i is a sequence of symbols $a_i^1,a_i^2,\ldots,a_i^{j_i}$. The cyclic permutation induced by Ψ over the disjoint sum of the symbols of the E_i's is the cyclic permutation defined by:*

$$\rhd a_1^1;a_1^2;\cdots;a_1^{j_1};a_2^1;a_2^2;\cdots;a_2^{j_2};\cdots;a_n^1;a_n^2;\cdots;a_n^{j_n} \rhd$$

In order to characterize the proof nets for the Lambek calculus we shall need the following proposition:

Proposition 6.36. *Let Ψ be a cyclic permutation over a finite set M of n sequences of symbols $M = E_1,\ldots,E_n$. Let ψ be the cyclic permutation induced on $E = \oplus E_i$, as in definition 6.35. Let σ be a 2-permutation of E, compatible with ψ, as in definition 6.34. Let Σ be the following (symmetric) relation over M: $E_i \Sigma E_j$ whenever there exists $x_i \in E_i$ such that $\sigma(x_i) \in E_j$. Let Σ^* be the transitive closure of Σ; if Σ^* has* exactly two equivalence classes *$\mathcal{G}$ and $\mathcal{D}$, then there exist $G \in \mathcal{G}$ and $D \in \mathcal{D}$ such that: $\mathcal{G} = [G,D[$ and $\mathcal{D} = [D,G[$.*

Proof. By induction on $\#E+n$.

If one of the class contains only one element, the result is obvious — this necessarily happens when a class has a single element, for instance when $n = 2$.

There exists z such that $\psi(z) = \sigma(z)$ Let z be a point such that $\#]z,\sigma(z)[$ has the smallest number of elements, and let us show that $\#]z,\sigma(z)[= 0$ — hence $\psi(z) = \sigma(z)$. Assume that there exists $y \in]z,\sigma(z)[$; since σ is compatible with ψ, $\sigma(y) \in]z,\sigma(z)[$. Thus one of the two intervals $]y,\sigma(y)[$ or $]\sigma(y),\sigma(\sigma(y)) = y[$ is a subset of $]z,\sigma(z)[$, and since none of them contains y, they have strictly less elements than $\#]z,\sigma(z)[$, contradiction.

Let z be an element such that $\psi(z) = \sigma(z)$ and let i be the index such that $z \in E_i$. Three cases can happen:

$\sigma(z) \in E_i$ and $E_i = z, \sigma(z)$ In this case, E_i is the only element in its equivalence class, and the result is clear.

$\sigma(z) \in E_i$ and $E_i = \ldots, z, \sigma(z), \ldots$ In this case, replace E_i with $E_i \setminus \{z, \sigma(z)\}$, restrict σ and ψ to $E \setminus \{z, \sigma(z)\}$. The induction hypothesis apply, and since Σ^* remains unchanged, the D and G provided by the induction hypothesis are solutions for the original problem.

$\sigma(z) \notin E_i$. In this case $\sigma(z)$ is the first symbol of $E_{i+1} = \Psi(E_i)$. Let us consider the following reduction problem:

let Ψ' be the cyclic permutation $\rhd E_1; \ldots; E_{i-1}; E_{i(i+1)}; E_{i+2}; \ldots; E_n \rhd$ where $E_{i(i+1)}$ is the sequence of symbols E_i, E_j

Observe that E, ψ and σ remains unchanged, and therefore σ is compatible with ψ. Since $E_i \Sigma E_{i+1}$ the equivalence relation Σ'^* for this reduction problem also has exactly two classes.

Hence we are faced with a similar problem with $\#M' = n-1$. The induction hypothesis yields G' and D' such that $\mathscr{G}' = [G', D'[$ and $\mathscr{D}' = [D, G'[$. A solution to the original problem is given by $G = G'$ and $D = D'$ — if G' (resp. D') is $E_{i(i+1)}$, then G (resp. D) should be E_i. □

Cyclic Proof Nets

Definition 6.37. *A cyclic prenet with conclusions $\Psi :\rhd A_1; \cdots; A_n \rhd$ is a prenet with conclusions $A_1, \ldots, A_n$ endowed with the cyclic permutation $\Psi_\Pi :\rhd A_1, \ldots, A_n \rhd$. We write Ψ_Π^{at} for the cyclic permutation induced by Ψ_Π on the atoms of Ψ — in the sense of the definition 6.35.*

Definition 6.38. *A cyclic prenet with conclusion $\Psi :\rhd A_1, \cdots, A_n \rhd$ is a cyclic proof net if and only if it is a proof net with conclusion $A_1, \ldots, A_n$ (the conditions ØÆ and* SAT*are satisfied) and:*

NC: *σ_Π is compatible with Ψ_Π^{at}*

For instance the example 6.19 is not a cyclic proof net. Indeed, $\Psi_\Pi = \rhd n_1^\perp \otimes n_2; n_3 \wp n_4^\perp \rhd$ (there are only two conclusions, so there is only one possible cyclic permutation), and $\Psi_\Pi^{at} = \rhd n_1^\perp; n_2; n_3; n_4^\perp \rhd$, while the 2-permutation σ_Π of its axiom links, given in example 6.31, is not compatible with Ψ_Π^{at} — as we have seen after the definition 6.34.

The proof nets of the examples 6.20, 6.21, 6.22 and 6.23 are cyclic proof nets.

Theorem 6.39. *Every sequent calculus proof of $\vdash A_1, \ldots, A_n$ in NC-MLL translates into a cyclic proof net with conclusions $\rhd A_1; \cdots; A_n \rhd$.*

Conversely, every cyclic proof net with conclusion $n \rhd A_1; \cdots; A_n \rhd$ is the translation of at least a sequent calculus proof of $\vdash A_1, \ldots, A_n$ in NC-MLL.

Proof. The first part is rather simple to establish by induction on the sequent calculus proof. Nevertheless one should take care of the compatibility of Ψ_Π^{at} with σ_Π; to do so, one should place atoms on a circle, and draw axiom links as chords of this circle, and draw R&B subformula trees outside the circle. Observe that the cyclic exchange (cx) corresponds to the equality of the proof nets.

The converse is proved by induction on the number of links of the proof net Π. As it is a proof net, Proposition 6.29 applies.

If Π is an axiom $\rhd \alpha, \alpha^\perp \rhd = \rhd \alpha, \alpha^\perp \rhd$ a sequentialisation is provided by the axiom $\vdash \alpha, \alpha^\perp$ of NC-MLL.

If Π has a final *Par* link $A_i = A \wp A'$, let us consider Π' the proof net obtained from Π by suppressing this final *Par* link and endowed with the cyclic permutation $\rhd A_1;\ldots;A_{i-1};A;A';A_{i+1};\cdots A_n \rhd$. The proof net Π' is a cyclic proof net as well, since $\Psi^{at}_{\Pi'} = \Psi^{at}_{\Pi}$ and $\sigma_{\Pi'} = \sigma_\Pi$. By induction hypothesis there exists a sequent calculus proof in NC-MLL corresponding to Π', and applying a $\wp$ rule to this proof yields a sequentialisation of Π.

Otherwise, by Lemma 6.29, Π has a splitting *Times*, say $A_i = A \otimes A'$. Suppressing this final link yields two proof nets Π_A and $\Pi_{A'}$ with conclusions $\Gamma_A = A_{i_1},\ldots,A_{i_p},A$ and $\Gamma_{A'} = A_{j_1},\ldots,A_{j_q},A'$ with $\{i_1,\ldots,i_p,j_1,\ldots,j_q\} = [1,n] \setminus \{i\}$. Consider the prenet $\Pi' = \Pi_A \cup \Pi_{A'}$ and endow its conclusions with the cyclic permutation $\rhd A_1;\cdots;A_{i-1};A;A';A_{i+1};\cdots;A_n \rhd$. Since $\Psi^{at}_{\Pi'} = \Psi^{at}_{\Pi}$ and $\sigma_{\Pi'} = \sigma_\Pi$, the 2-permutation $\sigma_{\Pi'}$ is compatible with $\Psi^{at}_{\Pi'}$. Let Σ be the (symmetric) relation between the conclusions of Π' defined by: $\exists x \in C\ \sigma_\Pi(x) \in C'$ — in other words, this relation holds whenever Π contains an axiom with a conclusion in C and the other in C'. The link $A \otimes B$ is splitting in Π, means that Σ^* has exactly two equivalence classes Γ_A and $\Gamma_{A'}$. Because of Proposition 6.36 the cyclic permutation of the conclusions of Π' can be written as $\rhd A_{i_1};\cdots,A_{i_p};A;A',A_{j_1};\cdots;A_{j_q} \rhd$. Thus Π_A (resp. $\Pi_{A'}$) endowed with the cyclic permutation $\rhd A_{i_1};\cdots,A_{i_p};A \rhd$ (resp. $\rhd A',A_{j_1};\cdots;A_{j_q} \rhd$) is a cyclic proof net. Indeed Π_A is a proof net and since σ_{Π_A} and $\Psi^{at}_{\Pi_A}$ are the restrictions to Γ_A of σ_Π and Ψ^{at}_Π compatibility is preserved — the same argument works for $\Pi_{A'}$.

Therefore, by induction hypothesis we have two sequent calculus proofs in NC-MLL with conclusions $\vdash A_{i_1};\cdots,A_{i_p};A$ and $\vdash A';A_{j_1};\cdots;A_{j_q}$ corresponding to Π_A and Π'_A. Applying the rule $\otimes$ of NC-MLL yields a proof with conclusion $\vdash \Gamma_A, A \otimes B, \Gamma_B$ corresponding to Π. □

For instance the proofs of the examples 6.10 and 6.11 correspond to the cyclic proof net of the example 6.22, which is *equal* to the proof net of the example 6.23. Indeed expressions $\rhd n^\perp \otimes n; (n^\perp \wp n) \otimes (n^\perp \otimes n); n^\perp; n \otimes np^\perp; np \rhd$ and $\rhd n^\perp; n \otimes np^\perp; np; n^\perp \otimes n; (n^\perp \wp n) \otimes (n^\perp \otimes n) \rhd$ denotes the same cyclic permutation.

6.4.5 Proof Nets for the Lambek Calculus — With or Without Empty Antecedent

In order to characterize the proof nets of the Lambek calculus L, which exclude sequents with empty antecedents, we need the following proposition. It involves the notion of a sub-prenet and subproof net: a sub-prenet (sub proof net) is a subgraph of a prenet (proof net) which is itself a prenet (proof net). A sub-prenet of a proof net is not always a proof net: it is possible that SAT does not hold in the sub-prenet (but ØÆ holds).

Proposition 6.40. *Let Π be a proof net; the following statements are all equivalent:*

1. *Every sub-prenet of Π has at least two conclusions. (ε-FREE)*
2. *Every sub proof net of Π has at least two conclusions.*
3. *Every sequentialisation of Π contains only sequents with at least two conclusions.*
4. *There exists a sequentialisation of Π which contains only sequents with at least two conclusions.*

Proof. Implications $1 \Rightarrow 2$, $2 \Rightarrow 3$ and $3 \Rightarrow 4$ are straightforward.

$4 \Rightarrow 1$ is shown by induction on the number of links in Π, which is equal to the number of axioms and logical rules of every sequentialisation of Π. Let us consider a sequentialisation Π^* of Π, such that every sequent of it contains at least two formulae. We can assume the last rule of Π^* is not an exchange rule: indeed the same proof without this exchange rule is also a sequentialisation of Π, with all sequents having at least two formulae.

If the last rule of Π^* is an axiom, Π^* consists of this axiom, which contains two formulae. In this case Π is an axiom, whose only sub-prenet is itself, which has two conclusions.

If that rule of Π^* is a two premise rule, applied to two proofs Π'^* and Π''^*, the corresponding link of Π is a splitting *Times* link: Π is obtained from two smaller proof nets Π' and Π'' connected by this *Times* link. The two proofs Π'^* and Π''^* are possible sequentialisations for Π' and Π'' and these proofs also have sequents with at least two formulae. Thus the induction hypothesis can be applied to Π' and Π'': every sub-prenet of Π' or of Π'' has at least two conclusions. The intersection of a sub-prenet $s\Pi$ of Π, with Π' (resp. Π'') is a sub-prenet of Π' (resp. Π'') which has $p > 1$ (resp. $q > 1$) conclusions. If the *Times* link is part of $s\Pi$ then the number of conclusions of $s\Pi$ is $p+q-1 > 1$, and otherwise the number of conclusion of $s\Pi$ is $p+q > 1$. Thus, in any case Π satisfies ε-FREE.

If the last rule of Π^* is a one premise rule applied to some proof Π'^*, the corresponding link of Π is a final *Par* link. Let Π' be the proof net obtained from Π by removing this final *Par* link; it is a proof net with strictly less links, which has a sequentialisation Π'^* with sequents with more than one conclusions. Hence, by induction hypothesis every sub-prenet of Π' has at least two conclusions. Given a sub-prenet $s\Pi$ of Π, its intersection $s\Pi'$ with Π' has at least two conclusions. It is impossible that $s\Pi'$ has only the two conclusions X and Y. Indeed we know that Π has at least two conclusions, hence it has another conclusion Z in addition to $X \wp Y$. Since Π is a proof net it is connected, and there exists a path joining $s\Pi'$ to Z conclusion, and this path can be assumed to lie outside $s\Pi'$ — by cutting the part inside $s\Pi'$. So there exists an edge of Π, incident to $s\Pi'$ starting this path. This edge can neither be the R edge below X, nor the one below Y, since any path starting by one of these edges has to enter again $s\Pi'$. But the only way to leave a sub-prenet is from one of its conclusions: therefore $s\Pi'$ has a conclusion which is neither X nor Y. Let p be the number of conclusions of $s\Pi'$. If X and Y are among the p conclusions of $s\Pi'$, then $s\Pi'$ has another conclusion and $p > 2$. Therefore, either $s\Pi$ has $p > 2$

conclusions (when $X \wp Y$ is not one of its conclusions), or $s\Pi$ has $p-1>1$ conclusions (when $X \wp Y$ is one of its conclusions). If X or Y is not a conclusion of $s\Pi'$, then $X \wp Y$ is not a conclusion of $s\Pi$, and $s\Pi$ and $s\Pi'$ have the same number of conclusions $p>1$.

In any case $s\Pi$ has at least two conclusions. □

Definition 6.41. *A Lambek proof net of conclusion $\Psi_\Pi = \rhd F_1;\cdots;F_n \rhd$ is an intuitionistic cyclic proof net, i.e. a prenet satisfying*

ØÆ: *there is no æ cycle* alternate elementary cycle.
SAT: *There always exists an æ path between any two vertices.*
INTUI: *Every conclusion F_i is in* $\mathsf{Li}^\bullet \cup \mathsf{Li}^\circ$.
NC: *σ_Π is compatible with Ψ_Π^{at} — the axioms of Π do not intersect.*

A Lambek proof net is said to be without empty antecedent if, moreover:

ε-FREE: *Every subprenet of Π has at least two conclusions.*

Among the four equivalent statements given above, we have chosen the first one, because subprenets are easier to define. It is enough to chose a set of vertices of the proof net, and to close it by subformula and axiom links, without verifying SAT or ØÆ. When NC and ØÆ hold, this amounts to the following fact: for every subformula G of a conclusion, the first and last atom of G are never linked by an axiom. If $G = H \otimes H'$ then this holds, and if $G = H \wp H'$, this exactly means that there is no sub-net with a single conclusion.

Theorem 6.42. *Every sequent calculus proof with conclusion $A_1,\ldots,A_n \vdash B$ in L_ε (resp. L) translates into a Lambek proof net (resp. a Lambek proof net without empty antecedent) with conclusions $\rhd -A_n;\cdots,-A_1;+B \rhd$.*

Conversely, let Π be a Lambek proof net (resp. a Lambek proof net without empty antecedent) with conclusions $\rhd F_1;\ldots;F_n \rhd$. and let i_0 be the unique index in $[1,n]$ such that $F_{i_0} \in \mathsf{Li}^\circ$ and $F_i \in \mathsf{Li}^\bullet$, for $i \neq i_0$. The proof net Π is the translation of at least a sequent calculus proof in L_ε (resp. L) of

$$(F_{i_0-1})^\bullet_L,(F_{i_0-2})^\bullet_{\mathsf{Lp}},\ldots,(F_1)^\bullet_{\mathsf{Lp}},(F_n)^\bullet_{\mathsf{Lp}},\ldots,(F_{i_0+1})^\bullet_{\mathsf{Lp}} \vdash (F_{i_0})^\circ_{\mathsf{Lp}}$$

Proof. The first part is a straightforward induction on the sequent calculus proof in L_ε (resp. L).

For the second part, we know from Proposition 6.39 that there is a sequentialisation corresponding to Π in NC-MLL, with conclusion $\vdash F_1,\cdots,F_n$. Because of Proposition 6.9, this sequent calculus proof in NC-MLL corresponds to a proof of

$$(F_{i_0-1})^\bullet_L,(F_{i_0-2})^\bullet_{\mathsf{Lp}},\ldots,(F_1)^\bullet_{\mathsf{Lp}},(F_n)^\bullet_{\mathsf{Lp}},\ldots,(F_{i_0+1})^\bullet_{\mathsf{Lp}} \vdash (F_{i_0})^\circ_{\mathsf{Lp}}$$

in L_ε. Using $1 \Rightarrow 3$ of Proposition 6.40, it is easily seen that whenever Π is a Lambek proof net without empty antecedent, the sequentialisation in L_ε is in fact in L, i.e. it does not contain sequents with only one formula. □

Among our proof net examples, only Examples 6.21, 6.22 and 6.23 are Lambek proof nets. Example 6.22 corresponds to the parse structures 6.5 and 6.6: we thus got rid of *spurious ambiguity* — a classical drawback of sequent proof search for categorial grammars, which provides too many proofs/parse structures for a single analysis. One advantage of working with cyclic permutation is that Examples 6.22 and 6.23 are *equal*. Example 6.21 is not a Lambek proof net without empty antecedent: indeed it contains a sub-net whose only conclusion is $n^{\perp} \wp n$. It corresponds to the Example 6.4 in L_{ε}.

6.4.6 Cut Elimination for Proof Nets

We have deliberately excluded the cut links from our discussion of proof nets so far. We will present two versions of the cut link, the first is a simple connection of a formula with its negation. The second is a special kind of tensor link with the constant "Cut" as its conclusion. This second formulation has the advantage that we can treat the cut link just as a tensor link in sequentialisation proofs.

Name	Graph	Premises	Conclusions
Cut		A and $A^{\perp}$	none
$\text{Cut}_{\otimes}$	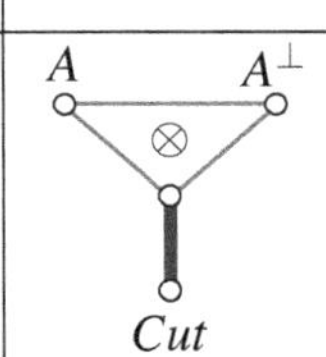	A and $A^{\perp}$	the special constant "Cut"

Cut elimination for proof nets is very simple — at least in the commutative case, we will discuss why cut elimination is more diffcult for non-commutative calculi in the next section. The base case occurs when the cut link is connected to an axiom link. In this case we are in the following situation.

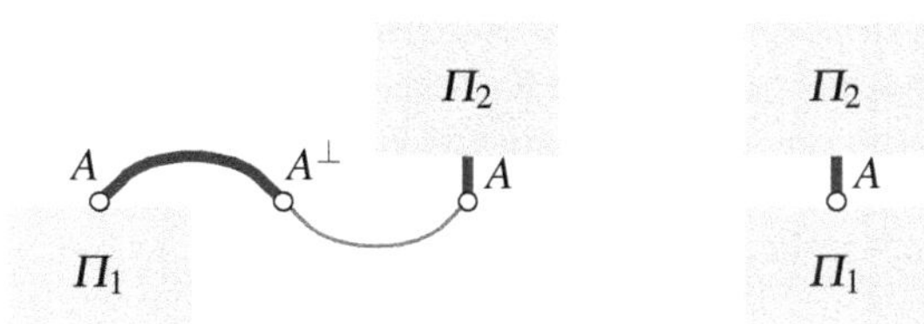

We know that Π_1 and Π_2 must be disjoint, since the vertices of Π_1 and Π_2 are connected by the path passing through the cut and the axiom link and we know that the complete proof net does not have an alternate elementary cycle. We can eliminate the cut as shown in the figure above on the right: we remove both the cut and the axiom link and the resulting structure satisfies ØÆ and SAT because the unreduced structure did.

In case the cut link is connected to a complex formula, we must be in the situation shown below on the left.

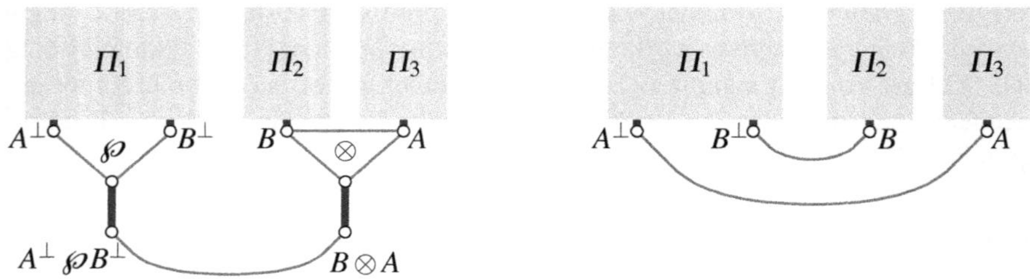

That is, given that we have a proof net on the left hand side of the figure, Π_2 and Π_3 are connected by the tensor link which is shown in the figure and therefore not connected elsewhere and Π_1 is connected to Π_2 and Π_3 by means of the paths shown in the figure and therefore not by any other paths.

We can replace the cut link by two cut links on the subformulas as shown in the figure above on the right. It is easy to see that the resulting structure is again a proof net. The path from A (and the formulas of Π_3) to B (and the formulas of Π_2) which used to be connected directly through the *Times* link now goes through Π_1, but all other paths have been shortened.

It is also easy to see that cut elimination is confluent.

Lemma 6.43. *Let Π be a proof net with cuts with possible cuts K_i, $K_i^\perp$ such that all conclusions are polarized and with one output conclusion, then all K_i, $K_i^\perp$ are polarized.*

Proof. This lemma is an easy corollary of Proposition 6.30 when we treat cut links as tensor links $Cut_\otimes$. We assume, without loss of generality, that Π has only atomic axiom links. Since Π is a proof net, we sequentialise as before.

If Π contains conclusions which are par links, then we can remove the par link and the result will be a proof net, it is easy to verify that this new proof net is still polarized since all polarized par links reduce the number of negative formulas, but keep the number of positive formulas constant. In case there are no terminal par links, by Proposition 6.30 there is a hereditary splitting *Times* link, possibly a cut. In case it is not a cut, removing the $n > 0$ hereditary splitting tensors will produce $n+1$ disjoint proof nets $\Pi_1, \ldots, \Pi_n$. We only need to verify that all Π_i are polarized. If the conclusion of the hereditary splitting link is a postive *Times* link with conclusion $A \otimes B$, then all n *Times* links are positive and each Π_i will have a positive conclusion after removal of all the *Times* links. If it is a negative *Times* link, then extactly one of the Π_i, say Π_k, has a positive conclusion which is already a conclusion of Π and is therefore connected to the hereditary splitting *Times* by an input conclusion of Π_k. Given the form of the polarized *Times* links, this means that all Π_i for $i \neq k$ have a single positive conclusion.

The interesting case is when there hereditary splitting *Times* link is a cut link between A and $A^\perp$ where A and $A^\perp$ are not necessarily polarized. However, we know by induction hypothesis that all conclusions of Π are polarized. In case A is an atomic formula, this means that A is connected by an axiom link to a formula $A^\perp$ which is a conclusion of Π and therefore both A and $A^\perp$ are polarized. Now suppose $A = B \wp C$ and $A^\perp = C^\perp \otimes B^\perp$: this is the only combination which is not polarized and we show it leads to a contradiction. One step of cut elimination connects B to $B^\perp$ and C to $C^\perp$. We also know that Π_1, the subnet with conclusion $B^\perp$, and Π_2, the subnet with conclusion $C^\perp$, are both a proof nets and the the proof net Π is polarized, that is Π has a single polarized output conclusion and all other conclusions of Π a polarized input conclusions. Since $B^\perp$ and $C^\perp$ are both input conclusions of their respective proof nets Π_1 and Π_2 and all other conclusions of both proof nets were conclusions of Π this means that one of Π_1 and Π_2 does not have a positive conclusion and therefore is not a polarized proof net.

6.4.7 Cuts and Non-commutative Proof Nets

There are a variety of multiplicative proof nets criteria in the usual commutative case that are fully satisfying. But the non-commutative case is rather tricky when there are cuts. To the best of our knowledge, only the criterion by Paul-André Melliès is fully satisfactory (Melliès, 2004). What do we mean by "satisfactory" for a correctness criterion?

1. every sequent calculus proof should be mapped to a correct proof net, rules corresponding to links (in particular cut-free proofs should be mapped to cut-free proof nets)
2. every correct proof net should correspond to a sequent calculus proof links corresponding to rules (in particular cut-free proofs should be mapped to cut-free proof nets)
3. sequent calculus proofs that only differ up to rule permutations should be mapped to the same proof net
4. the criterion should be preserved under cut elimination in proof nets (not as obvious as it may seem)

The reason why criteria are trickier in the non commutative case mainly comes from the difference between cut links and times links. In a planar representation, a cut link is allowed to be included in an internal face, while a times link which is a conclusion of the proof net is not. Figure 6.3 (after Melliès, 2004, p. 294) shows an example.

As the reader can easily verify (Exercises 6.5 asks you to verify a number of properties of this proof structure), it is a proof structure of

$$\vdash (b^\perp \wp b) \otimes (a^\perp \wp a)$$

which, though this sequent *is* derivable, the proof structure shown in Figure 6.3 is not sequentialisable in a non-commutative calculus. However, it is planar and satisfies all other conditions for proof nets (at least for proof nets which allow empty antecedent derivations).

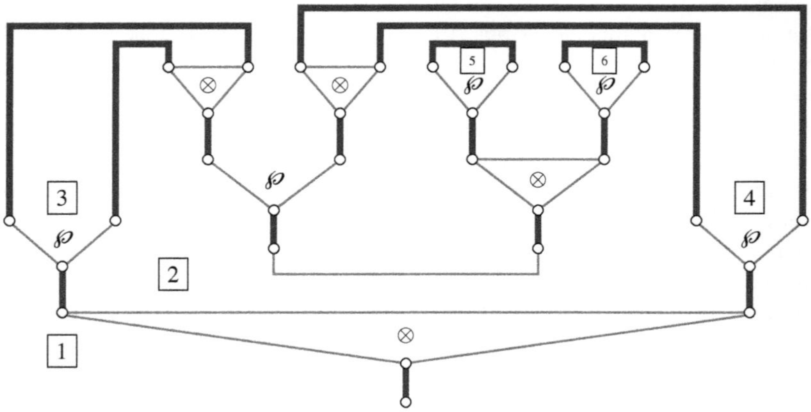

Fig. 6.3. Proof structure which does not correspond to a non-commutative sequent proof

To see that this proof net is not sequentialisable, the only conclusions of the proof net are the tensor link and the cut link (which, as before, we treat as a tensor link as well for the purpose of sequentialisation). Only the cut link splits the proof structure into two disconnected proof structures: one proof structure with conclusion $\vdash (a^{\perp} \wp a) \otimes (b^{\perp} \wp b)$ — which is both derivable and a substructure which is a proof net — but the second proof structure is a proof structure of

$$\vdash (b^{\perp} \otimes b) \wp (a^{\perp} \otimes a), (b^{\perp} \wp b) \otimes (a^{\perp} \wp a)$$

which is *not* derivable in a non-commutative logic.

The correctness condition of Melliès (2004) (though the terminology we use is closer to de Groote, 1999) formalizes this restriction on the conclusions. This condition is, to the best of our knowledge, the only correctness condition which works correctly for non-commutative proof nets with cut links.

Definition 6.44. *Let P be a planar drawing of a proof structure. A* face *f of P is a connected area enclosed by the edges of the proof structure such that the border of f contains at least one* B *edge.*

The faces of Figure 6.3 are shown as [n]. The R triangle of a *Times* link is not counted as a face, though face [5] and [6] show valid three-edge faces.

Definition 6.45. *A face f of a proof structure is an* internal *face iff it contains both* R *edges of at least one* Par *link. A face which is not internal is called* external.

Definition 6.46. *A proof structure Π is a proof net iff it satisfies* ØÆ, SAT *and all conclusions of Π are on the unique external face of Π.*

We can see that the external face of the proof structure in Figure 6.3 is face [2] (and not face [1] whose frontier contains both R edges of the par link connected to the cut link). As a conquence, the proof structure is not a proof net, since its conclusion is on the internal face [1].

6.4.8 Basic Properties of Graphs and Proof Nets

This section covers some basic properties of graphs and proof nets. Notably, it shows that, under certain conditions we can replace the acyclicity and connectedness condition of Danos and Regnier (1989) by either an acyclicity condition or a connectedness condition, using some basic properties of acyclic and connected graphs (Bondy and Murty, 1976; Diestel, 2010). Though people have been aware of many of these properties for a long time (eg. J. van de Wiele (1991, p.c.)), it is actually rather hard to find in print, though Guerrini (2011) gives a clear presentation of many of the results of this section and Morrill and Fadda (2008) independently prove that acyclicity implies connectedness (part of Corollary 6.56 in this section).

In this section, we will often say that a *prenet* Π is acyclic or connected, in the sense of Danos and Regnier (Definition 6.26) to indicate that all correction graphs of Π are acyclic or connected.

Definition 6.47. *Let G be a graph. We will use v to denote the number of its vertices and e to denote the number of its edges.*

Proposition 6.48. *If G is acyclic and connected, then* $e = v - 1$.

Proof. By induction on v. If $v = 1$ then $e = 0$, which is the only acyclic connected graph with a single vertex.

Suppose $v > 1$, then G contains at least one edge m and since G is acyclic $G - m$ has two components G_1 and G_2 which are acyclic and connected and which have less than v vertices. By induction hypothesis $e_1 = v_1 - 1$ and $e_2 = v_2 - 1$ and therefore, since $e = e_1 + e_2 + 1$, we have $e = (v_1 - 1) + (v_2 - 1) + 1 = v_1 + v_2 - 1$ and since $v = v_1 + v_2$, we have $e = v - 1$ as required. □

Proposition 6.49. *If G is a graph such that* $e = v - 1$ *then G is acyclic iff G is connected.*

Proof. If G is acyclic, then it consists of a number of connected components n, each of which, being acyclic and connected, satisfies $e_i = v_i - 1$ for $1 \leq i \leq n$, according to Proposition 6.48. Therefore, for the complete graph we have $e = v - n$, where n is the number of components. Since $e = v - 1$ by assumption, there is only a single connected component and therefore G is connected.

If G is connected then let G' be an acyclic, connected subgraph of G which contains all vertices of G (that is, a spanning tree: to obtain G', we delete the necessary number of edges from G to obtain an acyclic, connected graph). Since G' is acyclic and connected, according to Proposition 6.48 it has $v - 1$ edges. But since G has $v - 1$ edges by assumption, G is equal to G' and therefore acyclic. □

Definition 6.50. *Let* Π *be a (cut-free) prenet. We will use c to denote its number of conclusions, p to denote its number of par links, t to denote its number of tensor links and a to denote its number of axiom links.*

The following proposition, relating the number of tensor links, par links and conclusions of a proof net is also rather easy to show (Exercise 7.4 asks you to prove this proposition yourself). It was first noticed by in the early nineties and appears in Fleury (1996).

Proposition 6.51. *If Π is a proof net, then $c+p=t+2$.*

Proposition 6.52. *If a prenet Π is polarized and has a single positive conclusion, then $c+p=t+2=a+1$.*

Proof. When we look at the construction of a prenet from the axioms down, we see that a prenet with a axioms, without any tensor and par links, has a positive/output conclusions and a negative/input conclusions. If Π has a single positive conclusion, then it has $c-1$ negative conclusions.

- For each par link we add, we take two (possibly disconnected) conclusions of the prenet as its premises and introduce a new conclusion, increasing the number of par links by one, keeping the number of positive conclusions constant, but reducing the total number of negative conclusions by one — a polarized par link has either two negative premises and a negative conclusion or a positive and a negative premise and a positive conclusion.
- For each tensor link we add to the structure, on the other hand, we keep the number of negative conclusions constant, but reduce the total number of positive conclusions — a polarized tensor link has either two positive premises and a positive conclusion or a positive and a negative premise and a negative conclusion.

Therefore, the only way to obtain a prenet with $c-1$ negative conclusions is for the prenet to have $a-(c-1)$ par links and the only way to obtain a prenet with one positive conclusion is for the prenet to have $a-1$ tensor links. Therefore, every polarized prenet with a single output conclusion has $t=a-1$, $p=a-(c-1)$, which gives $c+p=t+2=a+1$. □

Proposition 6.53. *If Π is a prenet, then every correction graph G of Π has $2a+p+t$ vertices and $a+p+2t$ edges.*

Proof. Since each vertex is the conclusion of exactly one link, we can simply count the conclusions of the links in the proof nets: two for each axiom link and one for each tensor and par link. For the edges, each correction graph replaces a tensor link by two edges and a par link by a single edge (the axiom links stay single edges). □

Propositions 6.54 and 6.55 follow the simple and elegant proofs of Guerrini (2011).

In Proposition 6.52, we have seen that $a=t+1$ held for polarized prenets with a unique positive conclusion. The next proposition shows that $a=t+1$ holds in general for proof *nets*.

Proposition 6.54. *If Π is a proof net (not necessarily polarized) then $a = t + 1$.*

Proof. Since Π is a proof net, all its correction graphs are acyclic and connected. Therefore, according to Proposition 6.48, $e = v - 1$. By Proposition 6.53 all correction graphs of Π have $v = 2a + p + t$ and $e = a + p + 2t$ giving $a + p + 2t = 2a + p + t - 1$, which simplifies to $a = t + 1$. □

We can now show that for any prenet Π such that $a = t + 1$, it suffices to check either acyclicity or connectedness to determine whether or not Π is a proof net: in other words a prenet with a cycle (and satisfying $a = t + 1$) will necessarily be disconnected and a disconnected prenet satisfying $a = t + 1$ will necessarily contain a cycle.

Proposition 6.55. *Let Π be a prenet with $a = t + 1$, Π is a proof net iff Π is acyclic and Π is a proof net iff Π is connected.*

Proof. If Π is a proof net, then $a = t + 1$ by Proposition 6.54 and all correction graphs are both acyclic and connected by Definition 6.26.

For the other direction, suppose Π is a prenet such that $a = t + 1$, then, by the same reasoning as used for the proof of Proposition 6.54, $e = v - 1$, which, according to Proposition 6.49 means that acyclicity implies connectedness and vice versa. □

By the preceding propositions, we can do even better in the intuitionistic case, where all formulas are polarized and there is a conclusion of output polarity. In this case $a = t + 1$ is satisfied by Proposition 6.52.

Corollary 6.56. *If Π is polarized prenet with a single output conclusion then Π is a proof net iff Π is acyclic and Π is a proof net iff Π is connected.*

Proof. Immediate from Proposition 6.52 and Proposition 6.55 □

6.5 Parsing as Proof Net Construction

Assume we want to analyze the noun phrase '*un exemple très simple*', according to the lexicon provided in Example 6.4. We need a proof in L of

$$np \,/\, n, n, (n \setminus n) \,/\, (n \setminus n), n \setminus n \vdash np$$

Because of Proposition 6.42 this amounts to construct a Lambek proof net without empty antecedent with conclusions:

$$\triangleright\, n^{\perp} \otimes n; (n^{\perp} \wp n) \otimes (n^{\perp} \otimes n); n^{\perp}; n \otimes np^{\perp}; np \,\triangleright$$

— these "linear types" are automatically computed as we did in Example 6.1, and the order is inverted (see Proposition 6.9). So the lexicon automatically provides the R&B subformula trees of the proof net shown in Figure 6.4.

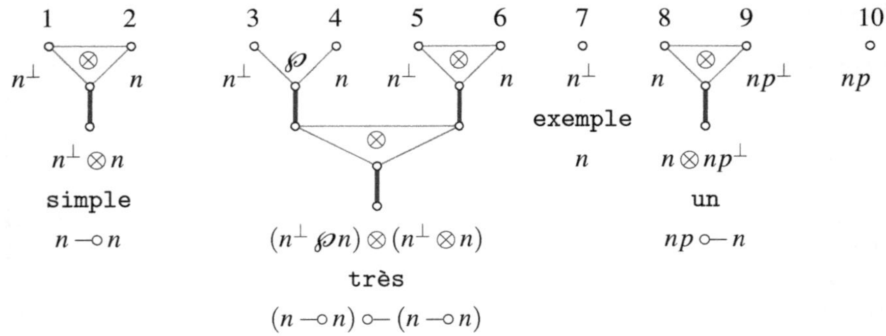

Fig. 6.4. Subformula trees of "un exemple très simple"

What is missing to obtain a proof net is σ_Π, the axiom links between the occurrences

$$n_1^\perp, n_2, n_3^\perp, n_4, n_5^\perp, n_6 n_7^\perp, n_8, np_9^\perp, np_{10}$$

They should be placed in such a way that the conditions ØÆ, SAT, INTUI, NC, ε-FREE are met. Of course, INTUI is automatically satisfied since all conclusions belong to $(\mathsf{Lp})^\perp$ and one (S) is in Lp

Because axioms link dual formulae there must be an axiom $(np_9^\perp, np_{10})$. One should then link the n and the $n^\perp$, and this makes 24 possibilities. However, thanks to the constraints expressed by ØÆ, SAT, NC and ε-FREE we almost have no choice:

$(n_4, n_5^\perp) \notin \sigma_\Pi$ — ØÆ, æ cycle with the *Times* link $(n_3^\perp \wp n_4) \otimes (n_5^\perp \otimes n_6)$.
$(n_5^\perp, n_6) \notin \sigma_\Pi$ — ØÆ, æ cycle with the *Times* link between these two atoms.
$(n_3^\perp, n_4) \notin \sigma_\Pi$ — ε-FREE, sub-prenet with a single conclusion.
$(n_4, n_7^\perp) \notin \sigma_\Pi$ — NC this would force $(n_5^\perp, n^6)$, which was shown to be impossible.
$(n_1^\perp, n_4) \in \sigma_\Pi$ — only possible choice for n_4.
$(n_2, n_3^\perp) \in \sigma_\Pi$ — NC, because of the previous line.
$(n_7^\perp, n_8) \notin \sigma_\Pi$ — SAT, yields a disconnected prenet, since we already have $(np_9^\perp, np_{10}) \in \sigma_\Pi$.
$(n_5^\perp, n_8), (n_6, n_7^\perp) \in \sigma_\Pi$ — only possible choice for these atoms, according to the above decisions.

Hence the only possible solution is the 2-permutation σ_Π given in the example 6.31: $(n_1^\perp, n_4), (n_2, n_3^\perp), (n_5^\perp, n_8), (n_6, n_7^\perp), (np_9^\perp, np_{10})$. It corresponds to the prenet 6.22.

Remark that though we have shown in Corollary 6.56 that it suffices to check either connectedness or acyclicity, enforcing the two conditions together allows us to disqualify more invalid axiom links directly (see (Moot, 2007) for discussion).

Next, one has to check that the result is a Lambek proof net, without empty antecedent, and this is straightforward and quick. It corresponds to the sequent calculus proofs given in examples 6.5 and 6.6. The identification of various sequent calculus proofs into a single proof net leads to less possibilities when constructing the proof.

A natural question is the algorithmic complexity of this parsing algorithm. For the less constrained calculus MLL (only satisfying ØÆ and SAT) it is known to be NP complete (Lincoln et al, 1992), but the notion of splitting *Times* leads to efficient heuristics using the fact that there never can be any axiom link between the two sides of a *Times* link (de Groote, 1995). This considerably reduces the search space. The intuitionistic restriction does not lead to any improvement.

For the non commutative calculi, and in particular for the Lambek calculus, the order constraint NC is so restrictive that one might be tempted to think that the complexity is polynomial. However, a recent paper of Mati Pentus (Pentus, 2006) shows that the Lambek calculus with product is NP complete as well, with the help of a variation of the proof nets studied in this chapter. In addition, Yuri Savateev (Savateev, 2009) has shown that NP completeness holds even for the Lambek calculus without product.

Interesting work has been done on using dynamic programming techniques for finding proof nets for the Lambek calculus. De Groote (1999) — who improves the tabulation techniques introduced in (Morrill, 1996) — uses dynamic programming for the placement of axiom links, defining them by a context-free grammar. Given the results by Pentus and Savateev cited above, these strategies evidently do not give polynomial algorithms, but they may be extended to find interesting polynomial fragments of the Lambek calculus.

We'll have more to say about parsing using proof nets in Section 7.2 in the next chapter, where we talk about parsing categorial grammars using *multimodal* proof nets.

6.6 Proof Nets and Human Processing

Starting with a study by Johnson (Johnson, 1998) for center embedded relatives and then improved and extended by Morrill (Morrill, 2000, 2011), proof nets happen to be interesting parse structure not only from a mathematical viewpoint, but also from a linguistic viewpoint. Indeed they are able to address various performance questions like garden paths, center embedding unacceptability, preference for lower attachment, and heavy noun phrase shift, that can be observed when we use proof net construction as a way to parse sentences.

We follow Morrill (Morrill, 2000) and consider the following examples:

Garden path sentences

1(a) The horse raced past the barn.
1(b) ?The horse raced past the barn fell.

2(a) The boat floated down the river.
2(b) ?The boat floated down the river sank.

3(a) The dog that knew the cat disappeared.
3(b) ?The dog that knew the cat disappeared was rescued.

The (b) sentences are correct but seem incorrect. Indeed there is a natural tendency to interpret the first part of the (b) sentences as their (a) counterparts. Hence the correct, alternative analysis, which is a paraphrase of "The horse *which was* raced past the barn fell" is difficult to obtain.

Quantifier-scope ambiguity

Here are some examples of quantifier-scope ambiguity, with the preferred reading:

I(a) Someone loves everyone. $\exists\forall$
I(b) Everyone is loved by someone. $\forall\exists$

II(a) Everyone loves someone. $\forall\exists$
II(b) Someone is loved by everyone. $\exists\forall$

So in fact the preference goes for the first quantifier having the wider scope.

Embedded relative clauses.

III(a) The dog that chased the rat barked.
III(b) The dog that chased the cat that saw the rat barked.
III(c) The dog that chased the cat that saw the rat that ate the cheese barked.

IV(a) The cheese that the rat ate stank.
IV(b) ? The cheese that the rat that the cat saw ate stank.
IV(c) ?? The cheese that the rat that the cat that the dog chased saw ate stank.

V(a) That two plus two equals four surprised Jack.
V(b) ?That that two plus two equals four surprised Jack astonished Ingrid.
V(c) ??That that that two plus two equals four surprised Jack astonished Ingrid bothered Frank.

VI(a) Jack was surprised that two plus two equals four.
VI(b) Ingrid was astonished that Jack was surprised that two plus two equals four.
VI(c) Frank was bothered that Ingrid was astonished that Jack was surprised that two plus two equals four.

In his paper (Morrill, 2000) Morrill provides an account of our processing preferences, based on our preference for a lower complexity profile. Given an analysis in Lambek calculus of a sentence depicted by a proof net, we have conclusions corresponding to the syntactic types of the words, and a single conclusion corresponding to S. All these conclusions are cyclically ordered. This cyclic order is easily turned into a linear order by choosing a conclusion and a rotation sense. Let us take the output conclusion S as the first conclusion, and let us choose the clockwise rotation with respect to the proof nets of the previous sections. According to the way proof nets are drawn we thus are moving from right to left, and we successively meet S, the type of the first word, the type of the second word, etc.

Now let us define *the complexity of a place in between two words* w_n *and* w_{n+1} (w_0 being a fake word corresponding to S) as the number of axioms $a - a^{\perp}$ which pass over this place, and such that the a belongs to a conclusion which is, in the linear order, before the conclusion containing $a^{\perp}$.

Observe that this measure relies on the fact that Lambek calculus is an intuitionistic or polarized calculus in which a and $a^{\perp}$ are of a different nature: indeed waiting for a category is not the same as providing a category. This measure also depends on the fact that we chose the output S to be the first conclusion: this corresponds to the fact that when someone starts speaking we are expecting a sentence (it could be another category as well, but we still expect some well-formed utterance).

Now we can associate to a sentence with n words a sequence of n integers (since S has been added there are n places) called its *complexity profile*.

In all examples above, the preferred reading always has the lower profile (that is a profile which is always lower, or at least does not go as high) and sentences that are difficult to parse have a high profile.

Here we only present one example, in Figures 6.5 and 6.6, as the others provide excellent exercises (and drawing proof nets on the computer is painful).

word	type u	$u^{\perp}$ for constructing the proof net
someone (subject)	$S/(np\backslash S)$	$(np^{\perp} \wp S) \otimes S^{\perp}$
(object)	$(S/np)\backslash S$	$S^{\perp} \otimes (S \wp np^{\perp})$
everyone (subject)	$S/(np\backslash S)$	$(np^{\perp} \wp S) \otimes S^{\perp}$
(object)	$(S/np)\backslash S$	$S^{\perp} \otimes (S \wp np^{\perp})$
loves:	$(np\backslash S)/np$	$np \otimes (S^{\perp} \otimes np)$

To complete the example, one should compute the semantics according to the algorithm given in Chapter 3.

6.7 Semantic Uses of Proof Nets

Once one is convinced of the relevance of proof nets for parsing, it is worth looking at what else can be achieved with proof nets, in order to avoid translating from one formalism into another, which can be unpleasant and algorithmically costly. A major advantage of categorial grammars is their relation to Montague semantics, and this link has been explored by many authors (Chapter 3 and the references therein provide an introduction to the subject).

As intuitionistic logic can be embedded into linear logic (Girard, 1987) the algorithm for computing semantic readings can be performed within linear logic. Indeed λ-terms can be depicted as proof nets, and β-reduction (or cut-elimination) for proof nets is extremely efficient. In particular the translation can limit the use of replication to its strict minimum. This has been explored by de Groote and Retoré (1996).

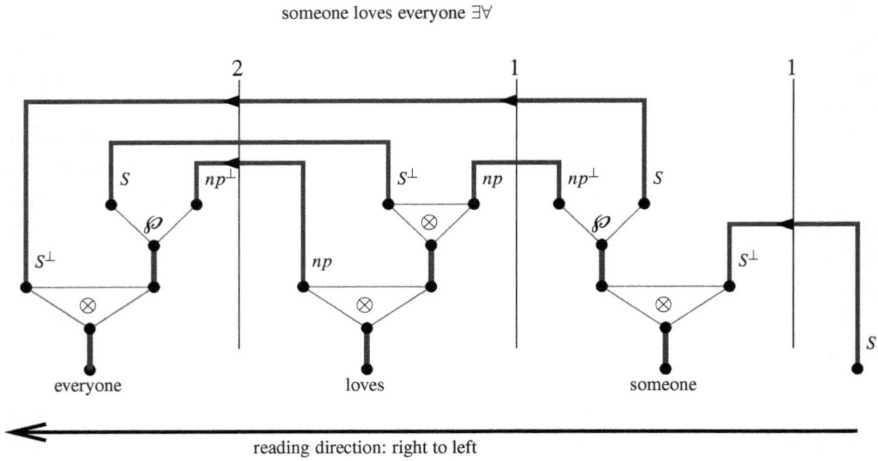

Fig. 6.5. "Someone loves everyone" with wide scope for someone. The complexity profile — read from right to left — is $1-1-2$.

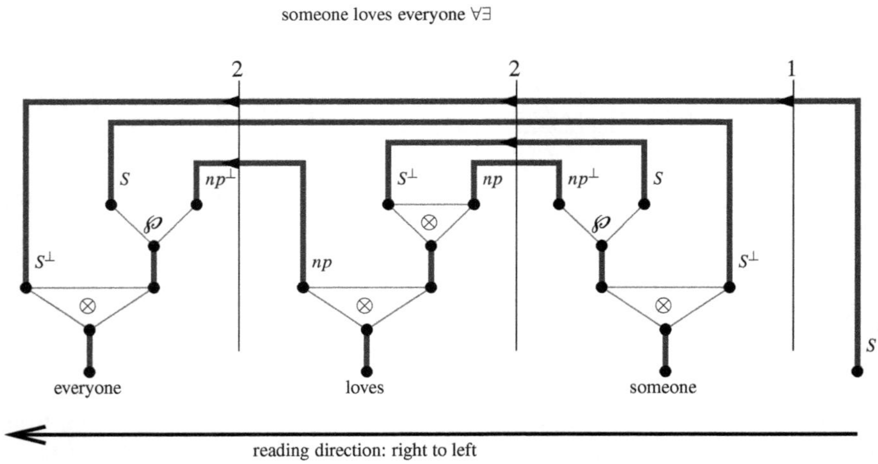

Fig. 6.6. "Someone loves everyone" with wide scope for everyone. The complexity profile — read from right to left — is $1-2-2$.

The correspondence between syntax and semantics with proof nets has been used for generation, firstly by Merenciano and Morrill (Merenciano and Morrill, 1996). Assuming that the semantics of a sentence is known, as well as the semantics of the words, the problem is to reconstruct a syntactic analysis out of this information. This mainly consists of reversing the process involved in the previous paragraph, which is essentially cut elimination. Using a representation of cut elimination by matrix computations (graphs can be viewed as matrices) Pogodalla has thus defined an efficient method for generation (Pogodalla, 2000c,a,b).

6.8 Concluding Remarks

This chapter has given a detailed treatment of proof nets for the associative Lambek calculus that have been discussed in Chapter 2. A central thesis has been that proof nets are not only interesting from a formal point of view, but also from the point of view of the nature of a parse. Indeed, proof nets identify proofs representing the same analysis thus avoiding the so-called spurious ambiguity problem. Therefore proof nets can be said to implement the very idea of *parsing-as-deduction*.

We have also touched upon several other aspects of proof nets: its connection to semantics and some suggestive evidence about proof net construction as a model for human sentence processing.

The proof nets presented here naturally suggest a further radicalisation: a formula can be depicted as a set of Red edges between its atoms, and the Blue edges are the atoms and a simple correctness criterion, "every alternate elementary cycle contains a chord", recognises exactly the proofs (Retoré, 2003). This way, the algebraic properties of the connectives, like associativity are interpreted by equality of the formulae, hence identifying even more proofs than usual proofnets with links. This was firstly done for commutative multiplicative linear logic, but it also works for non commutative logic like the Lambek calculus (Pogodalla and Retoré, 2004). In this setting as well, the cuts are a bit tricky to handle, and this is ongoing work.

Exercises for Chapter 6

Exercise 6.1. Using proof nets, show whether or not the following sequents are derivable in multiplicative linear logic.

$\vdash (a \otimes b) \otimes c, (c^{\perp} \wp b^{\perp}) \wp a^{\perp}$
$\vdash (a \wp b) \otimes c, (c^{\perp} \wp b^{\perp}) \otimes a^{\perp}$
$\vdash c^{\perp}, ((a \wp a^{\perp}) \otimes b) \otimes d^{\perp}, b^{\perp} \wp (c \otimes d)$

Exercise 6.2. Section 6.1.3 defines a translation of Lambek calculus formulae into linear logic formulae using polarities. For all of the following formulae F, give both the translation $+F$ and $-F$.

$(np \backslash S) / np$
$(n \backslash n) / (S / np)$
$S / (np \backslash S)$
$(S / np) \backslash S$
$((np \backslash S) / np) \backslash (np \backslash S)$

Exercise 6.3. Proposition 6.2 on page 196 defines the sets of formulae Li° and $\mathsf{Li}^{\bullet}$. For each of the following formulae, show if they are members Li°, members of $\mathsf{Li}^{\bullet}$ or if they are not a member of either set of formulae.

$(a \otimes b) \wp c^{\perp}$
$(a^{\perp} \wp b) \wp c^{\perp}$
$a^{\perp} \wp (b^{\perp} \wp c)$
$a \otimes (b^{\perp} \wp c)$
$a \otimes (b^{\perp} \otimes c)$

Exercise 6.4. Using Lambek calculus proof nets (refer to Definition 6.41 on page 220), show which of the following sequents are derivable.

$np \vdash S / (np \backslash S)$
$S / (np \backslash S) \vdash np$
$np, (np \backslash S) / np, np \vdash S$
$np, (np \backslash S) / np, S / (np \backslash S) \vdash S$
$np, (np \backslash S) / np, (S / np) \backslash S \vdash S$
$S / (np \backslash S), (np \backslash S) / np, ((np \backslash S) / np) \backslash (np \backslash S) \vdash S$

That is, translate each formula in the corresponding linear logic formula, compute the possible axiom linking σ_{Π} and verify that all conditions of Definition 6.41, including ε-FREE, are satisfied.

Exercise 6.5. Look back to Figure 6.3 on page 224.

1. Verify it is a proof structure of $\vdash (b^{\perp} \wp b) \otimes (a^{\perp} \wp a)$ by assigning formulas to each node in the proof structure.
2. Remove the cut link from the figure and compute the sequents corresponding to the two substructures. Are both of these structures derivable?

3. Perform cut elimination on the proof structure of Figure 6.3. Can you remark anything special about the result of cut elimination? If so, what does this mean?
4. Draw the proof structure in such a way the the external face is on the outside of the proof structure. What — if anything — is different about this proof structure?
5. Give a correct (planar) proof structure for $\vdash (b^{\perp} \wp b) \otimes (a^{\perp} \wp a)$ and verify it satisfies all constraints.

Exercise 6.6. Following Johnson and Morrill (Johnson, 1998; Morrill, 2000, 2011), Section 6.6 states that the acceptability of sentences is related the "nesting" of axiom links to the (Figures 6.5 and 6.6, page 232 gives an example comparison). Compute a similar complexity profile for each of the other phenomena discussed at the beginning of Section 6.6, by assigning each of them appropriate lexical formulas and constructing the proof nets.

References

Abrusci, V.M.: Phase semantics and sequent calculus for pure non-commutative classical linear logic. Journal of Symbolic Logic 56(4), 1403–1451 (1991)

Abrusci, V.M.: Non-commutative proof nets. In: Girard, et al., pp. 271–296 (1995)

Asperti, A.: A linguistic approach to dead-lock. Tech. Rep. LIENS 91-15, Dép. Maths et Info, Ecole Normale Supérieure, Paris (1991)

Asperti, A., Dore, G.: Yet Another Correctness Criterion for Multiplicative Linear Logic with Mix. In: Nerode, A., Matiyasevich, Y. (eds.) LFCS 1994. LNCS, vol. 813, pp. 34–46. Springer, Heidelberg (1994)

Bellin, G., Scott, P.J.: On the π-calculus and linear logic. Theoretical Computer Science 135, 11–65 (1994)

van Benthem, J.: Language in Action: Categories, Lambdas and Dynamic Logic. Studies in logic and the foundation of mathematics, vol. 130. North-Holland, Amsterdam (1991)

Bondy, J.A., Murty, U.S.R.: Graph Theory and Applications. Macmillan Press (1976)

Dalrymple, M., Lamping, J., Pereira, F., Saraswat, V.: Linear logic for meaning assembly. In: Morrill, G., Oehrle, R. (eds.) Formal Grammar, pp. 75–93. FoLLI, Barcelona (1995)

Danos, V.: La logique linéaire appliquée à l'étude de divers processus de normalisation et principalement du λ-calcul. Thèse de Doctorat, spécialité Mathématiques, Université Paris 7 (1990)

Danos, V., Regnier, L.: The structure of multiplicatives. Archive for Mathematical Logic 28, 181–203 (1989)

Diestel, R.: Graph Theory, 4th edn. Springer (2010)

Fleury, A.: La règle d'échange: logique linéaire multiplicative tréssée. Thèse de Doctorat, spécialité Mathématiques, Université Paris 7 (1996)

Fleury, A., Retoré, C.: The mix rule. Mathematical Structures in Computer Science 4(2), 273–285 (1994)

Girard, J.Y.: Linear logic. Theoretical Computer Science 50(1), 1–102 (1987)

Girard, J.Y.: Linear logic: its syntax and semantics. In: Girard, et al., pp. 1–42 (1995)

Girard, J.Y., Lafont, Y., Regnier, L. (eds.): Advances in Linear Logic. London Mathematical Society Lecture Notes, vol. 222. Cambridge University Press (1995)

de Groote, P.: Linear Logic with Isabelle: Pruning the Proof Search Tree. In: Baumgartner, P., Posegga, J., Hähnle, R. (eds.) TABLEAUX 1995. LNCS (LNAI), vol. 918, pp. 263–277. Springer, Heidelberg (1995)

de Groote, P.: A Dynamic Programming Approach to Categorial Deduction. In: Ganzinger, H. (ed.) CADE 1999. LNCS (LNAI), vol. 1632, pp. 1–15. Springer, Heidelberg (1999)

de Groote, P., Lamarche, F.: Classical non-associative Lambek calculus. Studia Logica 71(3), 355–388 (2002)

de Groote, P., Retoré, C.: Semantic readings of proof nets. In: Kruijff, G.J., Morrill, G., Oehrle, D. (eds.) Formal Grammar, pp. 57–70. FoLLI, Prague (1996)

Guerrini, S.: Correctness of multiplicative proof nets is linear. In: 14th Symposium on Logic in Computer Science (LICS 1999), pp. 454–463. IEEE (1999)

Guerrini, S.: A linear algorithm for MLL proof net correctness and sequentialization. Theoretical Computer Science 412(20), 1958–1978 (2011); Girard's Festschrift

Johnson, M.E.: Proof nets and the complexity of processing center-embedded constructions. Journal of Logic Language and Information Special Issue on Recent Advances in Logical and Algebraic Approaches to Grammar 7(4), 433–447 (1998)

Lamarche, F.: Proof nets for intuitionistic linear logic: Essential nets. 35 page technical report available by FTP from the Imperial College archives (1994)

Lambek, J.: From categorial grammar to bilinear logic. In: Došen, K., Schröder-Heister, P. (eds.) Substructural Logics, pp. 207–237. Oxford University Press, Oxford (1993)
Lincoln, P., Mitchell, J., Scedrov, A., Shankar, N.: Decision problems for propositional linear logic. Annals of Pure and Applied Logic 56(1-3), 239–311 (1992)
Melliès, P.A.: A topological correctness criterion for multiplicative non-commutative logic. In: Ehrhard, T., Girard, J.Y., Ruet, P., Scott, P. (eds.) Linear Logic in Computer Science. London Mathematical Society Lecture Note, vol. 316, ch. 8, pp. 283–321. Cambridge University Press (2004)
Merenciano, J.M., Morrill, G.: Generation as Deduction on Labelled Proof Nets. In: Retoré, C. (ed.) LACL 1996. LNCS (LNAI), vol. 1328, pp. 310–328. Springer, Heidelberg (1997)
Métayer, F.: Homology of proof-nets. Prépublication 39, Equipe de Logique, Université Paris 7 (1993)
Moot, R.: Filtering axiom links for proof nets. In: Kallmeyer, L., Monachesi, P., Penn, G., Satta, G. (eds.) Proceedings of the 12th Conference on Formal Grammar (FG 2007). CSLI Publications, Dublin (2007) (to appear) ISSN 1935-1569
Morrill, G.: Memoisation of categorial proof nets: parallelism in categorial processing. In: Abrusci, V.M., Casadio, C. (eds.) Third Roma Workshop: Proofs and Linguistics Categories – Applications of Logic to the Analysis and Implementation of Natural Language. CLUEB, Bologna (1996)
Morrill, G.: Incremental processing and acceptability. Computational Linguistics 26(3), 319–338 (2000); preliminary version: UPC Report de Recerca LSI-98-46-R (1998)
Morrill, G.: Categorial Grammar: Logical Syntax, Semantics, and Processing. Oxford University Press (2011)
Morrill, G., Fadda, M.: Proof nets for basic discontinuous Lambek calculus. Journal of Logic and Computation 18(2), 239–256 (2008)
Murawski, A., Ong, C.H.: Dominator trees and fast verification of proof nets. In: 15th Symposium on Logic in Computer Science (LICS 2000), pp. 181–191. IEEE (2000)
Pentus, M.: Lambek calculus is NP-complete. Theoretical Computer Science 357(1), 186–201 (2006)
Pogodalla, S.: Generation in the Lambek calculus framework: an approach with semantic proof nets. In: Proceedings of NAACL 2000 (2000a)
Pogodalla, S.: Generation, Lambek calculus, montague's semantics and semantic proof nets. In: Proceedings of Coling 2000 (2000b)
Pogodalla, S.: Generation with semantic proof nets. Research Report 3878, INRIA (2000c), `http://hal.inria.fr/docs/00/07/27/75/PDF/RR-3878.pdf`
Pogodalla, S., Retoré, C.: Handsome non-commutative proof-nets: perfect matchings, series-parallel orders and hamiltonian circuits. Tech. Rep. RR-5409, INRIA, presented at Categorial Grammars (2004); to appear in the Journal of Applied Logic
Regnier, L.: Lambda calcul et réseaux. Thèse de doctorat, spécialité mathématiques, Université Paris 7 (1992)
Retoré, C.: Réseaux et séquents ordonnés. Thèse de Doctorat, spécialité Mathématiques, Université Paris 7 (1993)
Retoré, C.: Calcul de Lambek et logique linéaire. Traitement Automatique des Langues 37(2), 39–70 (1996)
Retoré, C.: Perfect matchings and series-parallel graphs: multiplicative proof nets as R&B-graphs. In: Girard, J.Y., Okada, M., Scedrov, A. (eds.) Linear 1996. Electronic Notes in Theoretical Science, vol. 3. Elsevier (1996)
Retoré, C.: A semantic characterisation of the correctness of a proof net. Mathematical Structures in Computer Science 7(5), 445–452 (1997)

Retoré, C.: Handsome proof-nets: perfect matchings and cographs. Theoretical Computer Science 294(3), 473–488 (2003), complete version RR-3652, `http://hal.inria.fr/docs/00/07/30/20/PDF/RR-3652.pdf`

Savateev, Y.: Product-Free Lambek Calculus Is NP-Complete. In: Artemov, S., Nerode, A. (eds.) LFCS 2009. LNCS, vol. 5407, pp. 380–394. Springer, Heidelberg (2008)

Yetter, D.N.: Quantales and (non-commutative) linear logic. Journal of Symbolic Logic 55, 41–64 (1990)

7

Proof Nets for the Multimodal Lambek Calculus: From Theory to a Wide-Coverage Categorial Parser

Summary. In this chapter, we will extend the proof nets of the previous chapter to the multimodal Lambek calculus discussed in Chapter 5. This means incorporating non-associativity, mode information, unary connectives and structural rules in the proof nets. Fortunately, we will see that there is a single proof net calculus which can handle all these extensions together. The resulting proof nets will be two sided and have a more procedural correctness condition in the form of graph contractions. This correctness condition will also have a modular nature, due to the fact that different grammars may allow different classes of structural rules. These multimodal proof nets form the basis of an implementation and we will discuss the Grail theorem prover in Section 7.2.

7.1 Multimodal Proof Nets

In Chapter 5 we have seen that multimodal categorial grammars extend the non-associative Lambek calculus NL with mode information: different modes of composition can be combined in one logic and different structural rules can apply to each of these modes and their combinations. In addition, the multimodal calculus introduces unary connectives (and corresponding structural rules).

These additions and restrictions to the Lambek calculus complicate the theory of proof nets for multimodal categorial grammars a bit: we need a modular correctness condition, which can adapt itself to the set of structural rules and modes which are available in the logic. We will arrive at a multimodal proof net calculus in two steps: first, following Danos (1990); Puite (1998, 2001), we will introduce two sided proof nets: proof nets which have hypotheses in addition to conclusions. This doubles the number of links a prenet can have: each connective now has a link for when it occurs as a hypothesis and a link for when it occurs as a conclusion.

Then, following Moot and Puite (2002), we will introduce links for multimodal prenets and introduce a correctness condition based on graph contractions (a special case of the graph contractions of Danos (1990) for multiplicative linear logic).

Contractions will permit us to eliminate an appropriately connected pairs of links — each pair consisting of one par link and one tensor link — whereas the structural rules will correspond to additional graph rewrite rules in the proof net calculus.

Having introduced multimodal proof nets, we will prove the calculus so defined is equivalent to the sequent calculus formulation of the multimodal Lambek calculus in Section 7.1.4. In addition to the theoretical interest of the calculus, this multimodal proof net calculus is also the basis of an implementation: the Grail theorem prover, which will be discussed in Section 7.2.

7.1.1 Two Sided Proof Nets

As we have seen in Sections 6.2 and 6.3, linear logic allows both a one sided and a two sided sequent calculus. Up till now, it has been convenient to use proof nets which corresponded to the one sided sequent calculus: it reduces the number of rules and therefore the number of cases to treat in the equivalence proofs.

In this section, we will briefly look at two sided proof nets for multiplicative intuitionistic linear logic. Though this will multiply the number of rules (and the number of different types of links in the proof structures), it will be a step closer to the multimodal proof nets of the following sections. We will also remark that these two sided proof nets are rather close to natural deduction proofs.

Figure 7.1 shows the links for one sided proof nets. These are the same links as those of Definition 6.14 on page 207, with the addition of the cut link discussed in Sections 6.4.6 and 6.4.7. The cut link simply allows us to connect — by means of a R (regular, red) link — a formula to its negation. Alternatively, we can see the cut link as a special sort of tensor link with a special symbol "cut" as its conclusion: this is convenient when we want to prove sequentialisation of a sequent with cut links.

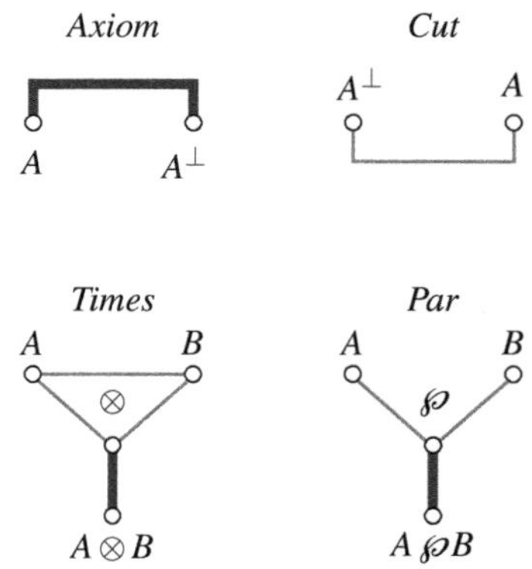

Fig. 7.1. Links for one sided proof nets

The cut links allow us to make the connection between the one sided and the two sided proof net calculus more clear. Figure 7.2 shows (in the middle of the figure) the two links for A / B: one which has A / B as its hypothesis and another which has

A / B as its conclusion. One the left hand side, we can see the partial prenets for $A \wp B^{\perp}$ which correspond to the links in the middle of the figure. As usual, the links are displayed with their hypotheses above the link and their conclusions beneath it and the left-to-right order of the hypotheses and the conclusions is important. As can be seen in the figure, hypotheses are connected to the rest of the graph by a R (regular or red) edge and conclusions are connected by a B (bold or blue) edge. We should immediately note two differences with the one sided links from Figure 7.1. First, a logical link can have multiple conclusions (in the one sided case only the axiom link has two conclusions). Second, the main formula is no longer necessarily the conclusion of the link: it can be one of its premises as well.

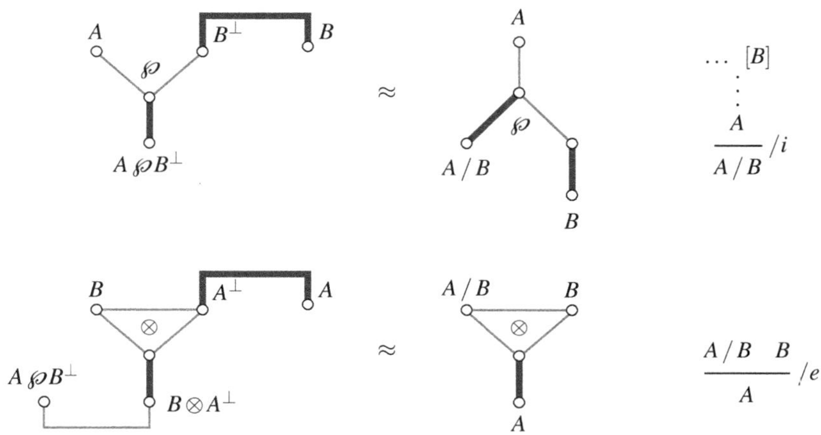

Fig. 7.2. Links for $A / B = A \wp B^{\perp}$

Inspecting the links of Figure 7.2, we see that in both cases, viewed from the outside, the new link in the middle of the figure and the partial prenet on the left are indistinguishable.

In the case for the link and the partial prenet which have A / B as their conclusion (both are on the top row of Figure 7.2), the two graphs are isomorphic: it is even the case that a tour along the outside of the graph will visit the nodes in the same cyclic order.

For the link and the partial prenet which have A / B as a hypothesis (both are on the bottom row of Figure 7.2), the link in the middle is slightly simpler. However, for the paths which pass *through* both structures, this doesn't make a difference: both graphs have an alternate elementary path from A / B to B which starts and ends at a R edge, both graphs have an alternate elementary path from A to A / B which starts with a B edge and ends in a R edge and both graphs have a path from A to B which starts with a B edge and ends in a R edge. In addition, a tour along the outside of both graphs will pass the external nodes in the same cyclic order.

For an $A \wp B^\perp$ conclusion, the link corresponds to a combination of a par link and an axiom link. However, since B is not necessarily an atomic formula, this can be a complex axiom, in which case B will be the main formula (and hypothesis) of another link.

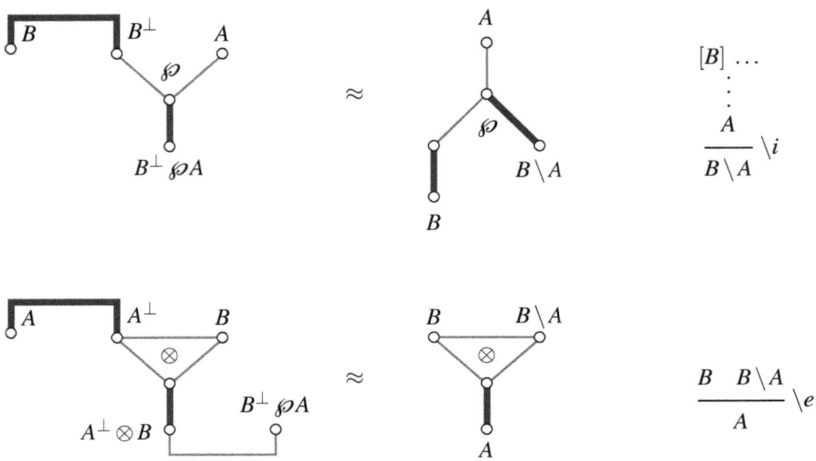

Fig. 7.3. Links for $B \backslash A = B^\perp \wp A$

Another thing which is worth pointing out is that this proof net representation is very close to natural deduction (compare Figure 7.2 to the similar figures in Appendix B.5 of Girard et al (1988)). Figure 7.2 shows the natural deduction rules next to the corresponding proof net links, in the rightmost columns. The correspondence between the introduction rule and the link for an A / B conclusion is more clear when we remind ourselves of the role which is played by the B formula in the natural deduction rule: we use a single B hypothesis to prove an A after which we withdraw this B to infer A / B. In a natural deduction proof the link between the B formula and the corresponding introduction rule is normally indicated in some way (by assigning a unique label to both the rule and the B hypothesis which is withdrawn, for example), in the proof net we simply connect the B hypothesis directly to B conclusion of the rule. This makes the up-down symmetry between the hypothesis and conclusion rule more evident than it is in natural deduction. We will see this symmetry even more clearly in the next section when we look at multimodal links.

Figure 7.3 shows the symmetric case for $B \backslash A = B^\perp \wp A$, with — as before — the partial prenets on the left of the figure, the new links in the middle and the corresponding natural deduction rules on the right.

The case for the product formula is slightly different. Figure 7.4 shows the links for the product formula. The link for the conclusion is identical to the link for the product formula in Figure 7.1. However, the link for $A \otimes B$ as a hypothesis does not correspond as clearly to the natural deduction rule shown to its left. The fact that a link in a proof net can have multiple conclusions in the current formulation helps to make the symmetry between the link for $A \otimes B$ as a hypothesis and $A \otimes B$ as a conclusion more clear than its asymmetric natural deduction counterpart allows. In addition, unlike the natural deduction rule (see Section 2.2.1), the hypotheses of a proof structure containing the hypothesis link for the product formula are in the right order: with the hypotheses Δ of $A \otimes B$ occurring between the hypotheses Γ_1 to the left of A and Γ_2 to the right of B.

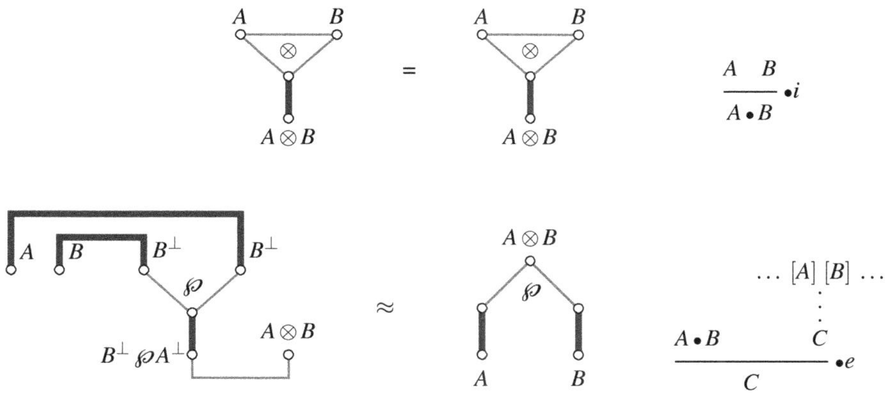

Fig. 7.4. Links for $A \otimes B$

We will finish this informal introduction with an example of a proof structure using the new links. Figure 7.5 revisits our Italian lexicon from Example 2.2 and shows the lexical formulae for "cosa guarda passare" along with the partial proof structures on the right.

The important point to make about this figure is that we no longer use axiom links to turn a partial proof structure into a proof structure, but that we identify nodes with are hypotheses with nodes which are conclusions. Another way to look at this is that the axiom links are already "incorporated" in Figure 7.5, but with each axiom link connected at only one of its ends; this means that we still have to attach the other end.

One solution to resolving this node identification problem correctly is by the S hypothesis from "cosa" with the S conclusion from "guarda", the *inf* hypothesis from "guarda" with the *inf* conclusion from "passare" and the np hypothesis from "passare" with the np conclusion from "cosa". This produces the proof structure on the right of Figure 7.5.

It is easy to verify that the resulting proof structure satisfies the following conditions of Definition 6.41 on page 220 (which defines proof nets for the Lambek calculus).

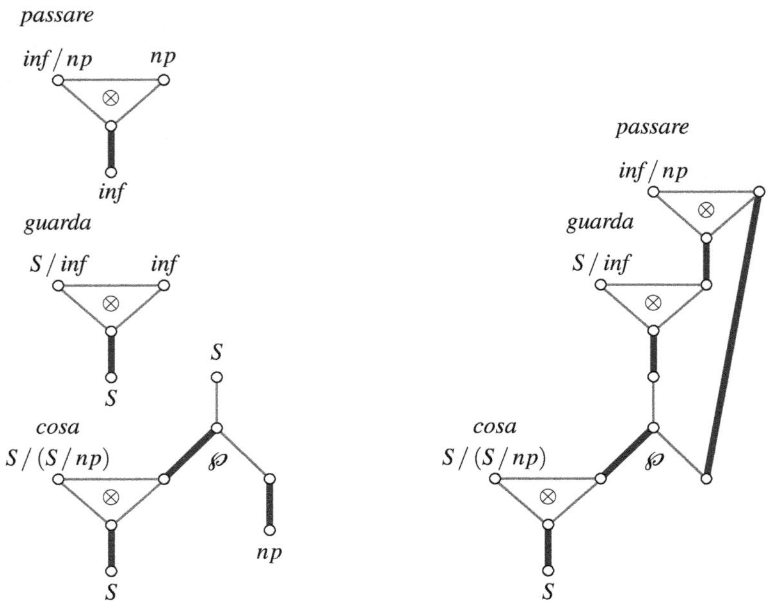

Fig. 7.5. "Cosa guarda passare" as a partial proof structure (left) and as a proof structure (right)

- ØÆ: it does not contain an alternate elementary cycle
- SAT: it contains an alternate elementary path between any two vertices
- INTUI: all links are intuitionistic (this is true by construction)

and is therefore a proof net in the sense of intuitionistic linear logic. The planarity condition NC is a bit harder to verify in the current proof net: it corresponds roughly to the fact that the left-to-right order of the words (the hypotheses corresponding to "cosa guarda passare") is correct, in combination with the fact that the np formula of the par link is connected to the rightmost formula of the tree which has the S formula of this same par link as its root, much as the natural deduction rule for $/_i$ would verify the position of the np formula.

Condition ε-FREE, which requires every subnet to have at least two conclusions, correspond in the two-side calculus to the condition that every subnet has at least one hypothesis.

Rather than proving these results more formally, we will move to the multimodal case, which generalizes the proof nets introduced here in such a way that we can incorporate non-associativity, unary connectives and structural rules in a single proof net calculus.

7.1.2 Multimodal Proof Structures and Abstract Proof Structures

Now that we have given an informal introduction to two sided proof nets and sketched its relation with the proof nets for linear logic we have seen before, we can introduce the proof nets for multimodal categorial grammars more formally.

The proof nets we have seen for linear logic and the Lambek calculus and proof search using proof nets are a three-step process, which operates schematically as follows.

1. We translate the formulae of the sequent into trees labeled with formulae and/or logical connectives. Some authors call the resulting structure a *proof frame*.
2. From this proof frame, we create a *proof structure* by linking atomic formulae of opposite polarity.
3. Finally, we check whether or not the proof structure is a *proof net* by looking only at properties of the underlying graph.

In this section we will follow the same three-step process for multimodal proof nets. We will give links and formula unfoldings for multimodal formulae, then turn these formula decomposition trees into proof structures and finally define proof nets for the multimodal calculus. These proof nets are essentially a notational variant of those used by Moot and Puite (2002) and those used by the Grail theorem prover we will discuss in Section 7.2.

Definition 7.1. *A* link *for a multimodal proof structure is a tuple* $\langle$*type, mode, Prem, Concl, Main*$\rangle$*, such that.*

[type] *is either* $\otimes$ *or* $\wp$*; if type* $= \otimes$ *we will call the link a* tensor link*, if type* $= \wp$ *we will call the link a* par link.

[mode] *is the* mode *of the link (either equal to an* $i \in I$ *or to a* $j \in J$*)*

[Prem] *is a list of vertices which are the* premises *of the link.*

[Concl] *is a list of vertices which are the* conclusions *of the link.*

[Main] *specifies the (optional)* main vertex *of the link, it is either the special constant nil — in which case we will say the link does not have a main vertex — or a member of the concatenation of Prem and Concl, in which case we will call this vertex the* main vertex *of the link and the other vertices the* active vertices *of the link.*

We will say the vertices *of the link are the union of the vertices in* Prem *and the vertices in* Concl.

Definition 7.2. *A* typed link *for a multimodal proof structure is a tuple* $\langle l, f\rangle$ *such that* l *is a link and* f *is a (total) function from the vertices of* l *(that is, the members of* Prem *and* Concl*) to multimodal formulae.*

The definition of links and typed links above are a natural extension of the links which we have seen for linear logic. Once we drop the restriction that links have a unique conclusion which coincides with the main formula of the link, then — since a link already had a sequence of premises — it is natural to allow a link to have a sequence of conclusions as well (we will only consides lists with one or two members here). In addition, the vertex of the link which corresponds to the main formula must now be explicitly designated. Finally, the addition of mode information to a link corresponds to multimodality in an very direct way. The possibility for links without main formulae and vertices which are not labeled by formulae are part of the definition in order to enable us to use the same definition of link for the underlying, more graph-like structures on which we will define our correctness condition.

Graphically, we will use the following conventions to portray a link:

- tensor links are displayed with an open circle in the center, par links with a filled circle,
- the hypotheses are displayed, from left to right, above the central circle of the link,
- the conclusions are displayed, from left to right, below the central circle of the link,
- the vertices v of typed links are displayed as the formula $A = f(v)$ which labels the vertex; when we want to talk about formula occurrences, we will display the formula as A_n where n is the unique identifier of the vertex in the graph (we will use integers in what follows) which is labeled by A,
- if the link is a par link, we will indicate the main vertex by an arrow from the central circle to the main vertex.

These graphical conventions are slightly different from those used in (Moot and Puite, 2002), where *all* links are drawn with an arrow arriving at the main vertex (not just the par links, but the tensor links as well) and with arrows *leaving* from the active vertices.

Definition 7.3. *A multimodal* proof structure *(or prenet) is a tuple* $\langle V, L, f \rangle$ *such that*

V *is the set of* vertices *of the proof structure.*
L *is the set of* links *of the proof structure.*
f *is a (total) function from* V *to multimodal formulae.*

- *each typed link* $\langle l, f_l \rangle$ *such that* $l \in L$ *and* f_l *corresponds to the restriction of* f *to the vertices of* l *corresponds to one of the typed links of Table 7.1,*
- *each vertex is the premise of at most one link,*
- *each vertex is the conclusion of at most one link.*

We will call a vertex h *which is not the conclusion of any link a* hypothesis *of the proof structure.*

We will call a vertex c *which is not the premise of any link a* conclusion *of the proof structure.*

When Γ *is the multiset of formulas which are assigned by* f *to the hypotheses of the proof structure and* Δ *is the multiset of formulas which are assigned by* f *to the conclusions of the proof structure, we will say it is a proof structure of* $\Gamma \vdash \Delta$.

Since a vertex in a proof structure is the premise of at most one link and the conclusion of at most one link, it is connected to at most two links and if it is connected to two links, it is the premise of one and the conclusion of another. If we take into account the fact that a formula is either a main formula of its link or an active formula, this gives us a natural division of the vertices (and the formulae by which they are labeled) into four classes:

1. *cut* vertices are the main vertex of two links,
2. *axiom* vertices are not the main vertex of any link,
3. *positive* vertices are the main vertex of exactly one link, and the conclusion of this link,
4. *negative* vertices are the main vertex of exactly one link, and the premise of this link.

Cut and axiom vertices will correspond to a cut or axiom rule on the formula with which this vertex is labeled. In a sense, axiom and cut vertices are *both* positive and negative: an axiom is positive to the link of which it is a premise and negative to the link of which it is a conclusion (ie. it is "negative at the top and positive at the bottom") and inversely for the cut vertices (which would be "positive at the top and negative at the bottom").

Definition 7.4. *Given a formula C and a polarity $O \in \{P,N\}$ we define its subformula tree, which is a proof structure with C as one of its conclusions (if its polarity is positive) or as one of its hypotheses (if its polarity is negative) according to Definition 7.3, inductively as follows.*

- *if C is an atomic formula, its subformula tree is C, that is a proof structure without links, with a single vertex v, with $f(v) = C$ and with C both a conclusion and a hypothesis of the proof structure (since there are no links, this is the only possibility).*
- *for a complex formula C of polarity O, Table 7.1 shows — for each connective and polarity — how to combine the proof structures which correspond to the direct subformula(s) of C into a subformula tree of C. In particular:*
  - *if C is of the form $\Diamond_j A$ or $\Box_j A$ and if $O(A)$ is the subformula tree corresponding to A, then $O(C)$ is obtained by extending the subformula tree with a unary branch as shown in Table 7.1. C is a conclusion (resp. hypothesis) of the new proof structure as required.*
  - *if $C = A \bullet B$ and if $O(A)$ is the subformula tree corresponding to A and $O(B)$ is the subformula tree corresponding to B, then $O(C)$ is subformula tree combining $O(A)$ and $O(B)$ by a binary branch for $\bullet_i$ of polarity O as shown in Table 7.1. Again C is a hypothesis (resp. conclusion) of the new proof structure as required.*
  - *if $C = B \backslash A$ or $C = A / B$, $O(A)$ is the subformula tree corresponding to A and $O'(B)$ is the subformula tree of the opposite polarity of O corresponding to B then we combine the two subtrees as shown the corresponding $/_i$ or $\backslash_i$ column in Table 7.1. Again, the fact that C is a hypothesis/conclusion of the new proof structure is easily verified.*

Table 7.1. Logical links for the multimodal Lambek calculus

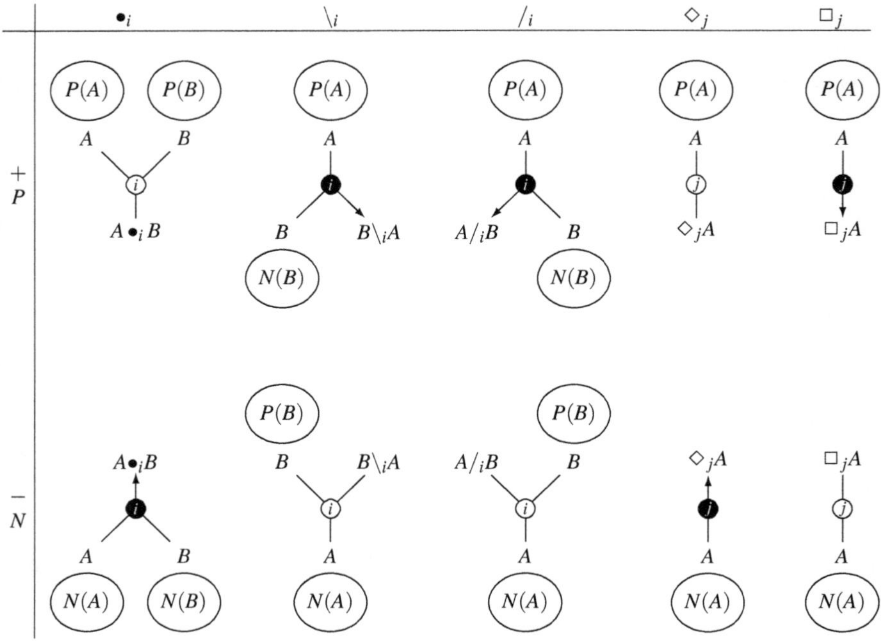

Table 7.1 summarizes the available links for all combinations of polarity–main connective and indicates the polarities of the active vertices of the link.

As can be seen from Table 7.1 positive formulae grow upwards (with an downward growing branch for the negative subformula in the case of the implications), whereas negative formulae grow downwards (with a upward growing branch for the positive subformula of an implication).

Example 7.5. Given that left/right and up/down have a specific meaning and given that the formula tree grows both upward and downward, we sometimes need to "stretch" the links a bit to avoid overlap when we draw the subformula tree graphically.

Figure 7.6 shows the negative unfoldings of the formula $S /_0 (np \backslash_0 S)$ and of the formula $(n \backslash_0 n) /_0 (S / \Diamond_1 \Box_1 np)$. For those interested in seeing the proof structure on the left of Figure 7.6 as it corresponds to the definitions, Figure 7.7 shows it in full formal detail. As the information in Figure 7.7 can be easily read off from the more graphical depiction of the proof structures, we will prefer the graphical representation in what follows.

Definition 7.4 guarantees that each formula decomposition tree is already a proof structure. For example, the formula decomposition tree of $S /_0 (np \backslash_0 S)$ in Figure 7.6 is a proof structure with hypotheses S and $S /_0 (np \backslash_0 S)$ and with conclusions S and

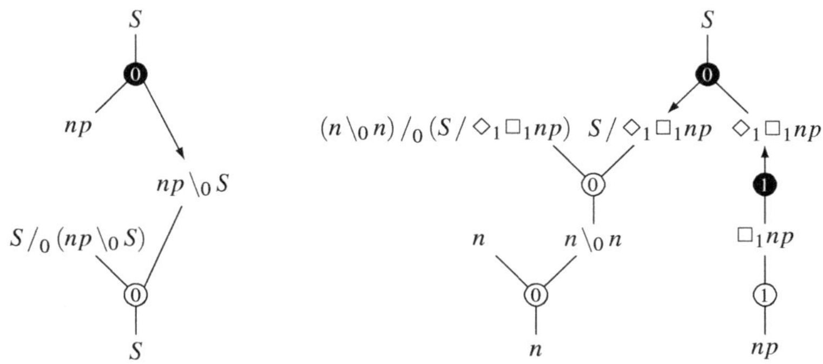

Fig. 7.6. Negative unfoldings of $S /_0 (np \backslash_0 S)$ and $(n \backslash_0 n) /_0 (S / \Diamond_1 \Box_1 np)$

$V = \{1,2,3,4,5\}$, $L = \{l_1, l_2\}$

The function f is defined as follows:

$f(1) = S$
$f(2) = np$
$f(3) = np \backslash_0 S$
$f(4) = S /_0 (np \backslash_0 S)$
$f(5) = S$

The links l_1 and l_2 are defined as follows:

$l_1 = \langle \wp, 0, [1], [2,3], 3 \rangle$
$l_2 = \langle \otimes, 0, [4,3], [5], 4 \rangle$

Fig. 7.7. Formal representation of the proof structure on the left of Figure 7.6

np. Similarly, the subformula tree of $(n \backslash_0 n) /_0 (S /_0 \Diamond_1 \Box_1 np)$ is a proof structure with hypotheses $(n \backslash_0 n) /_0 (S /_0 \Diamond_1 \Box_1 np)$, n and S and with conclusions n and np.

We want to establish a correspondence between sequents $\Gamma \vdash C$ and proof structures of $\Gamma \vdash C$, that is to say proof structures which have as their only hypotheses the formulae of Γ and as their only conclusion the formula C. This means that when we look at a proof structure which is the negative unfolding of a number of antecedent formulae $A_1, \ldots A_n$ and the positive unfolding of a formula C we need to construct a proof structure which has exactly the formulae $A_1, \ldots A_n$ as hypotheses and exactly the formula C as its conclusion. So in the formula unfolding of $S /_0 (np \backslash_0 S)$ of Figure 7.6, we want the S premise of the par link in the lexical proof structure to be the conclusion of another link of the proof structure we are constructing and we want the S and np conclusions of the lexical proof structure to be premises of other links in a larger proof structure.

To make this more clear, the negative unfolding of n, $(n \backslash_0 n) /_0 (S / \Diamond_1 \Box_1 np)$, np and $(np \backslash_0 S) /_0 np$ are shown together with the positive unfolding of n in Figure 7.8.

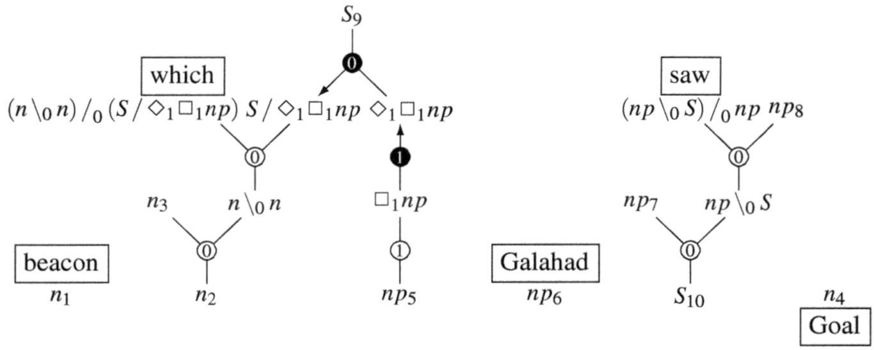

Fig. 7.8. Negative unfoldings of n, $(n\backslash_0 n)/_0(S/_0\Diamond_1\Box_1 np)$, np and $(np\backslash_0 S)/_0 np$, positive unfolding of n

The atomic formulae of the proof structure are subscripted with a unique integer to make it easier to refer to them in the following discussion.

Some words from the lexicon which correspond the given formulae are indicated in a square box above each of the unfolded formulae. In addition the succedent formula n is indicated by a square box marked "Goal" below it. The proof structures above correspond (up to the choice of lexical items) to the lexicon and example sentence of Example 5.9 on page 169, which motivates and explains the current formula assignments.

Using the lexical unfoldings above, we want to construct a proof structure of the following sequent.

$$n,(n\backslash_0 n)/_0(S/\Diamond_1\Box_1 np),np,(np\backslash_0 S)/_0 np \vdash n$$

Or, in other words we want to construct a proof structure which has exactly the hypotheses which have a word from the lexicon above them and which has as its conclusion the formula with "Goal" underneath it.

There are several possible ways to do this: we can connect the negative axiom n_1 (a conclusion of the proof structure of Figure 7.8) to either the positive atom n_3 or to the positive atom n_4 (both hypotheses of Figure 7.8), then connect the conclusion n_2 to n_4 (in case n_3 was chosen for n_1) or n_3 (in case n_4 was chosen for n_1). There is a similar choice for the negative atoms np_5 and np_6 and the positive atoms np_7 and np_8. For the S formulae, $S_9 - S_{10}$ is the only solution.

We will return to the issue of how to choose which atomic formulae to connect in Section 7.2, where we will discuss parsing with multimodal proof nets. For now, we will simply choose $n_1 - n_3$, $n_2 - n_4$, $np_6 - np_7$, $np_5 - np_8$ and $S_9 - S_{10}$ to obtain the proof structure shown in Figure 7.9.

This proof structure has only the formulae from the antecedent and succedent of the sequent as its hypotheses (resp. as its only conclusion). All other formulae are connected to two links, once as a premise (the link "above" it) and once as a

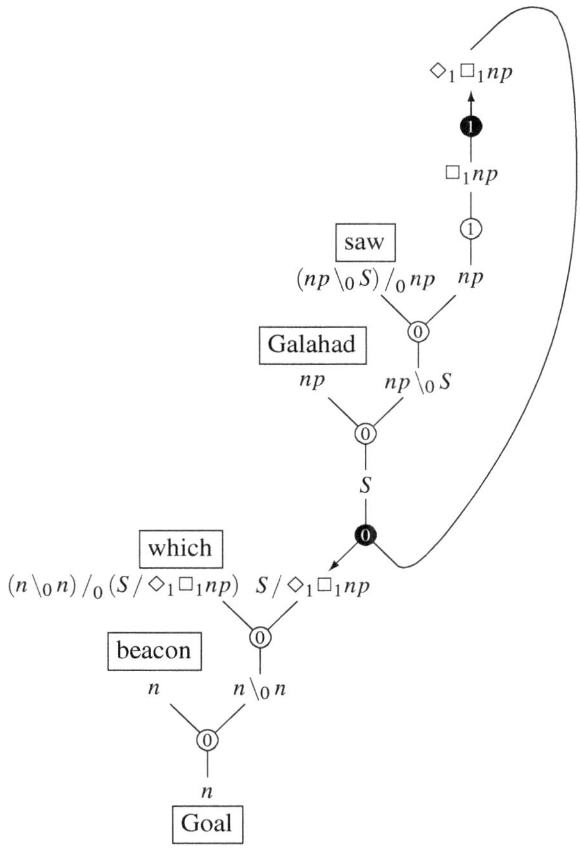

Fig. 7.9. Proof structure of $n, (n\backslash_0 n)/_0(S/_0\Diamond_1\Box_1 np), np, (np\backslash_0 S)/_0 np \vdash n$

conclusion (the link "below" it; note that the formula $\Diamond_1\Box_1 np$ is considered to be below par link for $\backslash$, which is why the line connecting it to the link is drawn as arriving at the formula from above).

As we have seen at the end of Section 7.1.1, the axiom links of the proof nets of one sided linear logic — which are links connecting positive and negative atomic formulae — have become identifications of vertices labeled by the same formula provided that one of the vertices is the conclusion of its link or the proof structure (i.e. is a positive atomic formula) and the other is the premise of its link or the proof structure (i.e. is a negative atomic formula). Formally, this corresponds to replacing all references to a vertex y by those of another vertex x: this makes sense only if $f(x) = f(y)$ (both vertices are labeled with the same formula) and, if the result of the substitution is to satisfy the conditions on proof structures, then at most one of x and y was the premise of a link before the substitution and at most one of x and y

(which, in addition, is not the premise of a link) was the conclusion of a link before the substitution.

Readers familiar with tree adjoining grammars will have remarked that the identification of atomic formulas looks a lot like the substitution operation used for tree adjoining grammars. This connection has been exploited to give an simple translation from tree adjoining grammars to multimodal proof nets which also models the adjunction operation (Moot, 2002, 2008b).

So we have seen the formula unfolding and the axiom linking step of proof nets for multimodal categorial grammars, the only thing we still need to do is explain how to decide which of the proof structures we can construct this way are actually proof *nets*.

As before, the correctness condition is stated graph-theoretically as conditions on the underlying graph of a proof structure. In the current case, the correctness condition is stated on a structure which removes some of the information in the proof structure and which we call an *abstract* proof structure.

Definition 7.6. *A multimodal* abstract proof structure *is a tuple* $\langle V, L, h, c\rangle$ *such that.*

V is the set of vertices *of the proof structure.*
L is the set of links *of the proof structure.*
h is a (partial) function from V to multimodal formulae, such that h assigns a formula to a vertex v iff v is a hypothesis of the abstract proof structure.
c is a (partial) function from V to multimodal formulae, such that c assigns a formula to a vertex v iff v is a conclusion of the abstract proof structure.

In addition, an abstract proof structure satisfies the following conditions.

- *each vertex is the premise of at most one link,*
- *each vertex is the conclusion of at most one link.*

We will call a vertex which is not the conclusion of any link a hypothesis *of the proof structure.*

We will call a vertex which is not the premise of any link a conclusion *of the proof structure.*

So an abstract proof structure assigns formulae only to its hypotheses and conclusions; these formula labels serve only to keep track of which formula occurrences correspond to the formulae of the antecedent and to the goal formulae of the corresponding sequent. In addition, as we will see below, tensor links no longer have a distinguished main vertex. We define a forgetful mapping from proof structures to abstract proof structures as follows.

Definition 7.7. *Let* Π *be a proof structure, the underlying abstract proof structure* $\alpha(\Pi)$ *is defined as follows.*

- *h is the restriction of* f *to the hypotheses of the proof structure.*
- *c is the restriction of* f *to the conclusions of the proof structure.*

- *if l is a par link of Π, then the same l is a par link of $\alpha(\Pi)$ as well, if l is a tensor link of Π then the corresponding tensor link l' of $\alpha(\Pi)$ is the link l, but with the main vertex replaced by the constant nil.*

The abstract proof structure which corresponds to the proof structure of Figure 7.9 is shown in Figure 7.10.

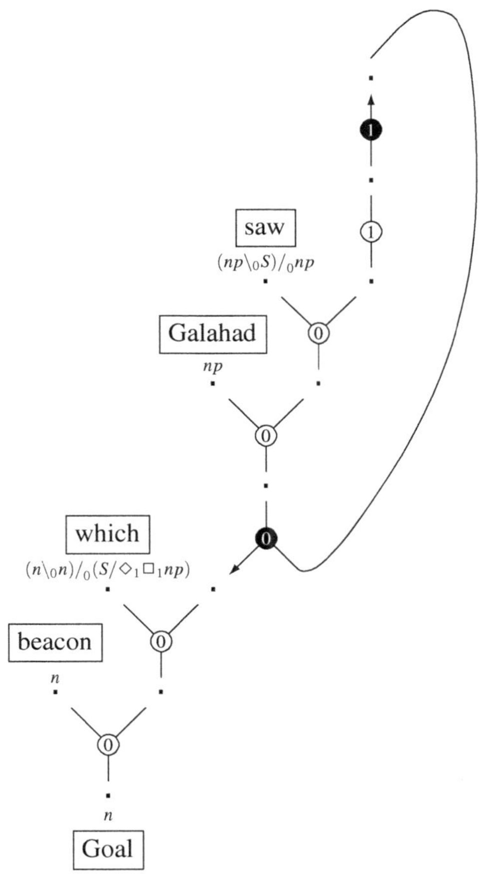

Fig. 7.10. Abstract proof structure of the proof structure of Figure 7.9

As the figure shows, abstract proof structures are displayed slightly differently than the corresponding proof structures. Vertices which do not have a formula assigned to them are displayed as a simple dot "**.**". A formula assigned as a hypothesis to a vertex is displayed above this dot, whereas a formula assigned as a conclusion to a vertex is displayed below it. The table below lists the possibilities.

$h(v)$	$=$ *undefined*	H	*undefined*	H
$c(v)$	$=$ *undefined*	*undefined*	C	C
vertex	·	H ·	· C	H · C

For a vertex which is assigned both a hypothesis and a conclusions, the two formulae can be distinct, as is shown in the rightmost column above.

Definition 7.8. *A* tensor tree *is a connected, acyclic abstract proof structure without par links.*

There is an obvious bijection between tensor trees and sequents of the multimodal Lambek calculus. As shown in Figure 7.11, we can translate an antecedent term Γ of a sequent $\Gamma \vdash C$ into a tensor tree without conclusion formula; labeling the conclusion vertex of the resulting tree with the formula C will give a tensor tree of $\Gamma \vdash C$.

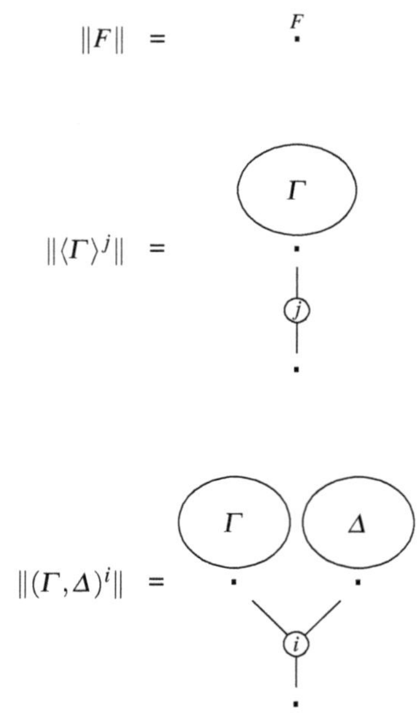

Fig. 7.11. Translating antecedent terms to tensor trees

It is easy to see too that this translation can be inverted and that we can transform a tensor tree into a sequent by the cases shown in Figure 7.11. The case without tensor links is trivial. Given that T is acyclic and connected and that each tensor link has only a single conclusion vertex, there is a unique node which is the conclusion vertex of T, we use the formula assigned as a conclusion to T as conclusion of the sequent, and use the cases shown on the right of Figure 7.11 to transform the tree without the conclusion label into an antecedent term.

In what follows, we will often implicitly convert a tensor tree into a sequent and vice versa to simplify our discussion.

7.1.3 Proof Nets and Contractions

Definition 7.9. *A proof structure for a multimodal categorial grammar without structural rules is a* proof net *if and only if its underlying abstract proof structure contracts to a tensor tree using the graph contractions of Figure 7.12.*

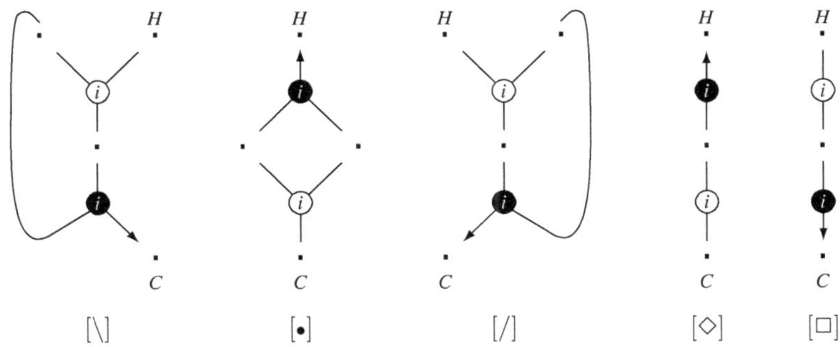

Fig. 7.12. Graph contractions

The contractions replace a connected pair of a tensor link and a par link by a single node.

Both links and the internal nodes (those which do not have a label H or C in the graph) are removed and the two external nodes labeled H and C are identified. In the figure, the labels H and C serve to distinguish the nodes. If node H is a hypothesis of the proof structure before contraction, it will still be a hypothesis of the proof structure after contraction and it will be labeled by the same formula. Similarly, C if it was a conclusion before the contraction, it will still be a conclusion and it will be labeled by the same formula after the contraction.

It is useful to see all contractions as instances of the same schema: a tensor and a par link are connected at all vertices except the vertex with the arrow, respecting left-right and inversing up-down.

Figure 7.13 shows how we can derive the sequent $np \vdash (S /_0 np) \backslash_0 S$ in the proof net calculus using the contractions of Figure 7.12. On the left, we see the only possible proof structure of $np \vdash (S /_0 np) \backslash_0 S$, with the corresponding abstract proof

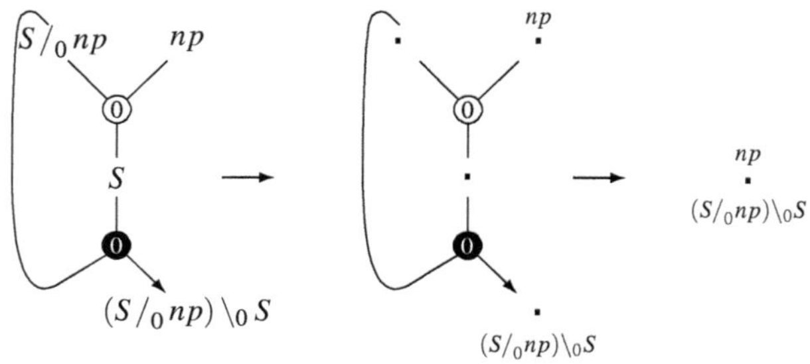

Fig. 7.13. Graph contractions: example

structure in the middle. This abstract proof structure is of the correct form for the $[\backslash]$ contraction, the result of which is shown on the right of the figure.

While proof nets according to Definition 7.9 are equivalent to the sequent calculus, we want to prove a stronger theorem, which allows us to add an arbitrary number of structural rules to the calculus.

However, we first observe a property of the contraction calculus. As shown in Figure 7.14, the contractions for the unary connectives can diverge: that is, some contractions can lead us into blind alleys. In other words, the contractions are not confluent. In the figure, the proof structure, which is simply the identity for $\Diamond_0\Box_0 A$, contracts to a tensor tree using the topmost branch but contracts to an abstract proof

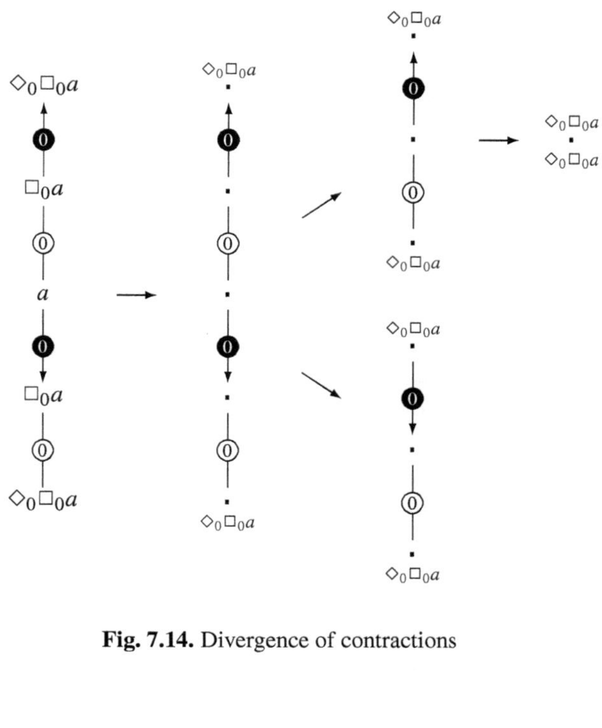

Fig. 7.14. Divergence of contractions

structure which is not a tensor tree (and to which no further contractions apply) in the bottom branch. So, though the existence of the first branch guarantees that the proof structure is a proof net, we need to explore *all* branches if we want to conclude a proof structure is *not* a proof net: discovering that one branch does not convert to a tensor tree does not allow us to conclude anything about whether or not the proof structure is a proof net.

In order to add structural rules to the system, we need to add, for each structural rule, a structural rewrite to the system. Remember that according to Section 5.2.4, each structural rule of the multimodal Lambek calculus is of the following form (where Ξ and Ξ' are contexts with n holes and π is a permutation of n).

$$\frac{\Gamma[\Xi[\Delta_1,\ldots,\Delta_n]]\vdash D}{\Gamma[\Xi'[\Delta_{\pi_1},\ldots,\Delta_{\pi_n}]]\vdash D}\,SR$$

which, given the correspondence between sequents and abstract proof structures, corresponds to the graph rewrite shown in Figure 7.15.

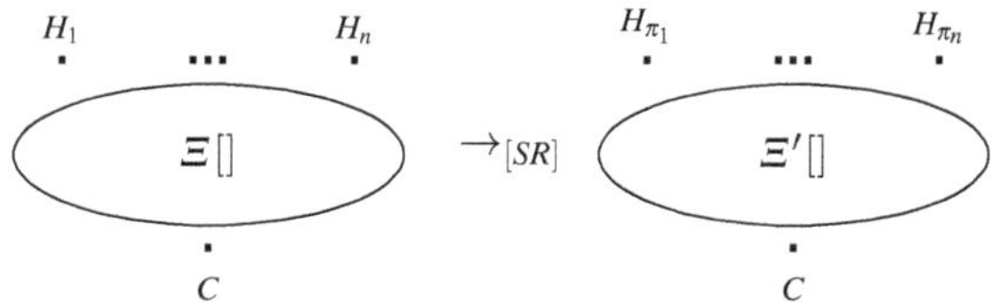

Fig. 7.15. General form of a structural rewrite from $\Gamma[\Xi[\Delta_1,\ldots,\Delta_n]\vdash C$ to $\Gamma[\Xi'[\Delta_{\pi_1},\ldots,\Delta_{\pi_n}]\vdash C$

That is to say, we remove tensor tree $\Xi[]$ from the abstract proof structure, keeping track of the n nodes $H_1,\ldots H_n$ which were the leaves of $\Xi[]$ and the node C which was the root of $\Xi[]$ and replace it by tensor tree $\Xi'[]$, attaching the root node of $\Xi'[]$ to the node which was the root of $\Xi[]$ and attaching the n leaves of the new tree to the leaves of the old tensor tree as indicated by the permutation π, which gives us a bijection between the n leaves of the two trees. So a structural rule will correspond to replacing a tensor tree by another tensor tree together with a correspondence between the leaves of the two trees.

To make the discussion more concrete, look at the two structural rules we have seen in Figure 5.12, repeated below.

$$\frac{\Gamma[(\Delta_1,(\Delta_2,\langle\Delta_3\rangle^1)^0)^0]\vdash D}{\Gamma[((\Delta_1,\Delta_2)^0,\langle\Delta_3\rangle^1)^0]\vdash D}\,MA\Diamond_1 \qquad \frac{\Gamma[((\Delta_1,\langle\Delta_3\rangle^1)^0,\Delta_2)^0]\vdash D}{\Gamma[((\Delta_1,\Delta_2)^0,\langle\Delta_3\rangle^1)^0]\vdash D}\,MC\Diamond_1$$

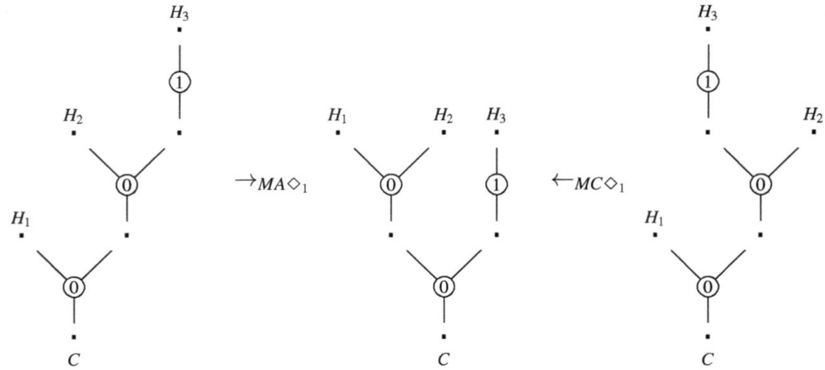

Fig. 7.16. The structural rules of Figure 5.12 as rewrites on abstract proof structures

Figure 7.16 shows these structural rules as rewrites on abstract proof structures. Compare this figure with Figure 5.12 on page 168 to make the correspondence more clear; the nodes labeled H_i correspond to the antecedent terms Δ_i of the structural rule.

Definition 7.10. *A proof structure for a multimodal categorial grammar with structural rules R is a* proof net *if and only if its underlying abstract proof structure contracts to a tensor tree using the graph contractions of Figure 7.12 and the graph conversions corresponding to R.*

Using Definition 7.10, we can show that the abstract proof structure of Figure 7.10 on page 253 is a proof net given the structural rules of Figure 7.16. Looking at the abstract proof structure, we see that two rules apply: either the unary contraction for $\diamond$ or the structural conversion for $MA\diamond_1$. Selecting the structural conversion — with H_1 being the np hypothesis of the abstract proof structure corresponding to "Galahad", H_2 corresponding to the $(np\backslash_0 S)/_0 np$ hypothesis and H_3 corresponding to the vertex which is the conclusion of the par link for $\diamond$ — will produce the abstract proof structure shown in Figure 7.17. The new abstract proof structure has the tensor tree shown in the middle of Figure 7.16 as a substructure.

Next, we apply the contraction for $\diamond$, which produces the abstract proof structure shown in Figure 7.18.

This abstract proof structure is in the right configuration for the $/$ contraction. When we apply it, the result is the tensor tree shown in Figure 7.19.

Given that there is a way to use the contractions and structural conversions to produce a tensor tree, we can conclude that the proof structure of Figure 7.9 (the proof structure we started with on page 251) is a proof net.

An important feature of the current calculus is that we *compute* the antecedent term which makes our sequence of formulae derivable. In addition, if the structural rules are non-increasing according to the definition in Section 5.2.4 then it is decidable — though PSPACE complete in the worst case — to determine whether or

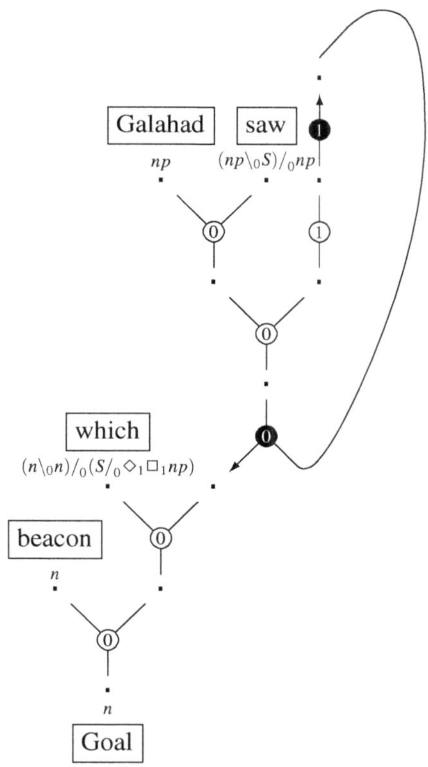

Fig. 7.17. The abstract proof structure of Figure 7.10 after the structural rule of mixed associativity

not a given proof structure is a proof net. In practice, though, for most of the structural rules we have seen in Chapter 5, applying the structural rules turns out to be relatively easy. Indeed, it is useful to see the current contraction condition with its separate structural rule component as a *generic* criterion, which we can specialize according to the available structural rules to get a more specific and more efficiently testable correctness condition. Moot and Puite (2002) give an example of such a specific correctness condition for the associative Lambek calculus. We give two other examples below.

Example 7.11. As an example of how we can specialize the correctness condition for specific packages of structural rules and specific grammars, Figure 7.20 shows a specialized contraction for the mixed associativity and mixed commutativity rule of Figure 7.16.

If the path from the root node of $\Delta'[]$ to its portrayed leaf node consists only of the binary mode 0, the we can — by successive applications of either the mixed associativity or the mixed commutativity rule — "move" $\Xi[]$ to Δ or, seen the other

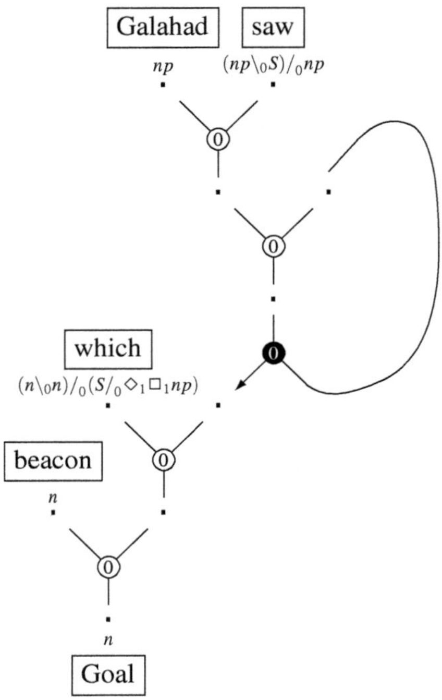

Fig. 7.18. The abstract proof structure of Figure 7.17 after the ◇ contraction

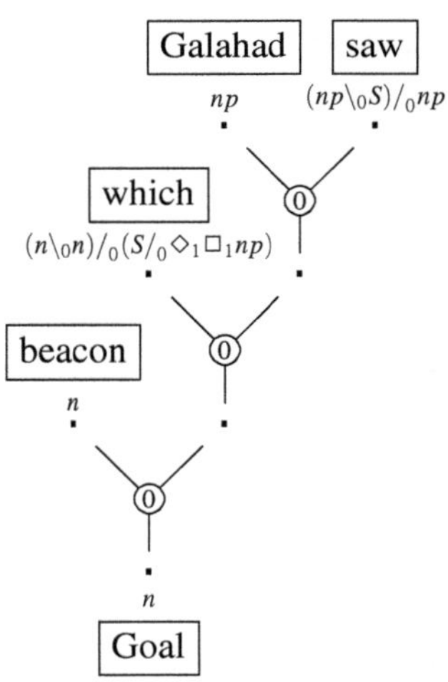

Fig. 7.19. The abstract proof structure of Figure 7.18 after the / contraction

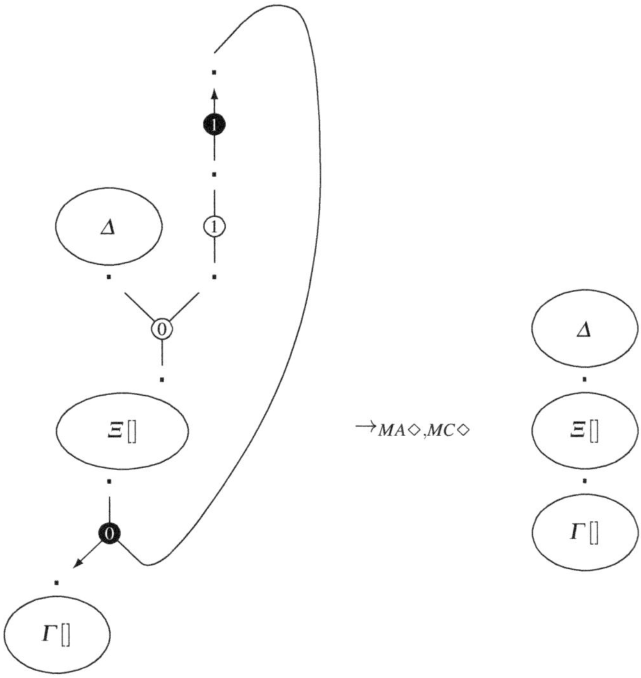

Fig. 7.20. Specialized contraction for a multimodal logic with both mixed associativity and mixed commutativity

way around "move up" the binary mode 0 and it unary branch 1 towards the root of $\Xi[\,]$. In the case where $\Xi[\,]$ is the empty context, we can simply apply the $[\Diamond]$ and the $[/]$ contractions directly.

It is important to note that this specialized contraction, while correct, can replace the corresponding structural rules only if the grammar requires the use of the structural rules only in the context shown above in the figure: if the grammar introduces the unary mode in other contexts (for example by assigning types of the form $\Box_1 A$ to formulas in the lexicon) then the specialized contraction shown in Figure 7.20 will not suffice to capture all possibilities.

Example 7.12. As a second example of a specialized contraction, Figure 7.21 on page 262 gives the $[\Box]$ contraction for a unary mode 1 to which the K structural rule applies (as in Example 5.8). The specialized contraction applies only if $\Xi[\,]$ contains only those binary modes over which the unary mode 1 distributes by means of the K structural rule (in Example 5.8, this was mode 4).

The standard contraction for $\Box$ is the special case where $\Xi[\,]$ is the empty context and there is only one leaf with a unary branch 1.

In short, it is often possible to give a simple characterization of the tensor tree (antecedent term) in which a par link must occur for the contraction to apply, which avoids the complexity of a generate-and-test algorithm (see Moot, 2008a, for more discussion).

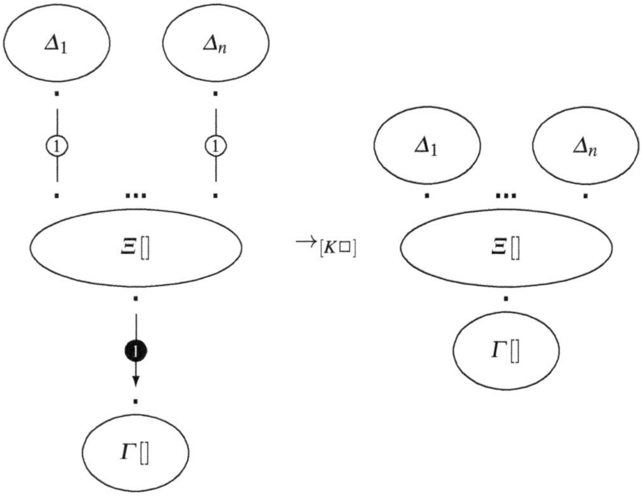

Fig. 7.21. Specialized contraction for $\Box$ and the K structural rule

Now that we have introduced the contractions and the structural conversions and used them to define proof nets for the multimodal Lambek calculus with a given set of structural rules R, we need the show that the proof nets so defined correspond exactly to the sequent proofs of the calculus.

7.1.4 Sequent Calculus and Multimodal Proof Nets

The following theorem, which shows that the proof nets for the multimodal Lambek calculus correspond exactly to the sequent proofs, was originally proved in (Moot and Puite, 2002). We will follow the structure of the proof quite closely.

Theorem 7.13. *If $\Gamma \vdash C$ is a sequent in a multimodal Lambek calculus with structural rules R, then $\Gamma \vdash C$ is derivable if and only if there is a proof net which converts to the tensor tree $\Gamma \vdash C$ using the contraction and the structural rules R.*

Proof. $\Longrightarrow$

We prove by induction on the length l of the sequent proof δ of $\Gamma \vdash C$ that we can construct a proof net contracting to tensor tree $\Gamma \vdash C$.

In case $l = 0$ we have an axiomatic sequent $A \vdash A$. The corresponding proof structure and abstract proof structure are shown below.

$$A \quad \to \quad \begin{matrix} A \\ \cdot \\ A \end{matrix}$$

If $l > 0$ we proceed by case analysis of the last rule.
If the last rule is $\backslash_h$

$$\frac{\Gamma[B] \vdash C \quad \Delta \vdash A}{\Gamma[(\Delta, A \backslash_i B)^i] \vdash C} \backslash_h$$

we know by induction hypothesis that there is a proof net Π_1, where the underlying abstract proof structure $A_1 = \alpha(\Pi_1)$ contracts to $\Gamma[B] \vdash C$ using conversion sequence ρ_1 and a proof net Π_2 where $A_2 = \alpha(\Pi_2)$ contracts to $\Delta \vdash A$ using conversion sequence ρ_2. Figure 7.22 shows the two proof nets we have by induction hypothesis.

Here and in what follows, we will use $\twoheadrightarrow$ to signify any number of conversions (structural conversions and/or contractions) and $\to$ to signify a single named conversion, or, in the current case, the function α which maps a proof structure to the underlying abstract proof structure.

Π_1 is a proof structure with hypothesis B (as well as all the other formulae in the context $\Gamma[]$) and conclusion C. Π_2 is a proof structure with conclusion A and the formulae of Δ as its hypotheses. Given that Π_1 and Π_2 are not just proof structures but also proof *nets*, we know that the underlying abstract proof structures A_1 and A_2 (in the middle of Figure 7.22) convert to the tensor trees $\Gamma[B] \vdash C$ and $\Delta \vdash A$, displayed on the right of the same figure.

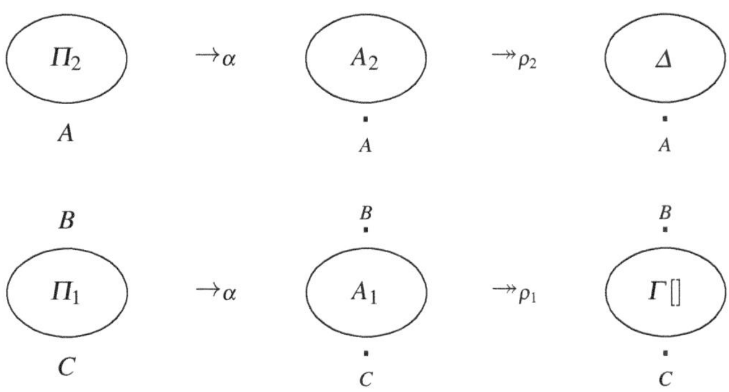

Fig. 7.22. Proof structures and abstract proof structures for $\Gamma[B] \vdash C$ and $\Delta \vdash A$

We need to show that we can combine these two proof nets into a single proof net of $\Gamma[(\Delta, A \backslash_i B)^i] \vdash C$.

Figure 7.23 shows how to do this. On the left we see the proof structures Π_1 and Π_2 of Figure 7.22 connected using a tensor link for $\backslash$. The corresponding abstract proof structure is again shown in the middle. Now it is easy to see that the resulting proof structure is a proof net as well, since the conversions ρ_1 which reduce A_1 to $\Gamma[]$ can apply to the new abstract proof structure, just as the conversions ρ_2 which reduce A_2 to Δ. This means that we can apply the concatenation of ρ_1 and ρ_2 to the resulting proof structure, which is therefore a proof net of $\Gamma[(\Delta, A \backslash_i B)^i] \vdash C$ as required.

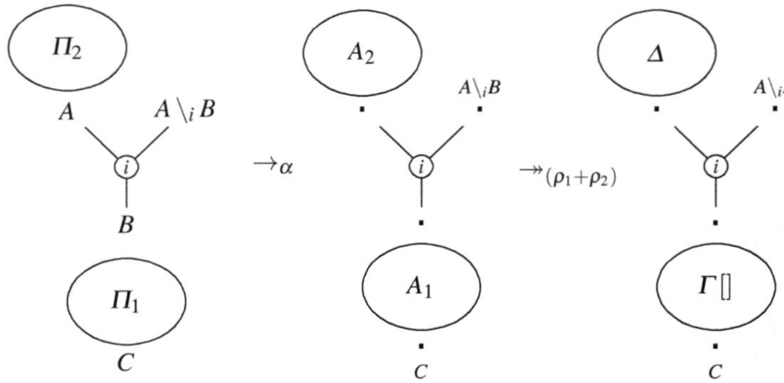

Fig. 7.23. Proof net of $\Gamma[(\Delta, A \backslash_i B)^i] \vdash C$, constructed using the proof nets of Figure 7.22

The other tensor rules ($/_h$, $\bullet_i$, $\Box_h$, $\Diamond_i$) are similar.

In what follows, we will no longer include the explicit conversion α of the initial proof structure into an abstract proof structure as we have done for the $\backslash_h$ case, since the conversion from proof structure to abstract proof structure is trivial. So when we appear to apply a conversion sequence to a proof structure, there is an implicit conversion of the proof structure into an abstract proof structure involved. This convention will simplify the figures and the arguments.

The case for $\backslash_i$ is more interesting, since it clearly illustrates how the par link and the contraction combine to function like the introduction rule.

$$\frac{(A, \Gamma)^i \vdash C}{\Gamma \vdash A \backslash_i C} \backslash_i$$

Figure 7.24 shows the proof net for $(A, \Gamma)^i \vdash C$, with the initial proof structure on the left and the abstract proof structure after contractions/conversions on the right.

We need to show we can turn the proof net of Figure 7.24 into a proof net of $\Gamma \vdash A \backslash_i C$. Figure 7.25 shows, on the left, the proof structure which results from connecting the par link for $\backslash$ to the proof structure on the left of Figure 7.24. We can apply the same conversion sequence ρ to the underlying abstract proof structure to obtain the abstract proof structure shown in the middle of the figure. This abstract proof structure is in exactly the right configuration to apply the $[\backslash]$ contraction, which produces the tensor tree shown on the right. As required, this is a tensor tree of $\Gamma \vdash A \backslash_i C$, making this a proof net of the same sequent.

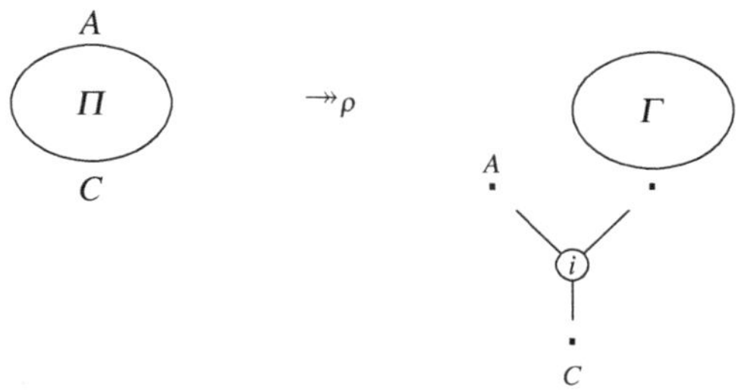

Fig. 7.24. Proof net for $(A,\Gamma)^i \vdash C$

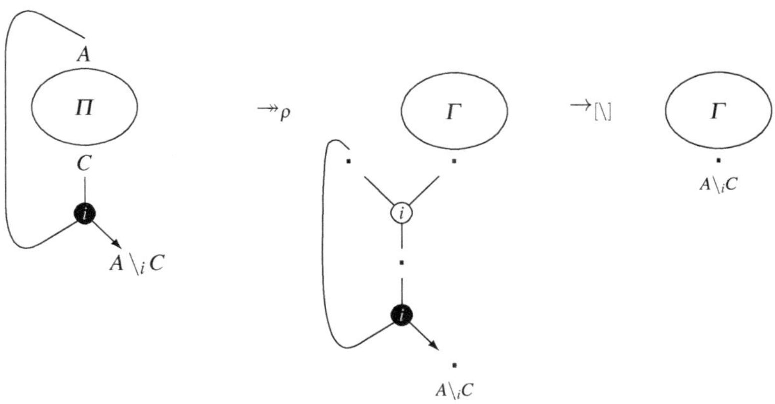

Fig. 7.25. Proof net for $\Gamma \vdash A \backslash_i C$ based on the proof net from Figure 7.24

Suppose the last rule of the sequent proof is the rule $\backslash_h$

$$\frac{\Gamma[(A,B)^i] \vdash C}{\Gamma[A \bullet_i B] \vdash C} \backslash_h$$

The induction hypothesis gives us the proof net shown in Figure 7.26. The proof structure with A and B (and the rest of the formulae of $\Gamma[]$ as hypotheses and C as conclusion converts to a tensor tree of $\Gamma[(A,B)^i] \vdash C$ using the conversion sequence ρ.

Figure 7.27 shows how connecting a par link for $\bullet$ to the A and B hypotheses of the proof structure allows us to convert the underlying abstract proof structure into a tensor tree of $\Gamma[A \bullet_i B] \vdash C$, making the resulting proof structure a proof net of $\Gamma[A \bullet_i B] \vdash C$ as required.

The other par rules ($/_i$, $\Diamond_h$, $\Box_i$) are similar; each involves adding the corresponding par link to the proof net we obtain by induction hypothesis and extending the conversion sequence by the contraction for the connective.

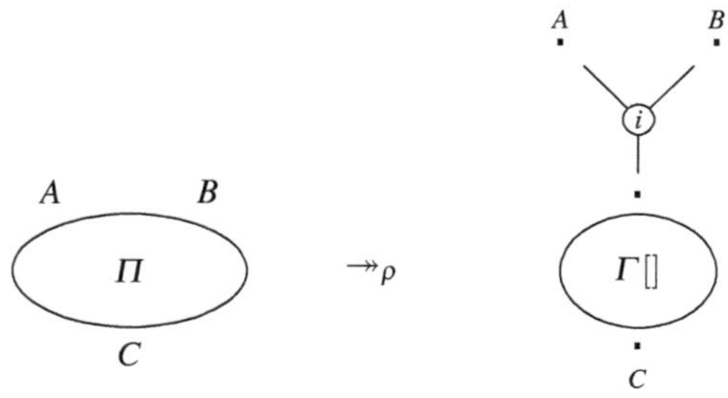

Fig. 7.26. Proof net for $\Gamma[(A,B)^i] \vdash C$

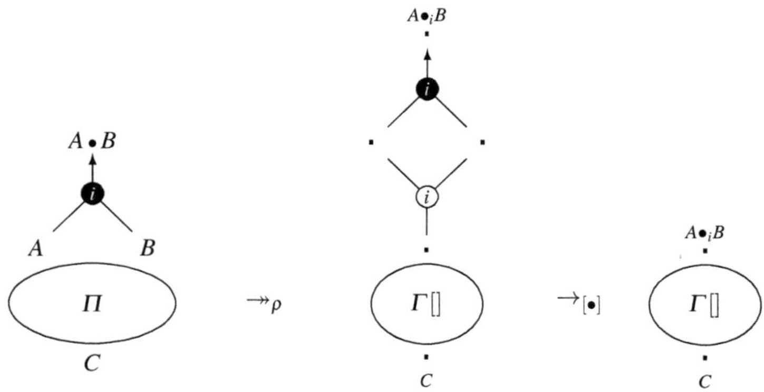

Fig. 7.27. Proof net for $\Gamma[A \bullet_i B] \vdash C$ based on the proof net of Figure 7.26

If the last rule of the proof is a structural rule,

$$\frac{\Gamma[\Xi[\Delta_1,\ldots,\Delta_n]] \vdash C}{\Gamma[\Xi'[\Delta_{\pi_1},\ldots,\Delta_{\pi_n}]] \vdash C}\,SR$$

then we know by induction hypothesis that we have a proof net of $\Gamma[\Xi[\Delta_1,\ldots,\Delta_n]] \vdash C$. We can simply extend it to a proof net of $\Gamma[\Xi'[\Delta_{\pi_1},\ldots,\Delta_{\pi_n}] \vdash C$ by extending the conversion sequence with the corresponding structural conversion as shown in Figure 7.1.4. Given that each structural conversion rewrites a tensor tree into another tensor tree, the right hand side of the figure is a tensor tree as required.

Finally, if the last rule is the *Cut* rule,

$$\frac{\Gamma[A] \vdash C \quad \Delta \vdash A}{\Gamma[\Delta] \vdash C}\,Cut$$

then we can simply connect the proof net of $\Gamma[A] \vdash C$ to the proof net of $\Delta \vdash A$ at the formula A (shown in Figure 7.29). Appending the two conversion sequences ρ_1 and ρ_2 gives us a conversion sequence to $\Gamma[\Delta] \vdash C$ as required.

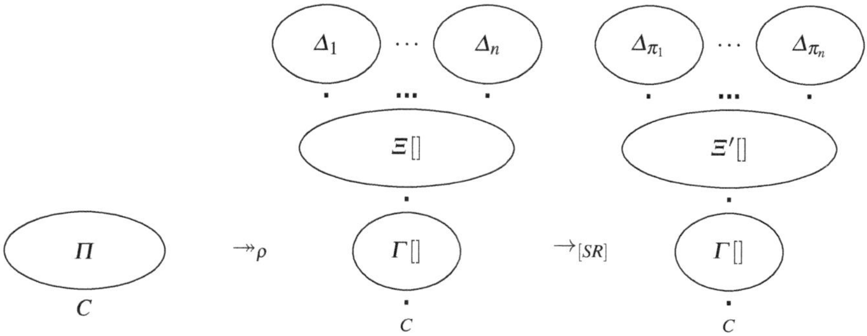

Fig. 7.28. Proof net for $\Gamma[\Xi[\Delta_1,\ldots,\Delta_n]] \vdash C$, extended to a proof net of $\Gamma[\Xi'[\Delta_{\pi_1},\ldots,\Delta_{\pi_n}] \vdash C$

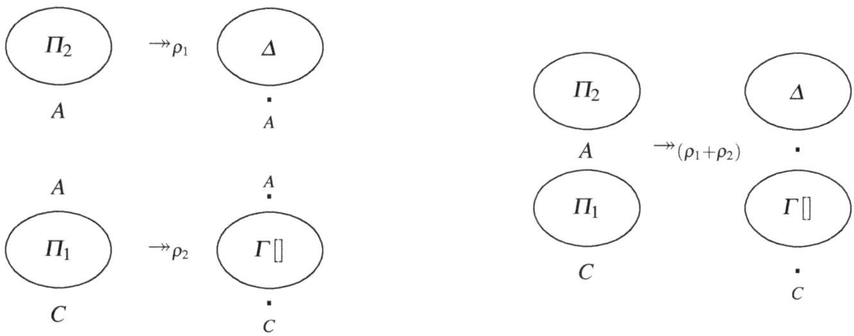

Fig. 7.29. Proof net for $\Gamma[A] \vdash C$, $\Delta \vdash A$ and $\Gamma[\Delta] \vdash C$

$\Longleftarrow$

To show that every proof net — that is, every proof structure of which the abstract proof structure converts to a tensor tree $\Gamma \vdash C$ — corresponds to a sequent proof of $\Gamma \vdash C$, we proceed by induction on the length l of the conversion sequence ρ.

If $l = 0$, then the abstract proof structure of the proof net Π is already a tensor tree. We proceed by induction on the number of tensor links t.

If $t = 0$, then our proof net consists of the single vertex A which corresponds to the axiom rule $A \vdash A$.

If $t > 1$, then

- Suppose at least one of the leaves of the tensor tree is the main formula of its link. Figure 7.30 shows the case of $\backslash_h$ (the case for $/_h$ is symmetric and the case for $\Box_h$ is obtained by removing Π_1 and the branch to it). The proof structure is shown on the left and the underlying abstract proof structure is shown on the right. Since the abstract proof structure is a tree, so is the proof structure. The sequent which corresponds to the abstract proof structure on the right is $\Gamma[(\Delta, A \backslash_i B)^i] \vdash C$.

  Since the complete proof structure is a tree, removing the tensor link and the formula $A \backslash_i B$ splits the proof structure into two smaller tensor trees: one which

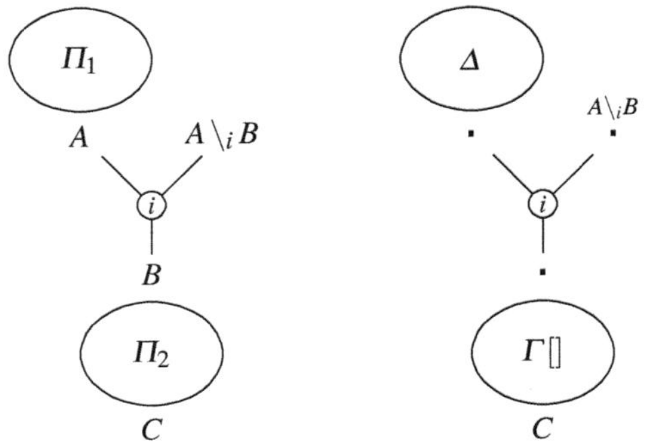

Fig. 7.30. Proof structure with $A \backslash_i B$ as main formula of its link and the underlying abstract proof structure

corresponds to $\Delta \vdash A$ and one which corresponds to $\Gamma[B] \vdash C$. By induction hypothesis, we have a sequent proof for each of these tensor trees. We can combine these two proofs into a proof of $\Gamma[(\Delta, A \backslash_i B)^i] \vdash C$ using the $\backslash_h$ rule.

$$\frac{\Gamma[B] \vdash C \quad \Delta \vdash A}{\Gamma[(\Delta, A \backslash_i B)^i] \vdash C} \backslash_h$$

- Otherwise, none of the leaves of the tensor tree is the main formula of its link. When we look at the possible forms of the tensor links in Table 7.1 this means that all links are either tensor links for $\bullet$ or tensor links for $\diamond$, including the link which has the conclusion of the tensor tree as its conclusion.

  Figure 7.31 shows the case where the tensor link for $\bullet$ is connected to the conclusion of the proof structure. Given the form of the proof structure, the underlying abstract proof structure is of the form shown in the picture on the

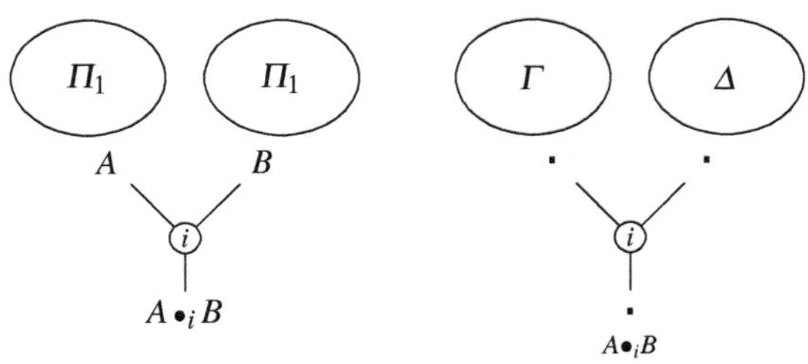

Fig. 7.31. Proof structure with $A \bullet_i B$ as main formula of its link and the underlying abstract proof structure

right. Note that this is a proof net of $(\Gamma,\Delta)^i \vdash A \bullet_i B$, so we need to find a sequent proof of $(\Gamma,\Delta)^i \vdash A \bullet_i B$.
Since the proof structure is a tree, Π_1 and Π_2 are disjoint and removing the tensor link connecting them gives two proof structures: Π_1 with conclusion A and Π_2 with conclusion B.
By induction hypothesis, we obtain a proof of $\Gamma \vdash A$ and of $\Delta \vdash B$ which we can combine into a proof of $(\Gamma,\Delta)^i \vdash A \bullet_i B$ using the $\bullet_i$ rule.

$$\frac{\Gamma \vdash A \quad \Delta \vdash B}{(\Gamma,\Delta)^i \vdash A \bullet_i B} \bullet_i$$

This completes the case for the empty conversion sequence.

Now, suppose $l > 1$. We look at the last structural conversion or contraction in the conversion sequence. We proceed by case analysis.

If the last contraction is a $[\backslash]$ contraction, we are in the situation shown in Figure 7.32.

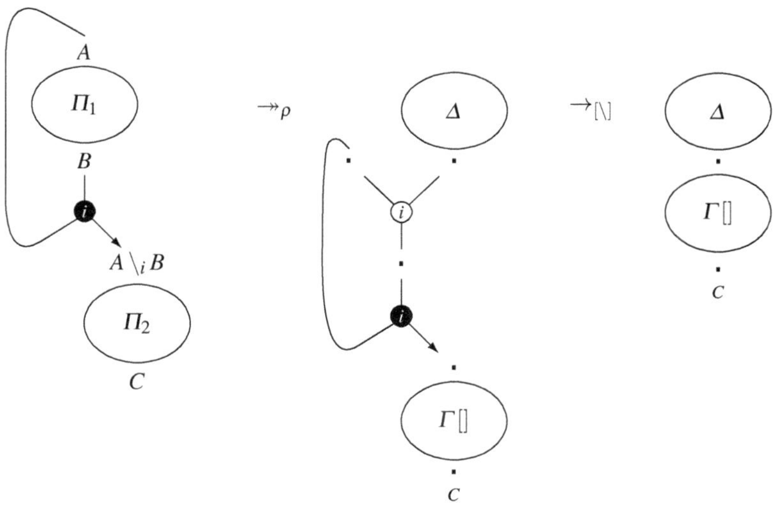

Fig. 7.32. Conversion sequence with $[\backslash]$ as its final contraction

Figure 7.33 shows the proof net split into two new proof structures. We need to show that both are proof nets as well, that is, we need to show that each of the conversions in ρ is applied entirely in Π_1 to form $(A,\Delta)^i$ or entirely in Π_2, where it will form $\Gamma[]$. Looking at the reduction sequence ρ in Figure 7.32 and reasoning backwards from the tensor tree $\Gamma[\Delta]$ to the initial proof structure, we see that each contraction replaces a point by a tensor and a par link and that each structural rule replaces a substructure which is a tensor tree by another tensor tree which has the same root and the same leaves (though not necessarily in the same order). Given that the par link forms a "bridge" between the two parts of the structure, a connected set of tensor links for a structural conversion must be completely on one of the two sides of it. Similarly, each point which "explodes" into a pair of links is on one of the two sides of the par link as well. So we conclude that it is possible to partition ρ into two conversion sequences ρ_1 and ρ_2 such that ρ_1 converts Π_1 to $(A,\Delta)^i \vdash B$ and ρ_2 converts Π_2 to $\Gamma[A \bullet_i B] \vdash C$, as shown in Figure 7.33.

Now, given that the length of both ρ_1 and ρ_2 is strictly less than l and both are proof nets, we can apply the induction hypothesis to obtain a sequent proof δ_1 of $(A,\Delta)^i \vdash B$ and a sequent proof δ_2 of $\Gamma[A \backslash_i B] \vdash C$. We need to combine these two proofs into a proof of $\Gamma[\Delta] \vdash C$, which is done as follows.

$$\dfrac{\dfrac{\dfrac{\vdots\ \delta_1}{(A,\Delta)^i \vdash B}}{\Delta \vdash A \backslash_i B}\backslash_i \qquad \dfrac{\vdots\ \delta_2}{\Gamma[A \backslash_i B] \vdash C}}{\Gamma[\Delta] \vdash C}\ Cut$$

If the last conversion of the conversion sequence is a structural conversion. Then the conversion sequence looks as shown in Figure 7.34.

Since the last rule is a structural conversion, the conversion still ends in a tensor tree even with the last conversion removed. Therefore we have to show that there is a proof net of $\Gamma[\Xi[\Delta_1,\ldots,\Delta_n]] \vdash C$ with a conversion sequence of length $l-1$. This allows us to apply the induction hypothesis to obtain a sequent proof δ of $\Gamma[\Xi[\Delta_1,\ldots,\Delta_n]] \vdash C$, which we can extend with the corresponding structural rule as follows.

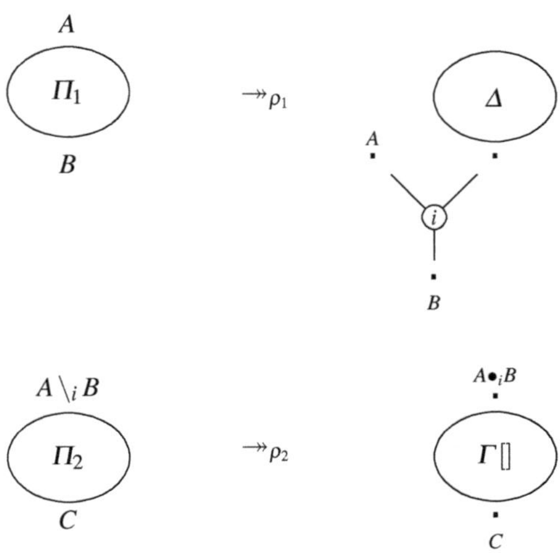

Fig. 7.33. The conversion sequence of Figure 7.32 divided among two proof nets

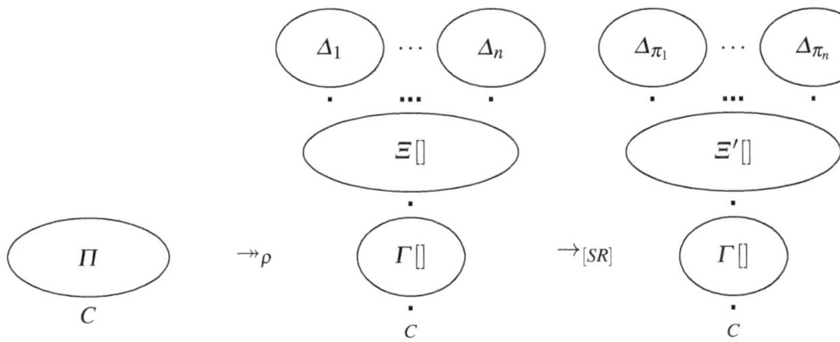

Fig. 7.34. Conversion sequence with a structural rule as its last conversion

$$\frac{\begin{array}{c}\vdots\,\delta\\ \Gamma[\Xi[\Delta_1,\ldots,\Delta_n]]\vdash C\end{array}}{\Gamma[\Xi'[\Delta_{\pi_1},\ldots,\Delta_{\pi_n}]\vdash C}\,SR \qquad \square$$

As shown in (Moot and Puite, 2002), we can modify the sequentialisation part of the proof to combine it with a cut-elimination result in such a way that the rule-to-rule correspondence is more direct. The proof above does not give cut-free sequent proofs, even if the initial proof net is a proof net without cut formulas. The modification allows us to give a sequentialisation proof in such a way that:

- each link (with its corresponding contraction if it is a par link) corresponds to a single proof rule of the same type in the sequent proof,
- each cut vertex in the proof net corresponds to a cut rule in the sequent proof,
- each axiom vertex in the proof net corresponds to an axiom rule in the sequent proof.
- each structural conversion in the conversion sequence of the proof net corresponds to a structural rule in the sequent proof.

In addition, (Moot and Puite, 2002) show that we can prove a cut elimination theorem directly for the multimodal proof net calculus: replacing cut formulas by cut formulas of strictly lesser degree and thereby giving a more direct proof than we have seen in Section 5.2.5.

7.2 Grail: Parsing with Multimodal Proof Nets

The proof nets discussed in the previous section form the basis of an implementation of a parser for multimodal categorial grammars (Moot, 2012a). Grail is a parser written in SWI Prolog, which uses the GraphViz macros for displaying the proof structures. It is distributed free of charge (under the GNU Lesser General Public

License). A step-by-step tutorial to installing and getting used to Grail can be found at (Moot, 2012b).

Grail allows you to design multimodal categorial grammars by specifying a lexicon, which assign a set of formulae (or pairs of formulae and lambda-terms, for those interested in semantics in the tradition of Montague as discussed in Chapter 3) to words and a set of structural rules for the grammar. New lexical entries can be produced either by editing the grammar file or by using the provided user interface (both are described in the tutorial, in the Sections *1 - Editing* and *3 - Grammar File Format* Moot, 2012b).

The modes I of the grammar are defined implicitly: all and only the modes i which occur either in the grammar or in the structural rules are members of I.

Grail uses abstract proof structures as its internal representation for its intermediate partial proofs during proof search. One of the particularities of Grail is that it has an interactive proof mode, where Grail and the user cooperate to find a proof of a given statement, with Grail handling bookkeeping details but also helping in filtering out branches of the search space which can never lead to a proof.

7.2.1 Interactive Parsing

Given a sentence $w_1,\dots,w_n$ and a goal formula C parsing this sentence corresponds to the following steps:

1. For each w_k, find a formula A_k such that $A_k \in \mathrm{Lex}(w_k)$,
2. Using the formula decomposition rules on the resulting sequent $A_1,\dots,A_n \vdash C$, unfold the sequent into a proof structure. As discussed, this proof structure will have a number of atomic formulae as hypotheses and as conclusions in addition to the formulae from $A_1,\dots,A_n \vdash C$.
3. Connect atomic formulae which are hypotheses of the resulting structure to those which are its conclusions, to obtain a proof structure which has as its hypotheses and conclusion only the formulae of $A_1,\dots,A_n \vdash C$.
4. Translate the resulting proof structure into an abstract proof structure.
5. Convert the abstract proof structure, using the structural rules and the contractions into a tensor tree with $A_1,\dots,A_n$ as its yield.

Steps 2 and 4 are trivial and are performed automatically by Grail. We'll have a bit more to say about how to handle Step 1 for very large lexicons in Section 7.2.3 below. For now, we will concentrate on how the user can interact with Grail to perform Step 3 — which connects the atomic formulae — and Step 5 — which performs the contractions and structural rules.

Figure 7.35 shows an example of the lexical unfolding for the complex noun "beacon which Galahad saw", which we have already seen in Figure 7.9 on page 251.

There are some differences with the proof structure of Figure 7.9. For example, only the atomic formulae are displayed in the structure. However, this does not result

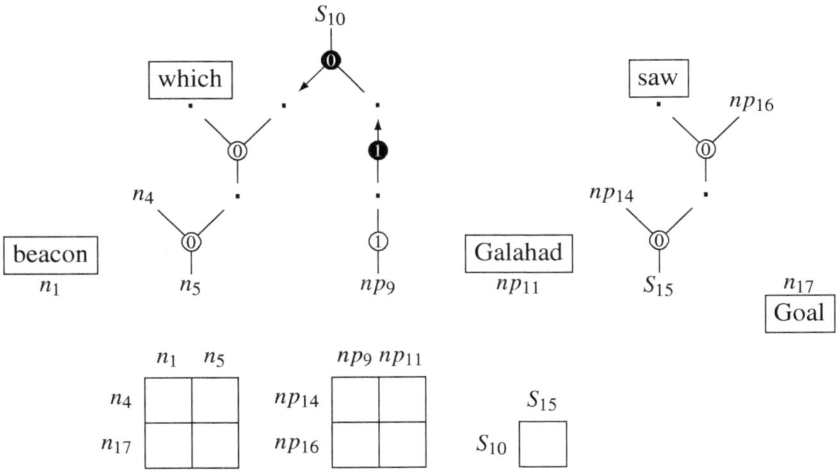

Fig. 7.35. Lexical unfolding and possible node identifications

in a loss of information: since the main formula of each link is always the vertex which is closest to the lexical leaf, we can use this information in combination with Table 7.1 on page 248 to compute the formulae for the unlabeled vertices. So we can deduce, for example, that the node above np_9 must be labeled $\Box_1 np$ and that the node above S_{15} and to the right of np_{14} must be $np \backslash_0 S$ and therefore that the node with the lexical leaf "saw" must be labeled $(np \backslash_0 S) /_0 np$.

There is also the addition of an explicit table containing the possibilities for identifying the atomic formulae. The negative atomic formulae — visually, we can identify them as "hanging" from the proof structure: attached from above, but not from below; np_9, which has the unfolding of "which" above it, and np_{11}, which has the lexical leaf "Galahad" above it, are negative atomic formulas — are written above the columns of this table, whereas the positive atomic formula are written left of its rows — the positive atomic formula are visually "on top" of the rest of the structure: attached from below, but not from above, n_4 and n_{17} are positive atomic formulas. All atomic formulae have a unique integer assigned to them to give a unique identifier to the different formula occurrences. A complete matching is a pairing of positive and negative atomic formulas: each negative atomic formula is matched with exactly one positive atomic formula and each positive atomic formula is matched with exactly one negative atomic formula.

Grail displays the table with the atomic formulae in a separate window and allows the user to select which atomic formulae to connect. Because of the count check (Proposition 2.6), only tables with as many rows as columns for each atomic formula make sense, since they have as many positive (hypothesis) as negative (conclusion)

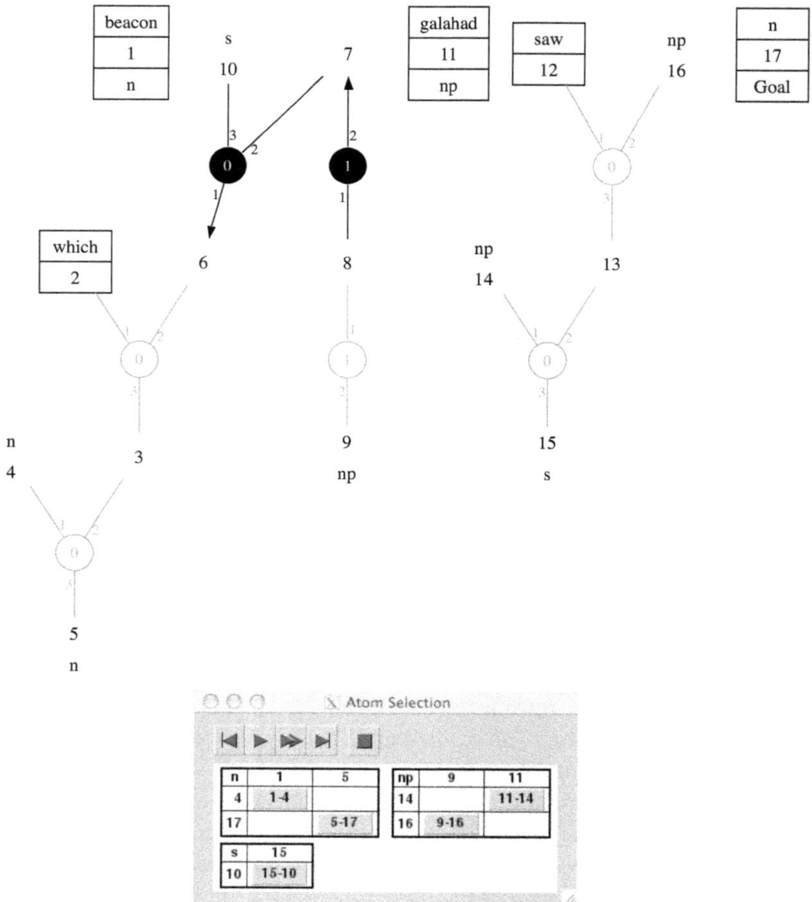

Fig. 7.36. Grail output corresponding to Figure 7.35: the proof structure produced by Graphviz (top) and the possible node identifications (bottom)

atomic formulae. Grail does not display proof structures which do not satisfy the count check. By default, Grail does not show a proof structure if it has demonstrated that — according to the specifications of the grammar — no total axiom linking is possible for the structure. Grail allows you to override this feature, which is useful when debugging grammars (see the Grail tutorial, Section *3 - Debugging Grammars* for details Moot, 2012b).

Figure 7.36 shows the Grail output for the example of Figure 7.35. The two figures are rather similar, but let's take some time to remark on the differences. First, Grail has already excluded a number of the possibilities for the axiom connections:

Fig. 7.37. One of the proof nets corresponding to the translation of $a \vee \neg a$ proposed by Savateev (2009)

in the window shown at the bottom of the figure, only the grey buttons marked with a pair of node numbers correspond to possible axiom connections, leaving only a single total linking as a possibility. We will discuss how Grail performs the calculations to arrive at this conclusion in the next section.

The Graphviz output shown at the top of Figure 7.36 does a good job of preserving the tree-like structure of the graph. However — for more cluttered graphs and for the different graph display algorithms provided by Graphviz which do not preserve the tree structure (see Figure 7.37 for an example) — there is possibly some confusion with respect to the left and right branches (and possible even the root) of a link. For this reason, Grail marks the left branch by 1, the right branch by 2 and the root node by 3.

In Grail, all nodes are given a unique integer as an identifier — it corresponds to the traversal of the formulas during the unfolding phase and has no other significance. The negative atomic formulas are written below this integer, and the positive atomic formulas above it. The words of the sentence are shown in rectangular boxes above the corresponding vertices, whereas the unique goal vertex has the word "Goal" below its vertex number.

Without the filtering strategies in place, there are two possibilities for the n formulae, two for the np formulae and one for the S formula ($S_{10} - S_{15}$). For the n formulae, the possibilities are $n_1 - n_4$, $n_5 - n_{17}$ and $n_1 - n_{17}$, $n_5 - n_4$. For the np formulae, the possibilities are $np_9 - np_{14}$, $np_{11} - np_{16}$ and $np_9 - np_{16}$, $np_{11} - np_{14}$. The user can select any pair of atomic formulae and continue selecting until all atomic formulae have been linked. In case any combination does not yield a proof net, Grail will present the unexplored alternatives — if any remain — and allow the user to select one of them. In case no other alternatives remain, the search space has been fully explored and Grail will output a message about the number of distinct proof nets found. Grail will also output the lambda-term semantics for the different parses of the sentence in a LaTeX file.

Since we have already shown in Section 7.1.3 that we can turn this proof structure into a proof net given the structural rule of mixed associativity, we will now show that this proof structure cannot be turned into a proof net without any structural rules.

First, suppose we choose $n_1 - n_{17}$, $n_4 - n_5$ for the n atomic formulae (in either order, whenever we choose the first, the second is the only possibility for the remaining formulae). This means that the n from "beacon" is the goal formula. However, this also means that the Goal node cannot be connected to the rest of the proof structure and therefore that there is no way to turn the resulting proof structure into a tree. Similarly, the $n_4 - n_5$ connection produces a cycle, which we cannot remove by further conversions and therefore, the resulting structure can never reduce to a tensor tree. This means that $n_1 - n_4$, $n_5 - n_{17}$ is the only possibility.

Second, suppose we choose $np_9 - np_{14}$, $np_{11} - np_{16}$ for the np nodes. This means that whatever we do, the final tensor tree will be one which contains a substring "saw Galahad", because of the $np_{11} - np_{16}$ connection and because we are considering the case without structural rules. Even *with* the mixed associativity and commutativity rules of Figure 7.16 (look back to page 258), the structure is not a proof net, because the structural rules cannot change the order of these two words with respect to each other: the mixed associativity and mixed commutativity can "move around" only material which is on a right branch and has the 1 unary mode. Since the $np_9 - np_{14}$ connection has put the only unary tensor branch on the left of a 0 binary branch and this means no structural rules can apply to the proof structure.

These arguments together show that the only remaining candidate for a proof net are the connections $n_1 - n_4$, $n_5 - n_{17}$, $np_9 - np_{16}$, $np_{11} - np_{14}$ and $S_{15} - S_{10}$, which corresponds to the possible connections listed at the bottom of Figure 7.36. Performing these connections will produce the proof structure we have seen in Figure 7.9 in Section 7.1.2.

After all atomic formulas have been connected and a total matching of the positive and negative atoms have been found, we still need to perform the contractions to obtain a tensor tree.

For the contractions, as we have seen in Figure 7.14, it can be possible that multiple contractions apply to an abstract proof structure and that some of these do not allow us to contract the abstract proof structure to a tree. Grail interacts with the user in these cases too. First, the contractions can be partially ordered, with some contractions necessarily taking place before others. The arrows of the par links give a natural order: if the arrow of a first par link "points towards" one of the tails (ie. non-arrows, the active vertices) of another par link, then this first par link can always be contracted before the second. In Figure 7.36 (and Figure 7.9), for example, we can (and even have to) apply the $\diamond$ contraction before the $/$ contraction. Grail computes which of the contractions can be applied first and colors the corresponding par link blue. We will call such par links *active*. When multiple active par links are connected to a same tensor tree, a conflict occurs and the user can give his input. A conflict is visible in the proof structure by a pair of par links connected to the same tensor tree, and with the arrows of both links pointing away from this tensor tree.

Figure 7.38 shows an example. The initial abstract proof structure is partially shown on the left of the figure. There are two active par links, the link with main vertex 8 and the link with main vertex 33 (remember that the main vertex of a par link has the arrow pointing towards it). Both par links are connected to the same tensor tree with their arrows pointing away from it. Grail allows you to decide which of the two contractions to perform. As a useful mnemonic to help us see when we are in the right configuration to perform a contraction: a par link and a tensor link must be connected to the same vertices by all connections which share the same number *except* for the connection with the arrow. In addition both links must share the same mode information.

As we have seen before, all connections from the central circle of the link to the vertices are numbered 1, 2, 3. For the par link with main vertex 8, its mode is 0 and the connection with the arrow towards vertex 8 is labeled 2. The other connection, labeled 1, which is important to see if we can contract must therefore be connected to a tensor link at the vertex connected to this tensor link by label 1. For both links, following the connection with label 1 means we arrive at node 9 and therefore we can perform the contraction of this par of links. Grail will present this choice by allowing you to select the main vertex of the par link, vertex 8.

The second possibility is to contract the par link with main vertex 33. The path labeled 2 leaving it arrives at vertex 34; a tensor link is connected to vertex 34 by a path labeled 2 and the tensor link and par link are both assigned mode 0, therefore there is another contraction possibility. Grail will present this choice by allowing you to select the main vertex of the par link, vertex 33.

Since the final par link in the figure, with the arrow arriving at Goal node 32, is connected to node 33, which is a single-node tensor tree with the arrow of par link 33 pointing towards it, we know that the contraction of par link 32 can (and in this case, *must*) be done after contraction 33.

Figure 7.38 shows two different contraction sequences starting from the same initial abstract proof structure: the first sequence starts with the 8 contraction, corresponding to a $\Box$ and the second sequence starts with the 33 contraction, corre-

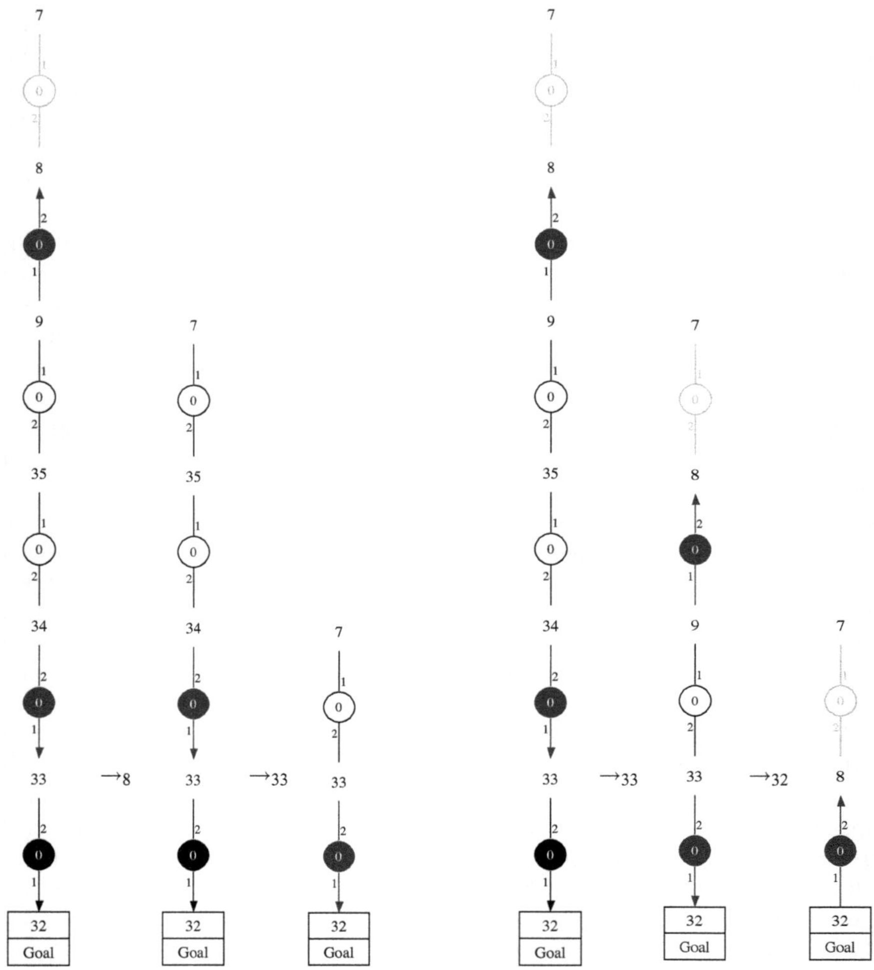

Fig. 7.38. Guiding the contractions in Grail: a successful sequence (left) and a failure (right)

sponding to a ◇. We can complete the first sequence by the 33 contraction, shown in the figure, following by the 32 contraction to obtain a single vertex.

The second sequence produces a second conflict after the 33 contraction: we can either apply the 8 contraction (this gives a permutation of the order of contractions of the first sequence) or the 32 contraction. The 32 contraction leads us to a dead end: there is one par link left, but it is on the wrong side of the remaining tensor link: following connection 1 from the par link leads us to the Goal vertex and not to a tensor link. As we have already discussed when presenting Figure 7.14, the order in which we perform the contractions is important. Trying a few of these contractions by hand is a good exercise. As we will see in the next section, Grail gives some help with this type of contractions.

After all contractions have been performed and the resulting tree contains only external modes, Grail presents a dialog window, allowing the user to try to find alternative proofs for the current sentence — Grail keeps track of the alternative axiom connections which have not been tried yet — or to abandon proof search.

7.2.2 Pruning the Search Space

Having given a brief overview of the interactive component of the Grail parser, which is a useful tool for writing and debugging grammars, we will now look at some of the algorithms employed by Grail which are used both to help the grammar writer but also to improve the efficiency of automatic proof search.

Grail uses several strategies to reduce the search space and to help prevent the user from making any choices which cannot possibly lead to a proof (Moot, 2004, 2007), and these will sound familiar if you look back to some of the arguments we have used for rejecting certain axiom connections. Figure 7.36 shows, at the bottom of the figure, how Grail has directly disqualified as potential axioms all but one total linking and in the previous section we discussed informally why this should be the case. In this section, we will explain how Grail arrives at the conclusion that this is the only possible linking.

The following definition lists an obvious translation from multimodal sequents into sequents of linear logic.

Definition 7.14. *Let $\Gamma \vdash C$ be a multimodal sequent, its translation into multiplicative intuitionistic linear logic $a(\Gamma) \vdash f(C)$ is defined as follows.*

$$\begin{aligned}
a(F) &= f(F)\\
a(\langle\Gamma\rangle^j) &= a(\Gamma)\\
a((\Gamma,\Delta)^i) &= a(\Gamma), a(\Delta)
\end{aligned}$$

$$\begin{aligned}
f(p) &= p\\
f(\Diamond_j A) &= f(A)\\
f(\Box_j A) &= f(A)\\
f(A \bullet_i B) &= f(A) \otimes f(B)\\
f(A \backslash_i B) &= f(A) \multimap f(B)\\
f(B /_i A) &= f(A) \multimap f(B)
\end{aligned}$$

Proposition 7.15. *If $\Gamma \vdash C$ is a derivable multimodal sequent, using structural rules of the form described in Section 5.2.4, then $a(\Gamma) \vdash f(C)$ is derivable in multiplicative intuitionistic linear logic.*

Proof. This is immediate from inspection of the rules. The rules for the unary connectives translate to the identity on proofs, given that both the unary punctuation and the unary connectives are erased. The structural rules are translated as (zero or more) applications of the structural rule of exchange (depending on the permutation π which is part of the definition of the structural rule). □

Therefore, by contraposition, if $a(\Gamma) \vdash f(C)$ is underivable we know that $\Gamma \vdash C$ is underivable as well. Since we know that it is easy to check whether a given proof structure is a proof net in multiplicative linear logic — by Proposition 6.27 and the results of Guerrini (1999); Murawski and Ong (2000), who give linear time algorithms — this gives us a simple but effective test which allows us to reject multimodal proof structures which do not correspond to linear logic proof nets according to the translation given above.

As a consequence, if we translate the proof structure of Figure 7.35 into its two-sided intuitionist proof structure, the proof structure shown in Figure 7.39 (it is possible to translate directly into the proof nets of Section 6.24 as well), then the conditions SAT, stating that any two vertices are connected by an alternate elementary path, and ØÆ, stating that there are no alternate elementary cycles, must hold of this proof structure.

From these conditions, we can deduce that an axiom connection between n_1 and n_{17} violates the SAT constraint, since there is no path from the Goal vertex to anywhere outside the path leading to the "beacon" vertex.

Similarly, an axiom connection between n_4 and n_5 would create a trivial alternate elementary cycle between the two vertices just connected and this connection must therefore be excluded as well.

A second strategy uses word order properties by assigning pairs of string positions to formulae, with the intended meaning of the leftmost position and the rightmost position of the corresponding formula (the Grail implementation allows any number of string positions, so that it would be possible for example to assign two pairs of positions to a single formula; for the sake of simplicity, we will only talk about single pairs of positions). The string positions for words are assigned from left to right, using $0,1$ for the first word, $1,2$ for the second and $n-1,n$ for the nth word, with the goal formula being assigned $0,n$. If the grammar does not contain any structural rules which change the order of words in the string, the string positions are propagated as shown in Figure 7.2. If $X - Y$ are the string positions assigned to the

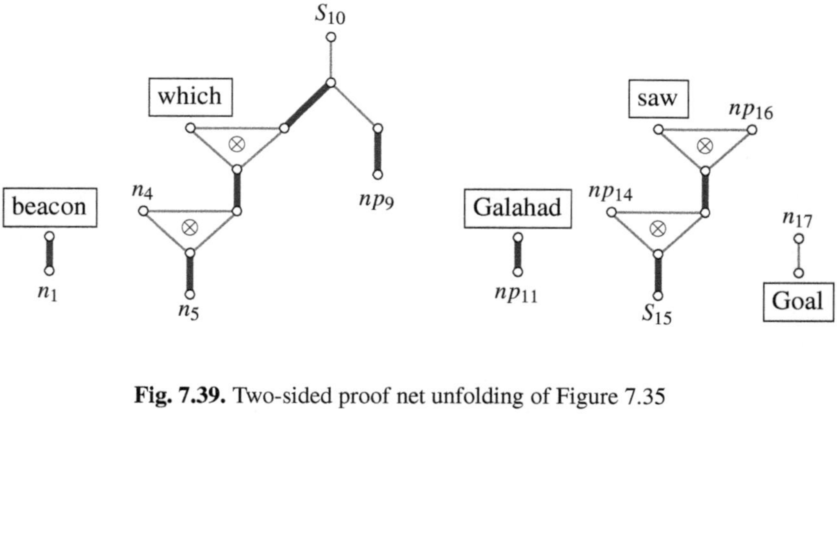

Fig. 7.39. Two-sided proof net unfolding of Figure 7.35

main formula of the link, then the string positions assigned to the active formulae are shown in the figure. Z is always an unused variable, it corresponds to a universal quantifier over positions, whereas f is an unused Skolem function taking all free variables which have scope over f as its arguments. The use of a Skolem function is necessary to get the correct interpretation of an existential quantifier over positions.

To illustrate the string position rules, consider the tensor rule for $A \,/\, B$. Suppose $A \,/\, B$ occupies the positions between X and Y and we combine it with an adjacent formula B. The leftmost position of B must be the rightmost position of $A \,/\, B$, that is Y, but the rightmost position can be any position, hence the free variable Z. The resulting A will therefore start at the leftmost position of $A \,/\, B$ and end at the rightmost position of B, which gives $X - Z$ as its string positions.

This is a classic strategy, variations of which have been used for categorial grammar parsing at least since (Morrill, 1995). The Grail implementation is based on the embedding of the Lambek calculus and some of its extensions into first-order linear logic of Moot and Piazza (2001). In particular, Grail allows a grammar to define its own string position rules for formulae, which means that Grail can assign string positions even in the case of some structural rules which change word order, a feature which will be illustrated in the example below.

Figure 7.40 shows an example of how the string positions are propagated to the atomic formulae. For example, "saw", according to its position in the string, has position $3,4$. Given that "saw" combines with an np to its right, this np must have 4 as its leftmost position, though we can say nothing about its rightmost position W except that it will be the rightmost position of the internal node of this lexical tree — which corresponds to $np \backslash_0 S$ — as well, so this $np \backslash_0 S$ node will have $3,W$ as its string range. Given that $np \backslash_0 S$ still needs to combine with an np to its left, the

Table 7.2. Logical links with string position labels for the multimodal Lambek calculus

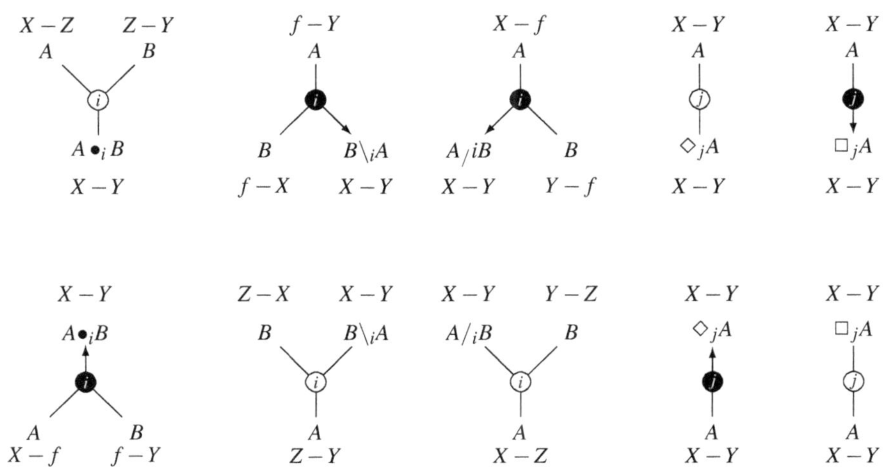

position of this np must be $V,3$, making the positions of the sentence S which is a conclusion of this lexical tree V,W.

Grail allows certain formulae, which are explicitly mentioned in the grammar, to get a non-standard translation. This is the case for the $S /_0 \Diamond_1 \Box_1 np$ formula, which, given a string position to the entire formula, passes this string position to its S subformula, while passing the empty sequence Z,Z to its np subformula. This models the intuition that this formula corresponds to a sentence which is missing an *np somewhere*. A slightly more clever encoding could also capture the fact that this np must occur on a right branch, however, we leave this as an exercise to the reader.

Taking al this into consideration, Figure 7.40 shows the string positions assigned to each of the atomic variables in the example proof structure.

Each atomic connection must unify the string positions of the two atoms which are connected. This constraint is quite powerful, and excludes, for example, a connection between n_1 and n_{17} (different rightmost positions) or a connection between np_{11} and np_{16} (different leftmost positions). In the figure, the axiom connections which are excluded by this condition are shown in grey.

A third filtering strategy exploits the fact that we need to find a perfect matching between the set of atomic formulae which are hypotheses and the set of atomic formulae which are conclusions. This means, for example, that when we look at Figure 7.40 and see that the connection between n_1 and n_{17} is excluded, then the connection between n_4 and n_5 should be excluded as well. Grail implements an algorithm by Régin (1994) to ensure that this constraint is satisfied in more complicated cases as well (see Moot, 2007, for more details).

A final strategy is specifically adapted for sequences of unary connectives as we have seen them in Section 5.2.2 (and some of the exercises). Though we have seen in Figures 7.14 and 7.38 that the unary contractions can diverge, a simple context-free grammar can decide whether or not a sequence of unary modes — we will call such a sequence a unary path — can contract in a calculus without structural rules for the unary modes. Figure 7.41 shows (on the left of the figure) how to assign a string to a unary branch in a proof structure, starting at the conclusion and going up until we arrive at an atom or a binary branch. The labels l_j, m_j and r_j represent (arrow pointing) left, mid (no arrow) and (arrow pointing) right. Turning the page 90 degrees clockwise will make the directions of the arrows of the par links correspond to the label, and this mnemonic helps when reading the string on the page: r_j should be to the immediate right of some m_j, which corresponds to the $\Diamond$ contraction, and l_j should be to the immediate left of some m_j, which correspond the the $\Box$ contraction (the contractions and the corresponding strings are shown on the right of Figure 7.41).

If no structural rules apply to the unary mode j and the unary mode is internal (that is, the abstract proof structure must contract to a tensor tree without unary j branches) we can verify whether or not the unary modes contract by means of the context-free grammar shown at the bottom of Figure 7.41.

The start symbol S of this grammar denotes the strings which correspond to an abstract proof structure which is a unary path and which contracts to a point. The correspondence between the contractions and the grammar is easy to see: the first

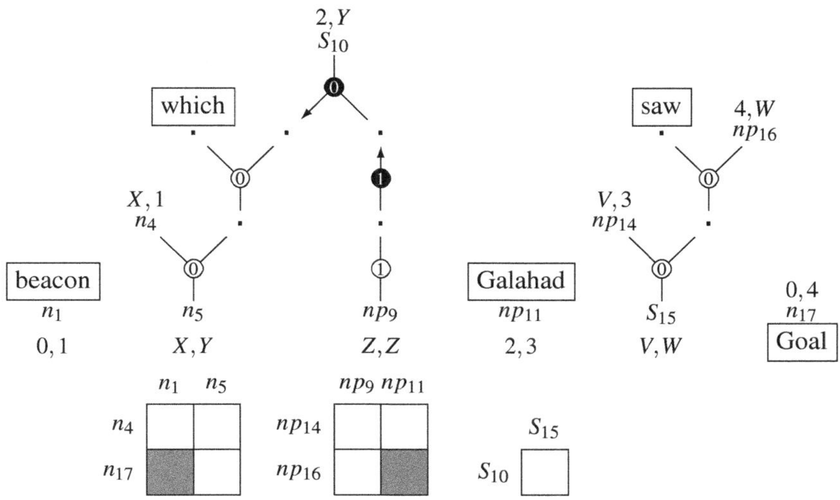

Fig. 7.40. Lexical unfolding with string position pairs for the atomic formulae

rule corresponds to no contractions, the second to the $\Diamond$ contraction and the third to the $\Box$ contraction.

Example 7.16. Calculating the string which corresponds to the abstract proof structure of Figure 7.38 on page 278 gives us $l_0 l_0 m_0 m_0 r_0 m_0$. Given that there is only a single mode we will not mention the mode subscripts in the derivation which follows.

We can parse this string using the context-free grammar of Figure 7.41 to produce the following context-free derivation.

$$
\begin{array}{l}
S \rightarrow \\
lSmS \rightarrow \\
llSmSmS \rightarrow \\
llmSmS \rightarrow \\
llmSm \rightarrow \\
llmmSrSm \rightarrow \\
llmmSrm \rightarrow \\
llmmrm
\end{array}
$$

To conclude, Grail implements a number of strategies to filter out (partial) proof structures which cannot be extended to a proof net. Taken together, these algorithms are very effective at reducing the total search space (see Moot, 2007, for an evaluation).

7.2.3 Wide-Coverage Parsing

Though we have seen, throughout this book, a number of small grammars illustrating different linguistic phenomena, modern computational linguistics has moved

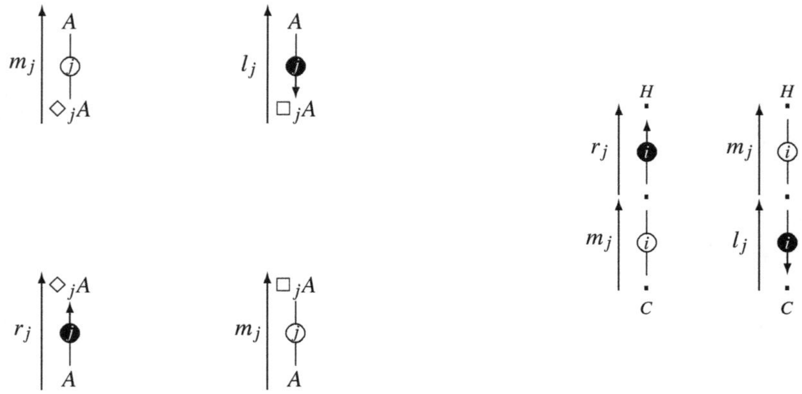

Context-free grammar

$S \rightarrow \varepsilon$
$S \rightarrow m_j S r_j S$ for each unary mode j
$S \rightarrow l_j S m_j S$ for each unary mode j

Fig. 7.41. Converting a unary branch into a string (left, the strings corresponding to the unary contractions (right) and the context-free grammar

beyond small grammars and has become very successful at using statistical estimation techniques to analyze unseen text: we will call this wide-coverage parsing (see Manning and Schütze, 1999, for an introduction to statistical natural language processing). Though wide-coverage parsing was first applied to parsing descriptively limited grammars (eg. context-free grammars), modern wide-coverage parsers have also been developed for linguistically more refined formalisms, including categorial grammars.

Given that it is very time-consuming to construct a large grammar (in the case of categorial grammars, the lexicon function Lex) by hand, it is often advantageous to start with an annotated resource (called a *treebank*, since it usually consists of an annotation tree for each sentence) and convert this resource into a grammar format we like — a categorial lexicon Lex in our case.

Since it is well-known that no matter how big our treebank and (as a consequence) the resulting extracted lexicon, there will always be a number of words which do not occur in the treebank: there are new proper names (for example, corresponding to a new generation of politicians, musicians and movie stars which were still unknown at the time of the construction of the corpus), some new verbs (like "to google") and some words which simply do not occur in the treebank because they are rare or because they are in a more colloquial style rarely adopted in newspapers (since treebanks are often based on newspaper articles). In short, even with a very large treebank and a resulting very large lexicon, we need some way of assigning lexical categories to unknown words. Another problem is that the size of the extracted lexicon becomes a bottleneck for parsing.

In the rest of this section, we will talk about how to convert a French annotated treebank into a categorial grammar, look at the entries in the extracted lexicon and use *supertagging* as a solution both to the parsing with a large lexicon but also for dealing with unknown words. We will conclude by giving an idea about how to use this extracted grammar for semantics.

The goal of this section is to illustrate that the logical view of categorial grammars is compatible with modern developments in computational linguistics and that the transparent syntax-semantics interface makes it possible to give *semantic* structures to unseen text as well.

Grammar Extraction

The French Treebank (Abeillé et al, 2003) is a set of newspaper articles from the newspaper "Le Monde" (from 1989 to 1994) which has been syntactically annotated at the Laboratoire de Linguistique Formelle at the university of Paris VII.

Figure 7.42 shows part of an annotated sentence a tree from the French Treebank, a segment of the longer phrase 7.1 below.

(7.1)	*La*	*cour*	*a,*	*d'*	*autre*	*part,*	*atténué*	*le*	*montant*	*des*	*amendes*	*que*
	The	court	has,	of	other	part,	reduced	the	amount	of the	fines	that

la	*11(sic)*	*chambre*	*avait*	*infligées*	*aux*	*autres*	*prévenus.*
the	11	chamber	had	inflicted	on the	other	defendants.

'The court has, on the other hand, reduced the height of the fines that the 11th chamber had inflicted on the other defendants.'

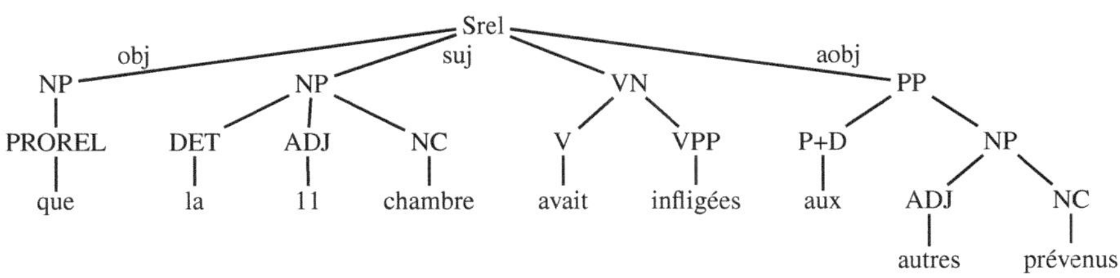

Fig. 7.42. Example tree from the Paris 7 treebank

The leaves of the annotation tree are the words from the sentence. The preterminals (ie. the nodes which have a leaf as daughter) are labeled with the Part-of-Speech tags of the corpus: *V* for verb, *VPP* for a past participle, *DET* for a determiner, *NC* for a common noun, $P+D$ for a combination of a preposition plus a determiner ("aux" in French corresponds to the combination of a proposition "à" (to) with a determiner "les" (the, plural), as indicated by the gloss of sentence 7.1 above).

The other nodes in the trees are labeled by the major syntactic groups: *NP* for noun phrase, *PP* for prepositional phrase and *VN* for the verb group. Finally, some of the connections in the tree are labeled by functional roles: "la 11 chambre" is the subject of the sentence (marked by *suj*), "que" is its object (*obj*) and "aux autres prévenus" is a prepositional complement headed by the preposition "à" (*aobj*).

When we want to transform the tree of Figure 7.42 into a categorial grammar derivation tree, we need to perform a number of steps.

1. The annotation of the trees in the corpus is rather flat, whereas we require a binary-branching tree. This is easily remedied by adding extra branches. For example, we group the adjective "11" and the common noun "chambre" together.
2. We need to decide, for each binary branch in the new tree, which will be the functor and which will be the argument: that is, do we assign the two nodes X/Y and Y or do we assign them Y and $Y\backslash X$? General rules help us make this decision: verbs, adjectives and determiners are functors, whereas object and subject noun phrases, and *aobj* prepositions are arguments.
3. We need to decide, for each argument type Y above what the corresponding formula is. Again, there are general rules to help us here: an argument labeled *NP* will correspond to the formula np, *PP* corresponds to pp, and the argument of a determiner DET or a preposition plus determiner $P+D$ will be n.
4. The annotation tree sometimes needs to be rebracketed in order to better reflect the argument structure. For example, as shown in Figure 7.42, the prepositional argument "aux autres prévenus" is annotated as being the argument of the verb *group*. Syntactically, it is an argument of the past participle "infligées", however, and the syntactic tree should be restructured to reflect this.
5. The annotation tree does not annotate "traces". This means in order to give a relativizer like "que" (that) its correct formula assignment (See Example 5.9 in Section 5.2.2 where we analyzed the English "which" as looking for a sentence missing a noun phrase to its right) we need to add the extracted np to the annotation tree. Since the position of the extracted np (or the np "trace") is neither annotated nor easily deduced from the annotation, this information needs to be added manually.
6. Finally, since the placement of adverbs can be rather free and we do not want to assign them a different formula for each of the different positions in which they can occur, we replaces these by a formulas with a mode permitting movement — much as we have done in Section 5.1.1 and Exercise 5.3.

Figure 7.43 shows the tree of Figure 7.42 with all the modifications listed above taken into account. Steps 1 and 2 correspond rather closely to the learning algorithms for AB grammars which we discussed in Section 1.6. Steps 5 and 6 move us beyond AB grammars and give us multimodal categorial grammars as we have seen them in Chapter 5 (there is a minor notational difference in that the "basic" continuous mode is not indicated in the assigned formulas: we use $\backslash$ and $/$ as a shorthand for $\backslash_0$ and $/_0$ respectively).

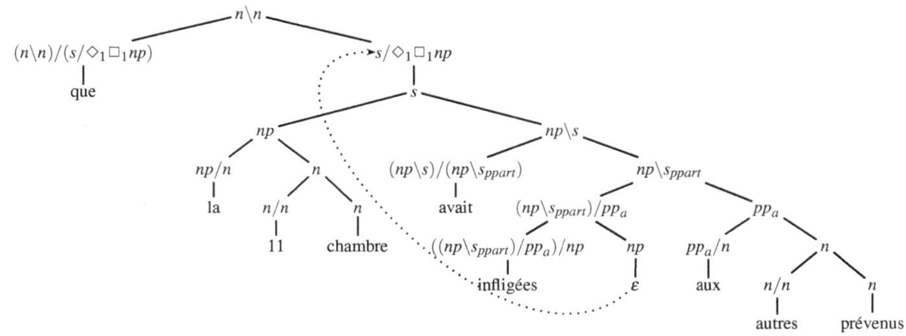

Fig. 7.43. The tree from Figure 7.42 binarized and with formula information added

Figure 7.43 is essentially a natural deduction tree (or an upside-down multimodal proof net!) and the leaves of the tree correspond to the entries of the lexicon. Compared to the atomic formulas we have seen before, there is some added detail: for example, the formula assigned to the preposition "à" is not $pp\,/\,np$ but $pp_a\,/\,np$, which allows us to distinguish between different types of prepositions when they occur as arguments: this allows us to distinguish between verbs taking different prepositional phrases as an argument. Similarly, in addition to the sentence type s, there is a separate type for s_{ppart} which is used for the past participle. This allows us to give the auxiliary verb "avait" the type $(np \setminus s)\,/\,(np \setminus s_{ppart})$, that is it selects a verb phrase headed by a past participle to its right to form a (simple) verb phrase.

Steps 4 and 5 require manual help: in total there is a rather large number of verb clusters and relative pronouns which have to be assigned their correct types and while some of these can be decided by means of general rules this is followed by a rather laborious step of manual verification.

The conversion and extraction algorithm sketched above is rather standard and versions of it have been applied to different styles of categorial grammar and to different treebanks and languages (Moortgat and Moot, 2001; Hockenmaier and Steedman, 2007; Moot, 2010a; Sandillon-Rezer and Moot, 2011).

The Extracted Lexicon

In order to give a better idea of the extracted lexicon, let's look at some of the extracted formulas from the French Treebank and several other resources: manually annotated sentences from the French Timebank (a corpus annotated with temporal structure, see Bittar, 2010), the Itipy corpus (a corpus of voyages in the Pyrenees mountain range, see Asher et al, 2008; Loustau, 2008; Moot et al, 2011) and some more recent newspaper articles from "Le Monde".[1] The total corpus consists of 13,337 sentences, 392,682 words and a total of 918 different formulas occurring in the lexicon.

[1] This choice of different corpora is rather eclectic but motivated by some of the applications with respect to spatio-temporal semantics for the Itipy project.

Perhaps surprisingly, one of the most difficult words in the corpus is the comma ",": though the comma is ignored most of the time (73% of the total number of occurrences) it often fulfills a role close to a coordination such as "et" (and), as can be seen from the many instances of the type scheme $(X \backslash X) / X$ occurring in the lexicon. The formula $(np \backslash np) / (n \backslash n)$ is used for appositive constructions such as those below.

(7.2) *Force Ouvrière, la CGT et la CFDT, qui avaient jugé*
Force Ouvrière, CGT and CFDT, who had judged
"inacceptable" (...)
"unacceptable" (...)

'Force Ouvrière, CGT and CFDT (French labour unions), who considered (...) "unacceptable"'

(7.3) *(...) le secrétaire américain au Trésor, M. Pierre Bérégovoy et*
the secretary American of-the Treasure, Mr Pierre Bérégovoy and
d' autres grand argentiers, soucieux de voir l' activité se ranimer
other big treasurers, concerned to see the activity itself reanimate
(...)

'(...) the US Secretary of the Treasury, Mr Pierre Bérégovoy and other finance ministers, concerned with seeing the (economic) activity pick up (...)'

In both of the sentences above, the first comma is part of a conjunction " X , Y et Z" which are fairly common. The second comma of type $(np \backslash np) / (n \backslash n)$ selects an adjective to its left (both "qui avaient jugé..." and "soucieux de ..." are of type $n \backslash n$) and a noun phrase on its left (the complex conjunction in both cases) to form a noun phrase.

Formulas for ","	
16,927	no formula
1,978	$(np \backslash np) / np$
728	$(n \backslash n) / n$
668	$(np \backslash np) / n$
441	$(s \backslash s) / s$
385	$(np \backslash np) / (n \backslash n)$
344	$((np \backslash s) \backslash (np \backslash s)) / (np \backslash s)$
336	$((np \backslash s) \backslash np) / np$
159	$(s \backslash s) / np$
110	$(pp \backslash pp) / pp$
107	$(n \backslash n) / np$
93	$(np \backslash s) / s$
69	$((np \backslash s) \backslash np) / np$
...	50 other formulas

Formulas for "est" (is)	
664	$(np \backslash s) / (np \backslash s_{ppart})$
633	$(np \backslash s) / np$
538	$(np \backslash s) / (n \backslash n)$
301	$(cl_r \backslash (np \backslash s)) / (cl_r \backslash (np \backslash s_{ppart}))$
222	$(np \backslash s) / pp$
55	$(np \backslash s) / (np \backslash s_{deinf})$
52	$((np \backslash s) / s_q) / (n \backslash n)$
50	$((np \backslash s) / s_q) / pp$
...	31 other formulas

Fig. 7.44. Most frequent formula assignments for "," and "est"

Looking at the types assigned to "est", the present tense of the verb "être" (to be) in Figure 7.44, we see it occurs most frequently as an auxiliary verb (selecting a past participle to its right, occurring 664 times, which is 24.4% of the total number occurrences for "est" as a present tense verb), closely followed by its use as a transitive verb and as a copula verb, selecting a subject and an adjective. The fourth type corresponds to the combination of a reflexive clitic "se" (himself/herself/themselves) which occurs before the auxiliary verb but is an argument of the past participle — sentence 7.4 gives an example: the lexical entry for the verb is "s'attaquer à" (to attack) and it occurs as a past participle. There is "est" with a locative *pp* and finally some less common constructions corresponding more ore less to the English sentences listed below.

(7.4) *M. Mayor s' est attaqué à (...)*
Mr. Mayor SE-himself has attacked at (...)
'Mister Mayor has attacked (...)' "est" = $(cl_r \setminus (np \setminus s)) / (cl_r \setminus (np \setminus s_{ppart}))$

(7.5) *(...) le siège est à Dallas (...)*
(...) the headquarters is at Dallas (...)
'(...) the headquarters is in Dallas (...)' "est" = $(np \setminus s) / pp$

(7.6) *Notre but est de résoudre (...)*
Our goal is to resolve (...)
'It is our goal to resolve (...)' "est" = $(np \setminus s) / (np \setminus s_{deinf})$

(7.7) *Il est très curieux que (...)*
It is very strange that (...)
'It is very strange that (...)' "est" = $((np \setminus s) / s_q) / (n \setminus n)$

(7.8) *C' est en juillet 1988 que (...)*
It is in July 1988 that (...)
' It was in July 1988 that (...)' "est" = $((np \setminus s) / s_q) / pp$

Supertagging

As we have seen, automatically extracting a lexicon — even after a significant amount of cleanup and manual corrections — produces a grammar which simply has too many formula assignments to each word to be useful for parsing anything but the most elementary sentences: many common words (and even interpunction symbols) are assigned a rather large number of formulas with the effect that lexical lookup becomes an important first obstacle to parsing with extracted grammars.

A solution to this problem has been proposed by Joshi and Srinivas (1994). Since the eighties, different methods have been used to successfully implement Part-of-Speech taggers, which reliably assign Part-of-Speech (POS) tags such as "adjective" and "verb" to texts using limited *local* information (word form, last characters of the word, previous word form) (see Manning and Schütze, 1999, Chapter 10 for an overview). Joshi and Srinivas (1994) propose to use these same techniques to assign richer structures than POS tags to words: hence *supertags*. This approach is especially interesting for lexicalized grammar formalisms — the original paper discusses Lexicalized Tree Adjoining Grammars, however it has since been applied to many other lexicalized grammar formalisms, including categorial grammars (see Bangalore and Joshi, 2010, Part III for an overview).

POS tagging is useful since it gives a preliminary disambiguation of the words in the lexicon: it would help us distinguish between the noun "est" (east) and the present tense verb form "est" (is) based only on the local context, for example. Supertagging is an extension of this strategy, where we disambiguate between the list of possible formulas for a word (if it occurs frequently enough) or for its Part-of-Speech tag (if we have seen it only a few times or not at all). Where POS-tagging is highly successful — depending on the detail of the tagset, scores of between 97-98% correctly assigned POS tags are common — supertagging, being inherently more difficult, has scores between 88-92% (again depending on the level of detail in the supertag set).

The Clark & Curran supertagger (Clark and Curran, 2004), trained on the categorial grammar extracted from the French Treebank, which has a total of 918 formulas (*supertags*), assigns 91.1% of unseen words their correct supertags and when assigning words all supertags within a factor of 0.01 of the best supertag, it assigns 98.4% of unseen words their correct supertag (Moot, 2010b). These scores are in a range comparable to the best supertaggers.

As an example, Figure 7.45 shows part of the graphical user interface (part of a set of tools downloadable from the Grail website) which connects Grail and the Clark & Curran tools.

(7.9) *Le gouvernement refuse tout commentaire sur cette proposition et*
The government refuses all comment about this proposal and
préfère avancer les chiffres positifs récoltés par la mesure.
prefers advance the numbers positive harvested by the mesure.

'The government refuses to comment on this proposal and prefers to advance the positive numbers obtained by the measure.'

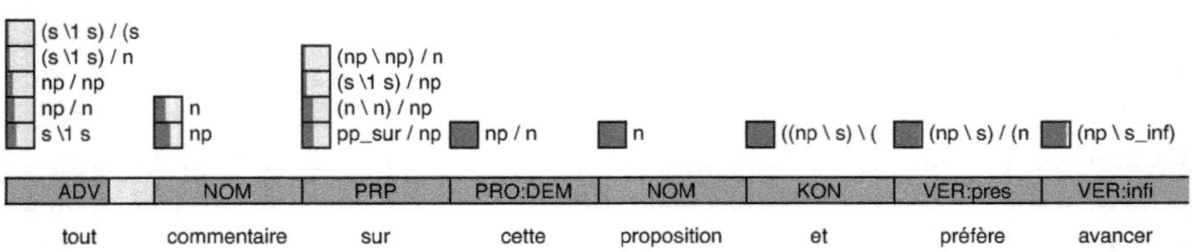

Fig. 7.45. Screenshot of the interface to the supertagger

The figure shows the words in the sentence on the bottom row, with the POS tags inside the rectangles above the corresponding word. So "préfère" (prefer) is assigned the tag VER:pres (for present tense verb), whereas "tout" (all/every) is assigned ADV (for adverb). This is actually an error of the POS tagger: "tout" should be a determiner in the current context, however, as we will see, the supertagger will be able to assign the correct formula $np\,/\,n$ in spite of this. Visually, the user interface indicates the confidence of the POS tagger in each tag, the percentage of the dark-colored background corresponding to the confidence of the tagger (more or less the probability with which the tagger estimates it has given the right response).

The supertagger results are shown above the words and POS tags, as a list of formulas with the likelihood decreasing towards the top of the figure. The confidence of the supertagger is indicated by the square on the left of the formula — the percentage with a darker background denoting (more or less) the probability with which the supertagger estimates the word and POS tag should be assigned this supertag. Many of the words are assigned only a single supertag: the demonstrative "cette" (this) is assigned only $np\,/\,n$ and even "et" (and) is assigned only the (correct) formula $((np\setminus s)\setminus(np\setminus s))\,/\,(np\setminus s)$ (possibly because it estimates this as highly likely for "et" followed by a verb).

The preposition "sur" (on/upon) is more difficult, with pp_{sur}/np — the type for a prepositional argument — being the first choice, followed at a short distance by the correct type $(n\setminus n)\,/\,np$.

The output of the supertagger serves as input to Grail, which performs proof search with the selection of formulas chosen by the supertagger.

Semantics

In the previous paragraphs, we have briefly described how to extract a multimodal categorial grammar from a treebank and how to use the resulting grammar for parsing.

As we have seen in Chapter 3 however, one of the advantages of using categorial grammar is its direct link with semantics in the style of Montague: we can transform a categorial parse, ie. a proof in the multimodal Lambek calculus, into an intuitionistic natural deduction proof, which corresponds to a lambda-term. In order for this to work, we need a lexicon which assigns lambda-terms of the appropriate type to all words. Here, we can use one of the points we criticized about Montague semantics at the beginning of Chapter 3 to our advantage: an unknown noun n is simply assigned the lambda-term $\lambda x.n(x)$ which is of the right type and this strategy extends to verbs and to proper names as well. So the semantic types of our lexicon fall into two general categories:

1. Words which require some special treatment: the auxiliary verbs "être" (to be) and "avoir" (to have), the conjunction "et" (and) etc. The working hypothesis is that these words can be listed in a lexicon,
2. A generic treatment for verbs, adjectives, nouns, etc. which assigns them a lambda-term based on their formula only — using the word only as a constant.

Bos et al (2004) were the first to combine categorial grammars and Montague-style Discourse Representation Theory as we have seen it in Section 3.6 to give wide-coverage semantics of English. In (Moot, 2010c), similar results are obtained using the multimodal lexicon of Moot (2010b), the Grail parser and a lexicon assigning DRSs to French words. The current version of this semantic lexicon contains lexical entries for over 300 words and over 200 default rules.

Figure 7.46 shows the Grail output for example sentence 7.9.

In the universe of the main DRS, there are two variables: z_{18} which is a discourse referent corresponding to the government (as indicated by the topmost condition) and d_2 which is an event variable (we have seen them before in Example 3.3) of an event which corresponds to the action "to prefer", performed by the government z_{18} (indicated as "agent" of the event of "preferring"). The thing which is being preferred is an embedded DRS labeled with x_4 ("advancing the positive numbers..."). The topmost condition is an implication — corresponding to "tout" (all) — stating that if z_{14} is a commentary about the proposition, then the government z_{18} refuses this commentary.

All in all, even though it has been calculated automatically by Grail, the resulting semantic representation is rather close to what we would assign by hand and we

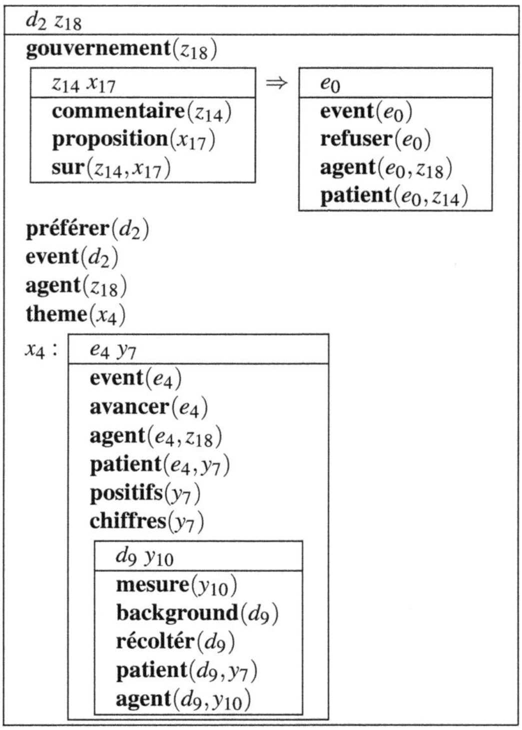

Fig. 7.46. Grail LaTeX output for sentence 7.9

leave it as an interesting open question how much detail we can add to these semantic representation: richer lexical semantics (as in Pustejovsky, 1995, for example) temporal structure (Partee, 1984; Kamp and Reyle, 1993), rhetorical structure (Asher and Lascarides, 1993), ...

7.3 Concluding Remarks

This chapter has given a detailed treatment of proof nets for the different logics based on the non-associative Lambek calculus that have been discussed in Chapter 5. A central thesis has been that proof nets are not only interesting from a formal point of view, but also from the point of view of implementation: given that proof nets are an "economic" way of representing proofs and partial proofs, we can see them as a way of constructing parses for the non-associative Lambek calculus and its multimodal extensions.

We have illustrated the use of proof nets as a tool for parsers by introducing a parser based on proof nets, the Grail theorem prover, and described some of its features.

Exercises for Chapter 7

Exercise 7.1. Look at Figure 7.13 on page 256. It gives a multimodal proof net of $np \vdash (S /_0 np) \backslash_0 S$. Give two other proof structures which convert to the same abstract proof structure shown in the middle ("the same" means that only the hypothesis and conclusion formula differ and that they are not instances of $X \vdash (Y /_0 X) \backslash_0 Y$ for any X or Y).

Exercise 7.2. Theorem 7.13 in Section 7.1.4 shows the equivalence of multimodal proof nets and the sequent calculus. The sequentialisation part of the proof, which shows that each multimodal proof net corresponds to a sequent proof, shows only the case for the $[\backslash]$ contraction explicitly (in Figure 7.32 and Figure 7.33). Give sequent proofs for proof nets whose conversion sequences end with the $[\bullet]$, $[/]$, $[\Diamond]$ and $[\Box]$ contractions.

Exercise 7.3. Draw both the positive and negative multimodal formula trees according to Definition 7.4 on page 247 for the following formulae.

$np /_0 n$
$np \backslash_0 S$
$(a \bullet_1 b) /_2 c$
$(S /_w np) \backslash_w S$
$((np \backslash_0 S) /_0 np) \backslash_0 (np \backslash_0 S)$

Exercise 7.4. Prove Proposition 6.51 on page 226.

Exercise 7.5. Reprove the sentences of Exercise 5.4 using multimodal proof nets: that is write down the negative multimodal formula trees for the lexical entries, give the rewrite rules corresponding to the structural rules and then show each of the proof structures can be contracted to a tensor tree.

Hint: suppose that the structural rules for the identity element of Figure 5.3 on page 153 correspond to the following rewrites.

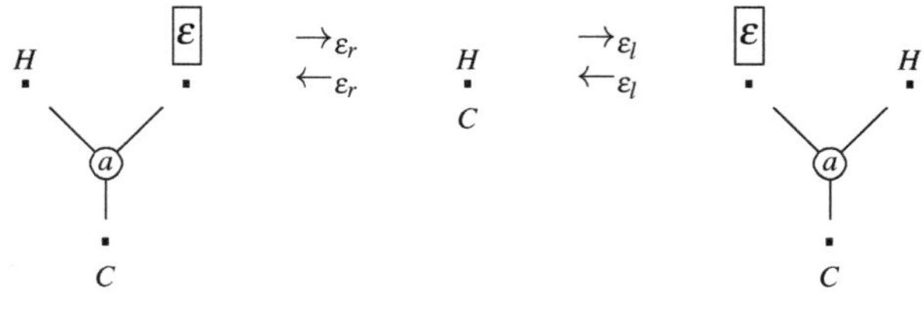

Exercise 7.6. Reprove the grammatical sentences of Exercise 5.6 using multimodal proof nets. In addition, prove that the sentences marked as ungrammatical are underivable by showing they do not contract to a tensor tree.

This exercise uses the same structural rules as shown in Figure 7.16 of Section 7.1.3.

Exercise 7.7. Reprove the following sentences (from Exercise 5.9) using multimodal proof nets.

(7.10) *(dat) ik Marie de nijlpaarden zag voeren.*
(that) I Mary the hippopotami saw feed.

'(that) I saw Mary feed the hippopotami.'

(7.11) *(dat) ik Henk Marie de nijlpaarden zag helpen voeren.*
(that) I Henk Mary the hippopotami saw help feed.

'(that) I saw Henk help Mary feed the hippopotami.'

Begin by writing down the proof structures corresponding to the lexical entries and the graph rewrites corresponding to the structural rules, both of which are repeated below.

$$\frac{\Gamma[(\Delta_1,(\Delta_2,\Delta_3)^0)^4]\vdash C}{\Gamma[((\Delta_1,\Delta_2)^4,\Delta_3)^0]\vdash C}MA \quad \frac{\Gamma[(\Delta_1,(\Delta_2,\Delta_3)^0)^4]\vdash C}{\Gamma[(\Delta_2,(\Delta_1,\Delta_3)^4)^0]\vdash C}MC$$

$$\frac{\Gamma[(\Delta_1,\langle\Delta_2\rangle^0)^0]\vdash C}{\Gamma[\langle(\Delta_1,\Delta_2)^0\rangle^0]\vdash C}K2 \quad \frac{\Gamma[(\langle\Delta_1\rangle^1,\langle\Delta_2\rangle^1)^4]\vdash C}{\Gamma[\langle(\Delta_1,\Delta_2)^4\rangle^1]\vdash C}K \quad \frac{\Gamma[\langle\Delta\rangle^1]\vdash C}{\Gamma[\langle\Delta\rangle^0]\vdash C}I$$

Word	Type(s)	Translation
ik	np	*I*
Marie	np	*Mary*
Henk	np	*Henk*
de	$np\,/\,n$	*the*
nijlpaarden	n	*hippopotami*
zag	$\Box_1((np\backslash_0(np\backslash_0 S))\,/_4\,inf)$	*saw*
helpen	$\Box_1((np\backslash_0 inf)\,/_4\,inf)$	*to help*
voeren	$\Box_1(np\backslash_0 inf)$	*to feed*

Exercise 7.8. Apply the specialized contraction shown in Figure 7.20 on page 261 to the abstract proof structure shown in Figure 7.10 on page 253. What is the resulting abstract proof structure?

Exercise 7.9. Give, following Example 7.12 on page 261, a specialized contraction for $[\Box]$ which corresponds to the $K1$ structural rule and a specialize contraction which corresponds to the $K2$ structural rules. Can you give a characterization of the $[\Box]$ contraction which corresponds to the combined K, I and $K2$ structural rules of Exercise 7.7 above?

Exercise 7.10. Using the translation from unary formulae to strings from Figure 7.41 (in Section 7.2.2), revisit Exercise 5.13 and reprove the relations among prefixes of the unary mode 0 using the context-free grammar at the bottom of Figure 7.41.

References

Abeillé, A., Clément, L., Toussenel, F.: Building a treebank for French. In: Abeillé, A. (ed.) Treebanks: Building and Using Parsed Corpora, ch. 10, pp. 165–187. Kluwer, Dordrecht (2003)

Asher, N., Lascarides, A.: Logics of Conversation. Cambridge University Press (1993)

Asher, N., Muller, P., Gaio, M.: Spatial entities are temporal entities too: The case of motion verbs. In: LREC 2008 Workshop on Methodologies and Resources for Processing Spatial Language, Marrakech (2008)

Bangalore, S., Joshi, A.K. (eds.): Supertagging: Using Complex Lexical Descriptions in Natural Language Processing. MIT Press (2010)

Bittar, A.: Building a Timebank for French: A reference corpus annotated according to the ISO-TimeML standard. PhD thesis, Ecole Doctoral de Sciences du Language Laboratoire Alpage (2010)

Bos, J., Clark, S., Steedman, M., Curran, J.R., Hockenmaier, J.: Wide-coverage semantic representation from a CCG parser. In: Proceedings of the 20th International Conference on Computational Linguistics (COLING 2004), Geneva, Switzerland, pp. 1240–1246 (2004)

Clark, S., Curran, J.R.: Parsing the WSJ using CCG and log-linear models. In: Proceedings of the 42nd Annual Meeting of the Association for Computational Linguistics (ACL 2004), Barcelona, Spain, pp. 104–111 (2004)

Danos, V.: La logique linéaire appliquée à l'étude de divers processus de normalisation et principalement du λ-calcul. Thèse de Doctorat, spécialité Mathématiques, Université Paris 7 (1990)

Girard, J.Y., Lafont, Y., Taylor, P.: Proofs and Types. Cambridge Tracts in Theoretical Computer Science, vol. 7. Cambridge University Press (1988)

Guerrini, S.: Correctness of multiplicative proof nets is linear. In: 14th Symposium on Logic in Computer Science (LICS 1999), pp. 454–463. IEEE (1999)

Hockenmaier, J., Steedman, M.: CCGbank, a corpus of CCG derivations and dependency structures extracted from the Penn Treebank. Computational Linguistics 33(3), 355–396 (2007)

Joshi, A., Srinivas, B.: Disambiguation of super parts of speech (or supertags): Almost parsing. In: Proceedings of the 17th International Conference on Computational Linguistics, Kyoto (1994)

Kamp, H., Reyle, U.: From Discourse to Logic. D. Reidel, Dordrecht (1993)

Loustau, P.: Interprétation automatique d'itinéraires dans des récits de voyages d'une information géographique du syntagme une information géographique du discours. PhD thesis, Université de Pau (2008)

Manning, C., Schütze, H.: Foundations of statistical natural language processing. MIT Press (1999)

Moortgat, M., Moot, R.: CGN to Grail: Extracting a type-logical lexicon from the CGN annotation. In: Daelemans, W. (ed.) Proceedings of CLIN 2000 (2001)

Moot, R.: Proof nets for linguistic analysis. PhD thesis, Utrecht Institute of Linguistics OTS, Utrecht University (2002)

Moot, R.: Graph algorithms for improving type-logical proof search. In: Proceedings Categorial Grammars 2004: an Efficient Tool for Natural Language Processing (2004)

Moot, R.: Filtering axiom links for proof nets. In: Kallmeyer, L., Monachesi, P., Penn, G., Satta, G. (eds.) Proceedings of the 12th Conference on Formal Grammar (FG 2007). CSLI Publications, Dublin (2007) (to appear) ISSN 1935-1569

Moot, R.: Lambek grammars and hyperedge replacement grammars. Tech. rep., LaBRI, CNRS (2008a)

Moot, R.: Lambek grammars, tree adjoining grammars and hyperedge replacement grammars. In: Gardent, C., Sarkar, A. (eds.) Proceedings of TAG+9, The Ninth International Workshop on Tree Adjoining Grammars and Related Formalisms, pp. 65–72 (2008b)

Moot, R.: Automated extraction of type-logical supertags from the Spoken Dutch Corpus. In: Bangalore, S., Joshi, A. (eds.) Supertagging: Using Complex Lexical Descriptions in Natural Language Processing. MIT Press (2010a)

Moot, R.: Semi-automated extraction of a wide-coverage type-logical grammar for french. In: Proceedings of Traitement Automatique des Langues Naturelles (TALN 2010), Montreal (2010b)

Moot, R.: Wide-coverage french syntax and semantics using Grail. In: Proceedings of Traitement Automatique des Langues Naturelles (TALN 2010), Montreal, System Demo (2010c)

Moot, R.: Main Grail website (2012a), `http://www.labri.fr/perso/moot/grail3.html`

Moot, R.: Grail tutorial (2012b), `http://www.labri.fr/perso/moot/tutorial/`

Moot, R., Piazza, M.: Linguistic applications of first order multiplicative linear logic. Journal of Logic, Language and Information 10(2), 211–232 (2001)

Moot, R., Puite, Q.: Proof nets for the multimodal Lambek calculus. Studia Logica 71(3), 415–442 (2002)

Moot, R., Retoré, C., Prévot, L.: Discursive analysis of itineraries in an historical and regional corpus of travels: syntax, semantics, and pragmatics in a unied type theoretical framework. In: Constraints in Discourse, Agay-Roches Rouges (2011)

Morrill, G.: Higher order linear logic programming for categorial deduction. In: Proceedings of the European Association for Computational Linguistics, EACL 1995, Dublin, pp. 133–140 (1995)

Murawski, A., Ong, C.H.: Dominator trees and fast verification of proof nets. In: 15th Symposium on Logic in Computer Science (LICS 2000), pp. 181–191. IEEE (2000)

Partee, B.: Nominal and temporal anaphora. Linguistics and Philosophy 7(3), 243–286 (1984)

Puite, Q.: Proof nets with explicit negation for multiplicative linear logic. Tech. rep., Department of Mathematics, Utrecht University (1998), preprint 1079

Puite, Q.: Sequents and link graphs: Contraction criteria for refinements of multiplicative linear logic. PhD thesis, Department of Mathematics, Utrecht University (2001)

Pustejovsky, J.: The generative lexicon. MIT Press (1995)

Régin, J.C.: A filtering algorithm for constraints of difference in CSPs. In: Proceedings of the Twelfth National Conference on Artificial Intelligence, pp. 362–367. AAAI, Seattle (1994)

Sandillon-Rezer, N.-F., Moot, R.: Using tree Using Tree Transducers for Grammatical Inference. In: Pogodalla, S., Prost, J.-P. (eds.) LACL 2011. LNCS (LNAI), vol. 6736, pp. 235–250. Springer, Heidelberg (2011)

Savateev, Y.: Product-Free Lambek Calculus Is NP-Complete. In: Artemov, S., Nerode, A. (eds.) LFCS 2009. LNCS, vol. 5407, pp. 380–394. Springer, Heidelberg (2008)

Index

www.ingramcontent.com/pod-product-compliance
Ingram Content Group UK Ltd.
Pitfield, Milton Keynes, MK11 3LW, UK
UKHW020953220926
475831UK00001B/57

* 9 7 8 3 6 4 2 3 1 5 5 4 1 *